Oncogenes

Kathy B. Burck
Edison T. Liu
James W. Larrick

Oncogenes

An Introduction to the Concept of Cancer Genes

With a Foreword by Joshua Lederberg

With 93 Figures, 8 in Full Color

Springer-Verlag
New York Berlin Heidelberg
London Paris Tokyo Hong Kong

Kathy B. Burck, M.D., Ph.D.
Fred Hutchinson Cancer Research Center
1124 Columbia Street
Seattle, Washington 98104
Division of Oncology
University of Washington
Seattle, Washington 98195, USA

James W. Larrick, M.D., Ph.D.
Director of Exploratory Research
Senior Research Scientist
GENELABS, Inc.
505 Penobscott
Redwood City, California 94063, USA

Edison T. Liu, M.D.
Assistant Professor of Medicine/Oncology
Curriculum in Genetics
Member, Lineberger Cancer Research Center
School of Medicine
University of North Carolina at Chapel Hill
Chapel Hill, North Carolina 27514, USA

Library of Congress Cataloging-in-Publication Data
Burck, Kathy B.
Oncogenes: an introduction to the concept of cancer genes.
Bibliography: p.
Includes index.
1. Oncogenes. I. Liu, Edison. II. Larrick, James W. III. Title. [DNLM: 1. Oncogenes. QZ 202 B9480]
RC268.42.B87 1988 616.99'4071 87-26634
ISBN 0-387-96423-1

Typeset by Publishers Service, Bozeman, Montana.
Printed and bound by Arcata Graphics/Halliday, West Hanover, Massachusetts.
Printed in the United States of America.

9 8 7 6 5 4 3 2 (Corrected second printing)

ISBN 0-387-96423-1 Springer-Verlag New York Berlin Heidelberg
ISBN 3-540-96423-1 Springer-Verlag Berlin Heidelberg New York

To our families
Clem and Winifred Bauman
Shih Chiu Liu, Grace Liu, Maggie Liu
William and Louise Larrick
and to Dianne Cassatt Jacobs, our trusted and tireless supporter.

Foreword

"Cancer viruses" have played a paradoxical role in the history of cancer research. Discovered in 1911 by Peyton Rous (1) at the Rockefeller Institute, they were largely ignored for several decades. Witness his eventual recognition for a Nobel Prize, but not until 1966 – setting an all time record for latency, and testimony to one more advantage of longevity.

In the 1950s, another Rockefeller Nobelist, Wendell Stanley, spearheaded a campaign to focus attention on viruses as etiological agents in cancer, his platform having been the chemical characterization of the tobacco mosaic virus as a pure protein – correction, ribonucleoprotein – in 1935 (2). This doctrine was a centerpiece of the U.S. National Cancer Crusade of 1971: if human cancers were caused by viruses, the central task was to isolate them and prepare vaccines for immunization. At that point, many observers felt that perhaps too much attention was being devoted to cancer viruses. It was problematic whether viruses played an etiological role in more than a handful of human cancers.

Nevertheless, some cancers were indubitably rooted in virus infection. What overarching concept could unify these observations with cancer induced by a range of environmental influences, ranging from hormones to poison gas to radiation?

The keys to this puzzle emerged out of a galaxy of microbial and cell biological studies that initially had no relationship to viruses. Alexander Haddow (1937) (3) and others formulated cancer as a somatic mutation, a change in the chromosomes (now we say DNA) of a somatic cell. Some contributory evidence that emerged was the overlap in biological activity of radiation and some carcinogenic chemicals: many of them could both cause tumors and induce mutations. And the cancer transformation was a hereditarily stable alteration of a cell clone. The theory could not, however, give a simple account of hormonal induction of cancers; not until the metabolic studies of the Millers (4) could we understand other discrepancies, like the carcinogenic activity of azo dyes and of polycyclic hydrocarbons. Nor did the theory explain viral cancers. In 1946, Lederberg (5) suggested that neoplasia resembled growth-restoring mutations in nutritionally regulated microbes, and speculated that viruses might mimic mutations by the import of extraneous genetic information. But there was no way to test the theory; absent were methods for the study of the genetics of somatic cells.

The first major breakthrough came from the study of pneumococcal transformation, and the epochal finding of Avery, MacLeod, and McCarty (1944) (6) that the agent of genetic transfer was DNA. The conflagration inspired by that work embraces almost all of contemporary biology. However, Avery did not have the luck to survive another 55 years, which would surely have won him recognition in Stockholm. It immediately impelled a rush of work on the genetics of bacteria, organisms that—like somatic cells—had been thought beyond the reach of genetic analysis. The discovery of recombination in *Escherichia coli* (1946) (7) was followed by that of virus-mediated transduction in *Salmonella* (1952) (8). This gave firm substantiation to the concept that genetic information could be transmitted from cell to cell by a virus. In the example of lysogenic conversion, information critical to the ecological functioning of the host was an integral part of the viral genome (9).

By the mid-1950s it was possible to declare a manifesto for somatic cell genetics (10). That the correct elucidation of the human karyotype, 2n = 46, took until 1956 (11) reminds us how primitive was our approach to somatic cells until the current generation. The advent of methods of somatic cell fusion opened the door to systematic mapping of somatic genomes in the 1960s (12), soon to be followed by the molecular genetic analyses of the current era.

The latter are the meat of this book. Oncogenes have become a central theme of contemporary molecular biology. Oncogene research addresses, of course, one of the most grievous of life's burdens, the disease of cancer. It is tautological that the understanding of cancer is inseparable from that of normal development. Rarely, if ever, is the oncogene an adventitious trick wholly invented by the predatory virus. Instead it is a subtle variant on an indigenous system of genic regulation. Cancer viruses have receded as a primary exogenous cause of human cancers. But we are in some measure lucky to have found them, for they have given us the most substantial clues to the occurrence of oncogenes within the human genome—the bits of DNA whose alteration by any means engenders cellular dysregulation and neoplasia.

No one interested in cancer, in viruses, in the cell biology of normal development—whom does that leave out?—can afford to be bereft of the overview of contemporary knowledge that is lucidly and comprehensively surveyed in this book.

Joshua Lederberg
The Rockefeller University

References

1. Rous P. A sarcoma of the fowl transmissible by an agent separable from the tumor cell. J Exp Med 13:397–411, 1911.
2. Stanley WM. Isolation of a crystalline protein possessing the properties of tobacco mosaic virus. Science 81:644–645, 1935.

3. Haddow A. Secondary colony formation in bacteria and an analogy with tumour production in higher forms. Acta Int Union Against Cancer 2:376–401, 1937.
4. Miller EC, Miller JA. Biochemistry of carcinogenesis. Annu Rev Biochem 28:291–320, 1955.
5. Lederberg J. A nutritional concept of cancer. Science 104:428, 1946. Where I quote my own work and thought it is for proximate familiarity, not any claim of uniqueness or priority. I appreciate the authors' asking me to contribute some archeological perspectives: une "recherche du temps perdu."
6. Avery OT, MacLeod CM, McCarty M. Studies on the chemical nature of the substance inducing transformation on pneumococcal types. J Exp Med 79:137–158, 1944.
7. See: Lederberg J. Genetic recombination in bacteria: a discovery account. Annu Rev Genet 21:23–46, 1987.
8. Zinder ND, Lederberg J. Genetic exchange in *Salmonella*. J Bacteriol 64:679–699, 1952.
9. Freeman VJ. Studies on the virulence of bacteriophage-infected strains of *Corynebacterium diphtheriae*. J Bacteriol 61:675–688, 1951.
10. Lederberg J. Prospects for the genetics and somatic and tumor cells. Ann NY Acad Sci 63:662–665, 1956.
11. Tjio JH, Levan A. The chromosome numbers of man. Hereditas 42:1–6, 1956.
12. See, i.e., Ruddle FH, Creagan RP. Parasexual approaches to the genetics of man. Annu Rev Genet 9:407–493, 1975.

Preface

Recent advances in molecular biology have allowed us to study cancer at the level of individual genes. Such work has led to a new conceptual framework now often used to explain the processes required for oncogenesis (cancer expression). The major paradigm within this framework is embodied in the *oncogene hypothesis*, which proposes that specific genes (oncogenes) can induce cancer, and that resident genes required for normal cellular function can be converted to oncogenes by genetic mutation. Like the germ theory that revolutionized our thinking about infectious disease, the oncogene hypothesis already has had a great impact on our understanding of cancer and of cell growth and differentiation, and on research directions.

Our purpose in writing this book is to explain recent findings about oncogenes to interested nonspecialists not immediately involved in the study of oncogenes, including physicians, graduate students, medical students, and advanced undergraduates. We believe that the book will be particularly useful as both an introduction to this important field and as a survey of current research.

The book begins with an outline of the experimental methods used in molecular biology and how they are applied to the study of oncogenesis. Next, a series of chapters discusses oncogenes in general, as well as the relationship of oncogenes to proto-oncogenes, viruses, and human cancers. This is followed by an in-depth examination of the major classes of oncogenes. We conclude with chapters on transgenic mice and the potential diagnostic and clinical uses of information derived from oncogene research. Though this text was written for the nonexpert, a basic understanding of biology and biochemistry would be helpful.

We hope that this book will provide the reader with a useful survey of an exciting field that is growing at an exponential pace.

Kathy B. Burck
Edison T. Liu
James W. Larrick

Acknowledgments

The tireless efforts of Ms. Joan Murphy, who typed and retyped the manuscript, is gratefully acknowledged; without her we would not have succeeded. Special assistance was provided by Dianne and Charles Jacobs and George Senyk. Several people reviewed the manuscript during preparation: Dr. Mat Kluger, University of Michigan; Dr. Robert Stern and Dr. Alan Wells, University of California, San Francisco; and Dr. Frank McCormick, Cetus Corporation, Emeryville, CA. Helpful suggestions were made by Drs. Elvin Kabat, Robin Clark, Mike Kriegler, David Ring, David Liu, and Yawen Chiang. Eric Ladner, Tim Culp, and Sharon Nilson helped with some of the illustrations.

Other illustrations were kindly provided by Dr. Steven L. Kunkel, Ms. Robin Kunkel, Dr. Larry Rohrschrieder, Dr. Dorthee Herlyn, Stefan Weiss, Ted Beals, Nancy Benson, Dr. Britt Marie Ljung, Dr. Ben Yen, Dr. David Ring, Sylvia Ma, Audrey Eaton, Dr. C. C. Lin, Dr. Marilyn Pearson, Dr. Chi Dang, Dr. William Lee, and Dr. Frank McCormick.

Contents

1 Introduction

During the past few years, we have witnessed an explosion in the understanding of basic cellular and molecular biology. With advances made possible by recombinant DNA and monoclonal antibody technology, some insight has been gained into how normal cells regulate their growth and differentiation and how abnormal cells, capable of multiplying and accumulating without the usual developmental constraints, are generated. Emerging from this work is the central concept of proto-oncogenes and oncogenes: normal cellular genes (proto-oncogenes) controlling growth, development, and differentiation that somehow become misdirected (i.e. converted to oncogenes) in the neoplastic cancer cell.

The seminal discovery leading to this concept was made in the laboratories of Drs. J. Michael Bishop and Harold Varmus at the University of California, San Francisco. These scientists showed that the transforming gene *src* of the Rous sarcoma virus shared sequences with normal cellular genes. Subsequently they proposed that other viral oncogenes were homologous to normal cellular "proto-oncogenes" and that when these normal genes are perturbed, neoplasia eventually results–hence the labeling of such genes as "enemies within."

The term "cancer" actually encompasses many different diseases, all of which are characterized by uncontrolled cell growth. Malignant cells are capable of spread or metastasis by both local invasion of adjacent normal tissue and lymphatic or hematogenous dissemination to distant sites. Cancer cells have somehow escaped constraints that confine normal cells to proliferate within closely defined patterns.

Our better understanding of cancer had been hampered by lack of a fundamental knowledge of how normal cells become transformed. Although the mechanism of transformation is not known with certainty for any cancer, certain etiologic and predisposing factors have been well defined. Chemical carcinogens are associated with skin, lung, and bladder tumors. Radiation exposure predisposes to hematological malignancies, leukemia and lymphoma, as well as to solid tumors such as thyroid carcinoma. Hormones influence cancer cell growth as in breast and uterine tumors. Viruses have been implicated in both human and nonhuman T cell malignancies as well as sarcomas and epithelial neoplasms in animals. Genetic predisposition has been noted with childhood tumors, e.g., retinoblastoma and Wilms' tumor, and associated with rare familial neoplasias such as polyposis coli, neurofibromatosis, and the multiple endocrine neoplasia syndromes. Familial clustering, implying underlying genetic factors, has been reported for nearly every common malignancy including breast and colon carcinoma and the acute and chronic leukemias. Finally, syndromes characterized by chromosomal instability and ineffectual DNA repair, including Bloom's syndrome and xeroderma pigmentosum, predispose to skin cancer and other malignancies. All of these agents share the common denominator of affecting DNA, the genetic material.

Concurrent with studies of carcinogenesis have been studies of normal cellular growth and differentiation. Normal genes can be divided roughly into two categories: structural and regulatory. Originally defined for simple prokaryotes and later shown to hold for even the most advanced eukaryotes, *structural genes* determine RNA or protein sequences used to build and run a cell, and *regulatory genes* govern the temporospatial expression of structural genes and other regulatory genes. In 1969 Huebner and Todaro proposed the concept that in multicellular organisms genes responsible for regulation of normal development could go awry. The result would be the misregulated growth typical of cancer. The term "oncogene" was introduced to describe such genes, and the concept became known as the oncogene hypothesis.

The oncogene hypothesis is attractive from several perspectives. First, it defines a genetic basis for cancer. Second, it links normal cellular functions of growth and differentiation (due to the actions of proto-oncogenes) with neoplastic transformation (due to the actions of oncogenes). And third, it provides the unifying theory whereby the ability of carcinogens and other genetic disturbances to contribute to oncogenesis can be explained.

This theory demands that perturbations of normal genes are necessary for cancer to emerge. Experimental evidence has confirmed the presence of a variety of genetic perturbations in human neoplasia. With human leukemias, and colon and bladder cancers, a single base pair substitution in a *ras* gene converts a normal allele to a transforming one. With chronic myelogenous leukemia, the shuffling of chromosomal material leads to the expression of a novel transforming protein which is a result of the fusion of two unrelated genes. Overexpression of an intact c-*myc* allele is seen in Burkitt's lymphomas due to an interrruption of normal elements that control gene expression. Lastly, the loss of an inhibitor of transformation has been described in retinoblastomas. Furthermore, an examination of the evidence will demonstrate that an abnormality of a single gene is often insufficient to induce the fully transformed phenotype, but that, two or more genetic lesions are necessary.

Though the oncogene hypothesis has contributed greatly to the conceptual framework used in the current study of cancer, many questions have since emerged. Oncogenes as *ras*, *src*, and *ski* transform certain cell types, but induce differentiation in others. Since these are opposite processes for a cell, one must speculate the presence of cellular factors that modulate the effects of these oncogenes. Thus oncogene-cellular gene interactions as well as oncogene-oncogene interactions may play important roles in determining the final phenotype of a cancer cell.

In the following chapters, we will explore how the study of oncogenes has both simplified and complicated our understanding of cancer. In most instances, however, the oncogene hypothesis has unified our thoughts and has helped mature our concepts concerning this disorder.

Bibliography

Alberts B, Bray D, Leust J, et al. The Molecular Biology of the Cell. New York: Garland Publishing, 1983.

Bishop JM. The molecular genetics of cancer. Science 235:305, 1987.

Darnell J, Lodish H, Baltimore D. Molecular Cell Biology. San Francisco: Freeman, 1986.

Freifelder D. Molecular Biology: A Comprehensive Introduction to Prokaryotes and Eukaryotes. Boston: Jones & Bartlett, 1983.

Hayes W. The Genetics of Bacteria and Their Viruses. 2nd Ed. Blackwell, Oxford, 1970.

Huebner RJ, Todaro GJ. Oncogenes of RNA tumor viruses as determinants of cancer. Proc Natl Acad Sci USA 64:1087–1094, 1969.

Langman J. Medical Embryology. 4th ed. Baltimore: Williams & Wilkins, 1981.

Lewin B. Gene Expression 2. 2nd ed. New York: Wiley, 1980.

Tooze J (ed). DNA Tumor Viruses: Molecular Biology of Tumor Viruses, 2nd ed., revised. Cold Spring Harbor, NY: Cold Spring Harbor Laboratory, 1981.

Watson JD, Tooze J, Kurtz DT. Recombinant DNA: A Short Course. San Francisco: Freeman, 1983.

2
Assays: Tools of the New Biology

Overview

Technical advances in molecular biology and tissue culture have permitted the detailed study of specific genes in the pathogenesis of cancer. Assays have been designed to examine the major macromolecules of a cell: DNA, RNA, and proteins. In addition, cell culture techniques and the use of genetically defined experimental animals allow for the identification of factors necessary for transformation.

In general, the power of these techniques lies first in their ability to sort and identify specific macromolecules among a large population of like species, and second in their capacity to expand and purify to a high degree a macromolecule of interest, such that structural and functional analyses can be performed. Methods for analyzing DNA require the ability to cleave DNA at specific sites, to sort the different fragments according to a particular characteristic (e.g., size), and to identify pertinent fragments that include DNA sequences of interest. This is accomplished by restriction endonuclease digestion, agarose gel electrophoresis, and Southern blot hybridization. DNA fragments can be isolated and expanded for more detailed studies by employing procedures whereby individual fragments are incorporated into separate bacteria or phage. A "library" of these fragments can be generated which may cover the entire genome and from this library, pertinent clones can be identified and expanded. Once significant quantities of cloned DNA are obtained, then the direct DNA sequence of the segment can be determined through methods developed by Sanger or Maxam and Gilbert. Similar techniques for sorting, probing, and isolating RNA (Northern blotting, cDNA cloning) and proteins (monoclonal and polyclonal antibodies, Western blotting) also have been adapted.

Cell culture techniques have been invaluable in determining those factors necessary for cell growth, and for defining a difference between normal and transformed cells. Immortalized cell lines, both transformed and non transformed, have permitted study of the cellular effects of oncogenes and their gene products. The ability to isolate or clone individual cells yielding a homogeneous population for experimentation, and the capacity to expand this cell population in order that sufficient material is made available for study are further benefits afforded by these techniques.

The ultimate test for carcinogenesis is induction of malignant growth in an intact organism. Animal assays are particularly helpful when the interplay of developmental factors and the effects of adjacent stromal cells are involved in the tumorigenic process.

The study of oncogenes depends, therefore, on the integration of knowledge obtained utilizing all these techniques.

In order to understand the molecular and cellular biology of normal cells and their cancerous counterparts, it is helpful to be familiar with some of the experimental procedures used to study them. It is also important to realize that what we "know" scientifically is based on procedural definitions and is dependent on experiments designed to prove or disprove particular ideas or hypotheses. The experiments that can be designed in the oncogene field depend, in turn, on the assays available to measure gene (and gene product) structure and function. It is these tools of molecular biology that are discussed in this chapter.

Assays are generally divided into two types: structural assays and functional assays. *Structural assays* determine the configuration of building blocks of large molecules, e.g., the arrangement of genes on a chromosome, the sequence of nucleotides in DNA, the primary amino acid sequence of a protein or its tertiary conformation. Structural assays can also determine which proteins are associated with cell membranes or the subcellular localizations of various macromolecules. *Functional assays* determine how things act, e.g., which genes are transcribed in particular cells, whether cells are "normal" or "malignant," how well a protein or its mutant counterpart can perform a particular task. Structural and functional assays are intimately related and frequently utilize quite similar techniques. We therefore discuss them together in relation to DNA, RNA, protein, cells, and organisms.

DNA Studies

Structure has implications for function as Watson and Crick so eloquently proved with their elucidation of the double helical structure of DNA. General mechanisms for both replication and transcription of DNA are implicit in this structure. However, because of the size of even small DNA molecules, few additional developments in molecular genetics would have been likely without the revolutionary advances made possible by recombinant DNA technology. Gene cloning, i.e., the isolation and propagation of fragments of DNA, has made it possible to study the organization of eukaryotic as well as prokaryotic genes on chromosomes, the relations of various genes to each other, and the relations among the several parts of a single gene. It is now possible to elucidate the order of nucleotides in a DNA molecule and thus have the primary sequences of, for example, genes, introns, promoters, repetitive elements, and splice sites.

Recombinant DNA technology encompasses restriction enzymes, Southern blotting, probes, cloning, restriction mapping, and sequencing.

Restriction Enzymes

Restriction endonucleases were found during the late 1960s to be present in bacteria such as the enteric bacillus *Escherichia coli*. They were named for their ability to "restrict," or prevent, DNA from a foreign bacterial strain from replicating or otherwise functioning in the original cell. To accomplish this purpose, these enzymes cut the foreign DNA into pieces and, moreover, at specific nucleotide sequences. The enzymes typically recognize four to six base pair sequences in the DNA and break the double-stranded molecule in or near that region. Often the recognized sequences are symmetrical. For example, the enzyme *Eco* R1 recognizes the sequence

5′ G↓AATTC 3′
3′ CTTAA↑G 5′ "sticky ends"

and breaks the bond between G and the 5′ A on each strand. *Hae* III recognizes

5′ GG↓CC 3′
3′ CC↑GG 5′ "blunt ends"

and cleaves between the G and C on each strand. Thus some enzymes leave overlapping or "sticky" ends, and other leave "blunt" ends (Table 2.1). Restriction enzymes are named for the bacterial strain of origin and order of discovery; for example, *Eco* R1 is the first enzyme from the R strain of *E. coli*, and *Hae* III is from *Hemophilus influenzae*, the third enzyme. Some 300 restriction enzymes have been isolated recognizing more than 100 different sites.

Restriction Mapping

Restriction enzymes form the basis of DNA technology. Once a length of DNA can be partitioned into a finite number of consistent fragments, study of the DNA is clearly facilitated. In the case of a viral chromosome or other small DNA of up to 20,000 to 50,000 base pairs,[1] maps of restriction fragments can be made by digesting the DNA with a restriction enzyme and separating the fragments by size on agarose gels (Fig. 2.1). The number of "recognition sequence" sites cleaved by the enzyme is directly proportional to the time of digestion; if sufficient time is allowed, all possible sites are cut (complete digest); shorter digestion times result in fewer cuts (partial digest). (Cuts are made in a quasirandom fashion at any of the properly recognized sites.) The rate at which fragments move on an agarose gel in an electric field is indirectly proportional to their size, with smaller fragments moving more rapidly than the larger ones. Fragments are visualized on gels by staining with ethidium bromide, a chemical that inserts

[1] Or 20 to 50 kilobases (kb); 1 kb = kilobase = 1000 bases (single-stranded molecule) or base pairs (double-stranded molecule).

TABLE 2.1. Restriction endonucleases.

Enzyme	Recognition sequence
Eco R1	G$^{\downarrow}$AATTCC CTTAA$_{\uparrow}$G
Bam H1	G$^{\downarrow}$GATCC CCTAG$_{\uparrow}$G
Hae III	GG$^{\downarrow}$CC CC$_{\uparrow}$GG
Hpa I	GTT$^{\downarrow}$AAC CAA$_{\uparrow}$TTG
Hind III	A$^{\downarrow}$AGCTT TTCGA$_{\uparrow}$A
Bgl II	A$^{\downarrow}$GATCT TCTAG$_{\uparrow}$A
Pst I	CTGCA$^{\downarrow}$G G$_{\uparrow}$ACGTC
Cla I	AT$^{\downarrow}$CGAT TAGC$_{\uparrow}$TA
Mbo I	$^{\downarrow}$GATC CTAG$_{\uparrow}$
Pvu I	CGAT$^{\downarrow}$CG GC$_{\uparrow}$TAGC
Sac I	GAGCT$^{\downarrow}$C C$_{\uparrow}$TCGAG
Sma I	CCC$^{\downarrow}$GGG GGG$_{\uparrow}$CCC

Restriction endonucleases are isolated from prokaryotic cells. They cleave double-stranded DNA molecules at specific nucleotide sequences. A number of common and representative enzymes are listed in this table.

between the base pairs and fluoresces under ultraviolet (UV) light. By comparing fragment sizes from the complete and partial digests, the order of fragments in relation to each other can be determined. Another method is to use combinations of restriction enzymes, comparing complete digestions with single enzymes to complete digestions with two or three enzymes. By comparing fragment sizes, it is possible to construct detailed maps.

As the size of the DNA increases, it becomes progressively more difficult to order restriction fragments (and even to determine the number of discrete fragments). Therefore it is not possible to generate "complete" restriction maps by this technique even for the genomes of large DNA viruses, let alone chromosomes of prokaryotic or eukaryotic cells. In these cases, smaller, more manageable pieces of DNA are generated by molecular cloning.

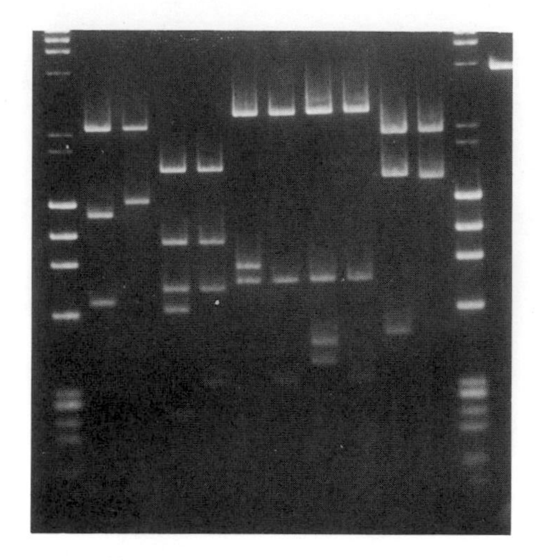

FIGURE 2.1. DNA fragments generated by restriction endonuclease cleavage electrophoresed on agarose gel. DNA molecules are negatively charged owing to their phosphate groups. When an electric field is applied across an agarose gel, DNA fragments migrate at a rate indirectly proportional to their size: The smallest fragments migrate most quickly. Ethidium bromide is a fluorescent dye with a flat planar structure. It intercalates between DNA base pairs and fluoresces orange under ultraviolet light, allowing visualization of the separated DNA fragments. DNA "marker" fragments of known size are frequently run in parallel with unknown DNA fragments in order to provide a standard for size estimation of the unknown fragments. In the gel pictured, the leftmost and second from right lanes represent marker fragments. The other lanes show clones of a feline retrovirus (FeLV) cut with various restriction enzymes.

DNA CLONING

The objective of cloning DNA is to produce large numbers of identical copies of particular DNA fragments. This goal is most easily accomplished by taking advantage of prokaryotic systems with their capacity for rapid reproduction, limited only by the availability of nutrients. There are essentially two methods of producing large copy numbers of foreign DNA fragments in bacteria: use of plasmids or use of bacteriophages as "cloning vectors."

Plasmids are of several varieties. Large ones tend to be limited to one or a few copies per cell ("stringent" control of replication) and are difficult to work with because of their size and low copy number. Thus the plasmids important for molecular cloning are the smaller ones that normally exist in multiple (10–30) copies per cell. These small plasmids replicate autonomously from the bacterial chromosome ("nonstringent" control of replication) and can be multiplied to even higher per cell copy numbers by blocking replication of the bacterial chromosome (e.g., by adding an antibiotic such as chloramphenicol to which the cell but not the plasmid is sensitive). Some of these plasmids contain antibiotic resistance genes that can be used as a naturally occurring selection marker. They are small enough to be easily mapped by the restriction technique outlined above. Fortuitously, the original "wild-type" (unmodified) plasmids used in cloning contain three essential features: autonomous replication, selection marker(s) (e.g., antibiotic resistance or ability to utilize an unusual substrate), and restriction sites for insertion of foreign DNA into these plasmids. By genetic engineering and appropriate selection techniques, it has been possible to construct "ideal" cloning plasmids trimmed to the essentials (small size → higher potential copy number) and containing one (or more) unique insertion site(s) for one (or more) of the most useful restriction enzymes.

Bacteriophage vectors are generally modified forms of the lysogenic phage lambda (λ). The lambda genome is organized such that its genes for replication and packaging of DNA are on its two ends, whereas the central portion contains the genes for lysogenic functions, i.e., for stable integration into the host bacterial chromosome (Fig. 2.2). The lysogenic functions are not required for the lytic life cycle. Lambda cloning vectors carry replication and packaging genes on the end fragments, while foreign DNA can be inserted into the nonessential middle region.

Cloning in plasmids is technically simpler than cloning in phages (Fig. 2.3). Both foreign and plasmid DNA are cleaved by a restriction enzyme; the plasmid is opened at one particular site, whereas the foreign DNA is cut into many linear fragments. Association of the pieces is facilitated by single-stranded regions, "sticky" or cohesive ends that enable the molecules to anneal to each other by base-pairing at their ends. Some enzymes, e.g., *Eco* R1 or *Bam* H1, naturally

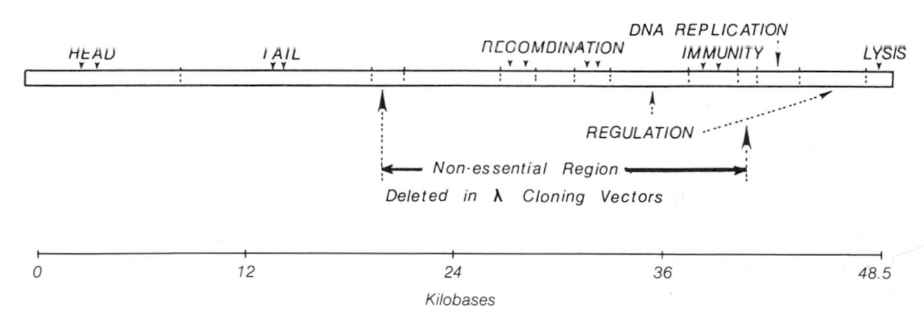

FIGURE 2.2. Genetic map of bacteriophage λ. The λ phage is approximately 48,500 bp in length. Its genome codes for structural proteins and enzymes necessary for a lytic life cycle and for regulatory/repressor functions necessary for a lysogenic life cycle. For the latter, the phage genome is incorporated stably into the chromosome of a bacterial host. The lytic functions are at the 5′ and 3′ ends of the genome, whereas the lysogenic genes are in the middle of the linear phage DNA. The lysogenic functions are not essential for the phage lytic life cycle and can be deleted and replaced by other sequences when the phage is used as a cloning vector (see Fig. 2.4).

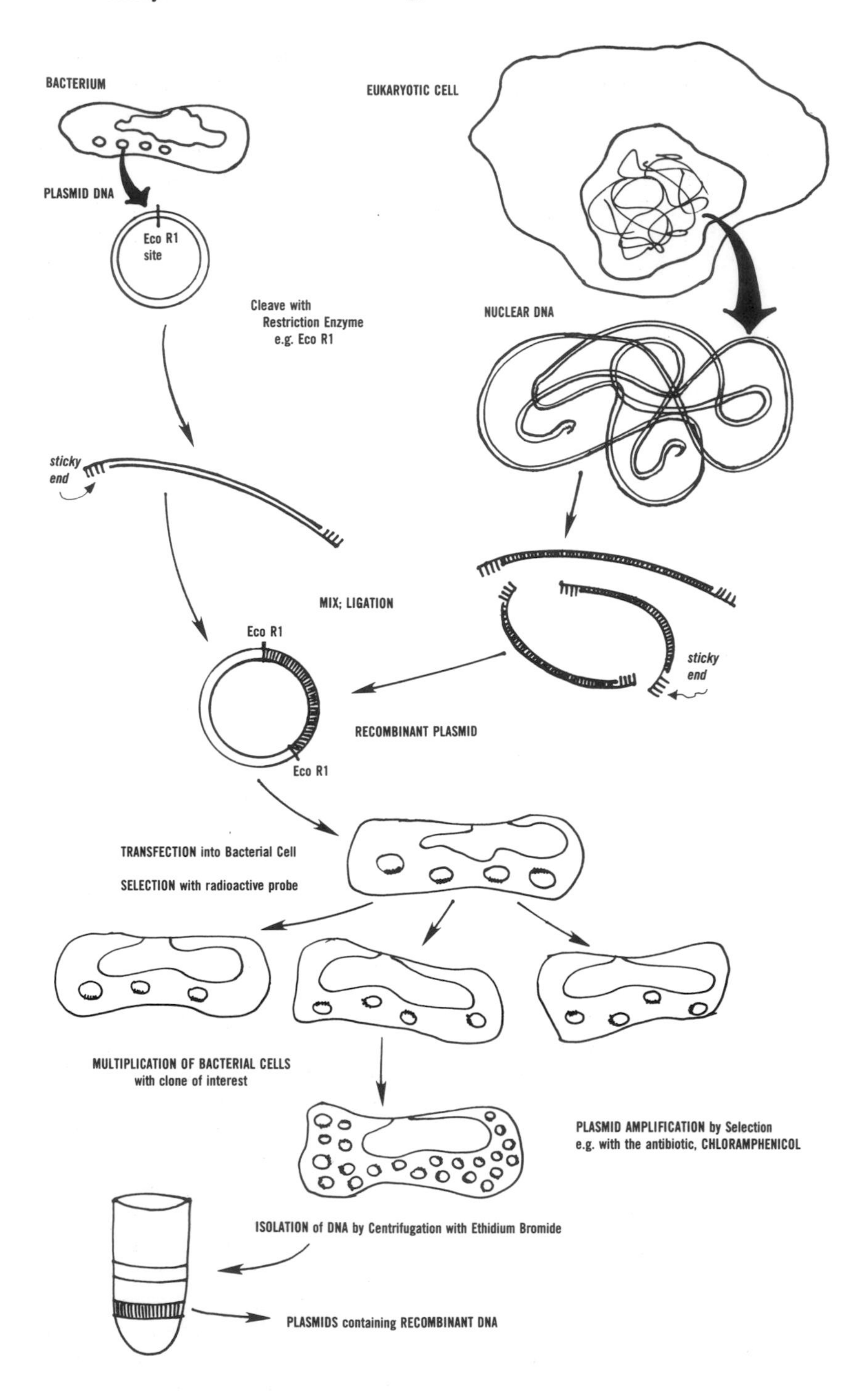
BACTERIUM
EUKARYOTIC CELL
PLASMID DNA
Eco R1
site
Cleave with
Restriction Enzyme
e.g. Eco R1
NUCLEAR DNA
sticky
end
MIX; LIGATION
Eco R1
sticky
end
RECOMBINANT PLASMID
Eco R1
TRANSFECTION into Bacterial Cell
SELECTION with radioactive probe
MULTIPLICATION OF BACTERIAL CELLS
with clone of interest
PLASMID AMPLIFICATION by Selection
e.g. with the antibiotic, CHLORAMPHENICOL
ISOLATION of DNA by Centrifugation with Ethidium Bromide
PLASMIDS containing RECOMBINANT DNA

leave single-stranded "sticky ends" (see above). When an enzyme leaves "blunt" ends (e.g., *Hae* III), linker fragments can be added by blunt ligation or annealing using the enzyme T_4 ligase before the DNAs are mixed. The plasmid DNA/foreign DNA ratio is adjusted so that the probability for one foreign fragment to stick to one plasmid is optimized. Once the foreign DNA has stuck to the plasmid, a "DNA ligase" enzyme is used to seal the cuts. When the foreign fragment has been sealed into the plasmid, the new hybrid plasmid is again circular. These hybrids can be reintroduced into bacteria by a transformation technique. Under selective conditions (e.g., growth in media containing an appropriate antibiotic), only bacteria containing plasmids with the antibiotic resistance gene grow, and the foreign DNA is replicated in the bacteria along with the plasmid DNA itself.

Plasmid-containing bacteria generally replicate more slowly than bacteria without plasmids, and the time required for replication is proportional to the size of the plasmid. The result is that a selective advantage accrues to bacteria with smaller plasmids, and a corollary is that plasmids containing large inserts tend to lose these plasmids under the selection pressure entailed by growth and replication itself. (The foreign insert of a plasmid is "extra" and can be lost without disadvantage to the bacteria so long as the plasmid retains the genes necessary for growth.) Therefore plasmids are best for cloning relatively small pieces of DNA: fewer than 5000 base pairs.

The particulars of lambda phage cloning (Fig. 2.4) make it complementary to plasmid cloning. Lambda cloning vectors are mutated, genetically engineered phages that contain specific restriction sites. To clone with lambda, the vector is cut with its particular enzyme to give two large end fragments, or "arms," containing the essential replication and packaging regions discussed above and multiple smaller fragments from the middle nonessential region. The "arms" are isolated and mixed with the foreign DNA to be cloned. The annealing and litigation reactions take place in essentially the same manner as for plasmid cloning. The

---

◄

FIGURE 2.3. Plasmid cloning. Plasmid vector DNA and the "foreign" DNA to be cloned are isolated by standard microbiological techniques. Both DNAs are cut with restriction enzymes; if the DNA is cut with an enzyme that leaves blunt ends or if the foreign DNA is cut with an enzyme other than the one used to cleave the vector DNA, DNA linker fragments frequently are added to facilitate annealing between vector and foreign DNA. The DNAs are mixed in a ratio that maximizes insertion of one foreign fragment into each plasmid. After annealing, the recombinant plasmids are ligated to "seal" single-stranded nicks resulting in typical closed, circular, supercoiled plasmids. Recombinant plasmids are transfected back into host cells and propagated by standard techniques. Bacteria containing recombinant plasmids are selected by restrictive growth conditions (e.g., antibiotic-containing medium) and then are plated to give single colonies. DNA from the colonies is transferred to filters that are hybridized with radioactive probes to select particular clones of interest. Once a clone is selected, recombinant plasmids are amplified by incubating bacteria in chloramphenicol; plasmid DNA is then isolated from bacterial DNA by banding with ethidium bromide in cesium chloride gradients.

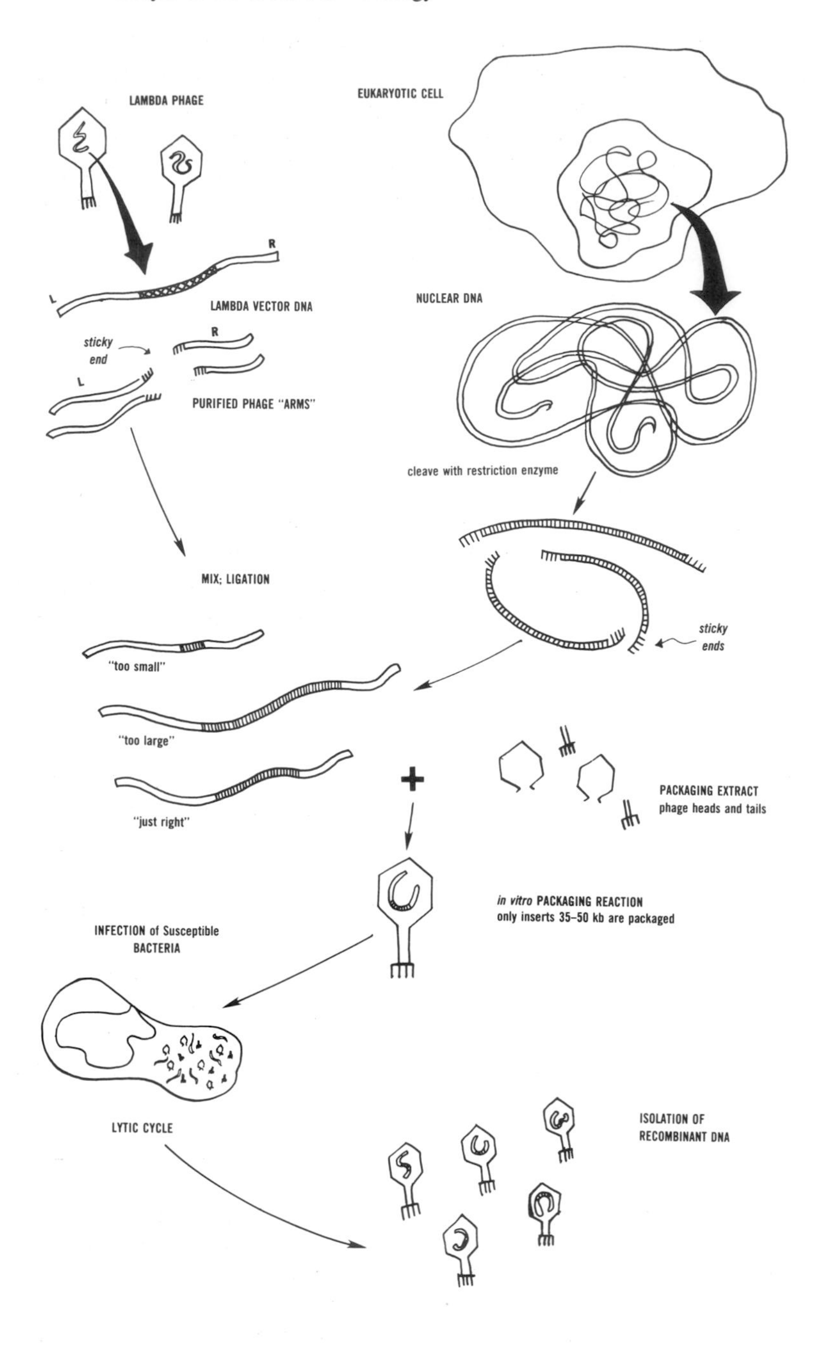
LAMBDA PHAGE
EUKARYOTIC CELL
R
L
LAMBDA VECTOR DNA
NUCLEAR DNA
sticky end
R
L
PURIFIED PHAGE "ARMS"
cleave with restriction enzyme
MIX; LIGATION
sticky ends
"too small"
"too large"
"just right"
+
PACKAGING EXTRACT
phage heads and tails
in vitro PACKAGING REACTION
only inserts 35–50 kb are packaged
INFECTION of Susceptible
BACTERIA
LYTIC CYCLE
ISOLATION OF
RECOMBINANT DNA

annealed DNA is added to a "packaging extract" of lambda phage heads and tails. In order to insert DNA into the limited-size phage head during packaging, the DNA must be a certain relatively uniform length (40–50 kb). Because the genes on the arms require approximately 30 kb of DNA, inserts cloned in lambda must be on the order of 15 to 20 kb. DNA is packaged in the head followed by addition of the tail and completion of virion assembly. To produce the foreign DNA, the finished particles infect fresh bacteria and complete a lytic cycle. Phages containing the insert then are isolated by standard microbiological techniques.

Probes

Using the cloning techniques outlined above, it is possible to isolate total DNA from complex eukaryotes such as human cells and to clone the entire genome. Such a set of clones is called a *genomic library.* The size of the genome, however, precludes the analysis of each and every clone to find any particular gene or DNA segment. It is thus necessary to have a means of identifying particular clones. One useful method is the use of nucleic acid probes, radioactive sequences of DNA or RNA complementary to the desired gene. Given the propensity of separated complementary DNA or RNA strands to reanneal with their partners, the radioactively labeled probe can be used to screen a series of bacterial colonies (in the case of plasmid clones) or phage "plaques" (in the case of lambda clones) for the clone of interest (Fig. 2.5). In either case, the DNA of the clone is fixed to a filter and the radioactive probe added in solution. The probe anneals with its complementary DNA, and the correct clone is revealed by exposure of the filter to x-ray film (autoradiography). Black spots on the film indicate exposed areas caused by radioactive particles from the probe and thus the location of complementary clones; the film replica of the filter then is compared to the original set of clones and the matching clones selected and grown.

The original probes were prepared from RNA: tRNA, rRNA, or mRNA. Each type can be isolated from the others in relatively pure form; but whereas tRNA and rRNA comprise a limited number of species, a different mRNA exists for each unique protein specified by a cell. Enrichment for particular mRNAs can be

---

◄

FIGURE 2.4. Bacteriophage λ cloning. Phage "arms" containing lytic function genes are mixed with restriction endonuclease-cleaved foreign DNA fragments. As for plasmid cloning (Fig. 2.3), linkers may be added to facilitate annealing. Inserts are ligated to the arms, and the recombinant DNA is mixed with a "packaging extract" containing phage structural proteins. After packaging, fresh bacterial host cells are infected with the recombinant phage; the latter undergo a lytic life cycle and are plated to give single "plaques" (clear spaces) in a "lawn" of bacteria representing lysed bacteria and released progeny phage particles. DNA from the plaques is transferred to filters that are hybridized with radioactive probes to select particular clones.

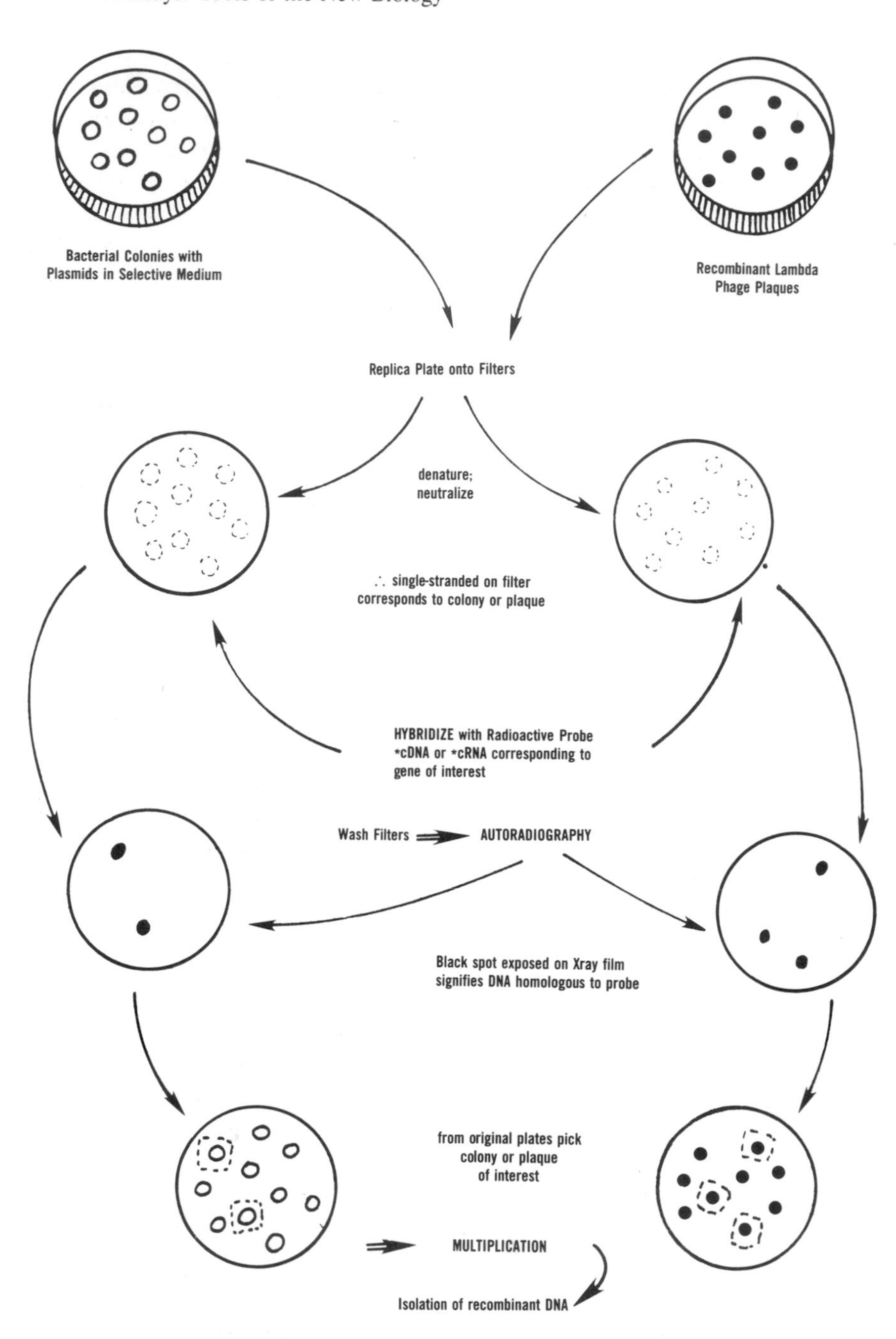
Bacterial Colonies with
Plasmids in Selective Medium
Recombinant Lambda
Phage Plaques
Replica Plate onto Filters
denature;
neutralize
∴ single-stranded on filter
corresponds to colony or plaque
HYBRIDIZE with Radioactive Probe
*cDNA or *cRNA corresponding to
gene of interest
Wash Filters
AUTORADIOGRAPHY
Black spot exposed on Xray film
signifies DNA homologous to probe
from original plates pick
colony or plaque
of interest
MULTIPLICATION
Isolation of recombinant DNA

accomplished by isolating mRNA from cells expressing a lot of the protein of interest (e.g., hemoglobin mRNA from red blood cell precursors); however, it is not possible to obtain a pure probe in this manner. This problem was overcome by using reverse transcriptase, the RNA-dependent DNA polymerase of the retroviruses (see Chapter 4). This enzyme permits synthesis of DNA copies of mRNA (called cDNA for "copy" DNA); cDNA can be cloned itself and then used as a probe. New cDNA probes can be characterized by using them to reisolate the mRNA from which they were made, which in turn is translated into protein. Because many proteins have previously been characterized biochemically, comparison of the translated protein with known proteins allows identification of the cDNA. Once clones have been isolated using cDNA probes, the clones themselves can be used as probes to allow identification of similar but nonidentical genes. [Variations on this method have been invented including *chromosome walking*, whereby a series of overlapping clones are used sequentially to obtain probes for DNA segments far (on the chromosome map) from the original probe.]

Southern Blotting

Characterization of eukaryotic genes has been greatly facilitated by a technique developed independently of other recombinant DNA technology by Southern (Fig. 2.6). DNA is cleaved with one or more restriction enzymes and run on an agarose gel where fragments are separated according to size, as discussed above. The DNA from the gel is denatured with a basic solution to give single strands, neutralized, then transferred by overlaying the gel with a filter sheet. The single-stranded fragments are transferred to the filter by setting up a buffer system such that the buffer passes through the gel and then the filter via a wick. The DNA fragments stick to the filter. Thus a replica or "blot" of the DNA fragments on the original gel is made on the filter. Once the replica is obtained, sizes of specific DNA fragments can be determined by hybridizing the filter with a radioactive probe in solution. The probe binds to any complementary DNA and can be visualized by autoradiography as described above (Fig. 2.7).

Southern blotting can be used to map restriction fragments in cloned DNA. The technique can also be used to map uncloned genes and provide an estimate

---

◄

FIGURE 2.5. Selection with radioactive probes. DNA from recombinant clones is transferred to filters, denatured to single strands, then neutralized and hybridized in solution with radioactive single-stranded probe DNA or RNA. The probe hybridizes specifically with complementary single strands on the filter. Filters are washed to remove nonspecifically bound radioactivity and placed in proximity to x-ray film. Radioactivity from probe nucleic acid exposes the film in a pattern corresponding to the desired clones, which thus are identified and selected for further propagation.

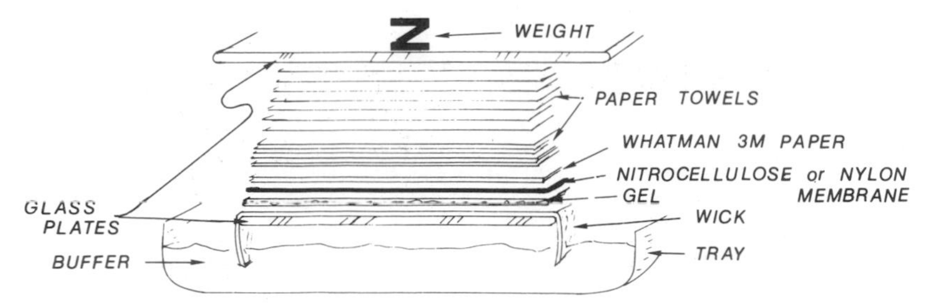

FIGURE 2.6. Southern blot apparatus. Separated DNA fragments on a gel are first denatured to single strands, then neutralized to prevent further degradation of the DNA. The treated gel is placed on a filter-paper wick spanning two buffer tanks containing a hypertonic saline–citrate solution. A hybridization filter is placed on top of the gel. Additional paper filters or towels are stacked on top of the hybridization filter and weighted slightly. Buffer is drawn from the wick through the gel and to the filter and paper above it. Over a period of some hours the DNA fragments are transferred to the hybridization filter, which then can be hybridized with radioactive probes as described in Figure 2.5. Transfer of RNA to filters (Northern blotting) and proteins to filters (Western blotting) is accomplished with a similar apparatus.

of gene abundance in total cellular DNA. If total cellular DNA is digested by restriction enzymes and electrophoresed on a typical agarose gel, large numbers of fragments in a continuum of sizes are generated. These fragments are not resolvable with ethidium bromide staining. (They appear as a stained smear on the gel.) However, if the fragments are transferred to a Southern blot and hybridized with a specific probe, only complementary DNA hybridizes and is made visible on the film after autoradiography. By comparing the number, sizes, and amounts of radioactivity in various DNA fragments, an estimate of relative gene abundance can be deduced. For example, repetitive DNA elements show multiple hybridizing bands of various sizes, whereas unique or low copy number genes such as hemoglobin show one or a few well defined bands. This technique is essential for studying proto-oncogenes in cellular DNA as well as the genetic predisposition to cancer.

DNA SEQUENCING

An enormous breakthrough in understanding the primary structure and organization of eukaryotic genes came with the advent of techniques for direct sequencing of DNA. Earlier nucleic acid sequencing techniques, developed by Frederick Sanger and co-workers in Cambridge, England, during the mid-1960s, required synthesis of an RNA copy of the DNA and sequencing of this cRNA in order to deduce the DNA sequence. This method, of course, was limited by the size of the complementary RNA that could be produced and the difficulties inherent in working with RNA (see below). Sanger developed the first direct DNA sequenc-

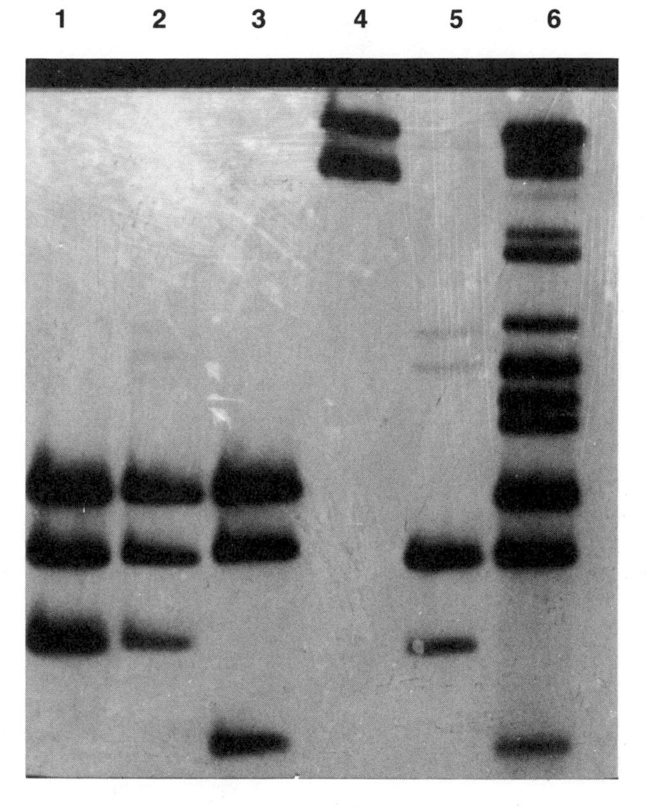

FIGURE 2.7. Autoradiograph of a Southern blot showing blackened (exposed) areas on the film corresponding to DNA fragments complementary to the hybridization probe.

ing method in 1975, and an alternative method was devised by Allan Maxam and Walter Gilbert at Harvard University in 1977. Sanger subsequently developed a third method of DNA sequencing, more powerful than his first, and the latter procedure and that of Maxam and Gilbert are currently the most widely used DNA sequencing techniques.

The Maxam–Gilbert method utilizes specific chemical cleavages to differentiate among the DNA bases: A double-stranded DNA fragment is labeled with radioactive phosphate at both 3′ or both 5′ ends by a 3′ or 5′ "kinase" (an enzyme that adds a phosphate residue to a macromolecule). The strands then are separated, and each single strand (labeled at one end) is divided into four reaction mixtures. A chemical that specifically destroys one of the bases is added to every mixture; reaction conditions are such that only a few bases are "hit" in any one DNA molecule. Another chemical, piperidine, then is added to each reaction, causing cleavage of the DNA at each site of the previously destroyed base. Every reaction mixture is run in a separate lane on an acrylamide sequencing gel that separates the fragments by size such that each one-base difference in molecular size is resolvable. After autoradiography, the film shows a "ladder" of bands corresponding to the original DNA sequence.

The newer Sanger method takes advantage of enzymatic synthesis of DNA by DNA polymerase (Fig. 2.8B). Here a single-stranded unlabeled DNA fragment (to be sequenced) is isolated, and a short-end-labeled complementary "primer" is attached by its 5′ end to the 3′ end of the longer fragment. Four reaction mixtures are then established, each containing the four normal deoxynucleotide precursors for DNA synthesis and one each of a different dideoxynucleotide. DNA is synthesized by linking nucleotides in a 5′ to 3′ direction. The dideoxynucleotide is missing the 3′ linkage site such that it can be incorporated into the growing strand at the 5′ site, but no additional nucleotide can be incorporated after it. Thus when DNA polymerase is added to the reaction mixture, nucleotides are added to the 3′ end of the primer until a dideoxynucleotide is incorporated; subsequently no further elongation of that chain occurs. By controlling the ratio of normal to dideoxynucleotides in the same reaction mixture, a series of strands are generated incorporating the dideoxy precursor at random and generating strands of different lengths. The four reaction mixtures are run in separate lanes of an acrylamide sequencing gel and again separated by size, resolving each one-base difference as for the Maxam-Gilbert method. Again a "ladder" of bands is generated corresponding to the affected base of each reaction mix, but in this case the sequence complementary to the original single strand is determined.

The dideoxy sequencing method has been augmented by the advent of single-stranded cloning using *E. coli* bacteriophage M13 (Fig. 2.8A). This phage packages an infectious single strand of DNA that is replicated to a double-stranded "replicative form" (RF). The RF is approximately 7200 base pairs and can easily be isolated and used as a cloning vector. By orienting the insert in one or the other direction and reinserting the RF into *E. coli*, as the phage replicates and packages its single infectious strands large amounts of single-stranded foreign DNA can be produced. Because the DNA is inserted into a particular known site, one primer complementary to phage DNA immediately upstream from that site can be synthesized and used for sequencing *any* foreign DNA cloned into that site. By saving the time necessary for primer synthesis as well as by allowing isolation of large quantities of defined single-stranded DNA, the combination of M13 cloning and Sanger dideoxy sequencing is powerful indeed. At present, a well trained technician can sequence up to 500 base pairs per day.

Sequencing techniques have illuminated primary gene structure, which in eukaryotes includes the sequences of both coding regions (exons) and intervening sequences (introns), as well as the number and arrangement of these regions (by comparison with final processed mRNA). In addition, these techniques have allowed elucidation of gene-controlling elements, e.g., promoter and enhancer regions, termination signals, and signals for mRNA processing: splice site, "cap" sites, and polyadenylation sites (see next section).

RNA Studies

Gene expression begins with the synthesis of RNA. Studies of eukaryotic gene expression therefore entail elucidation of which RNAs are produced in which cells at which times. Study of RNA in general is facilitated by its smaller size relative to DNA. The first methods for nucleotide mapping and sequencing were developed for RNA precisely because of this property. In addition, study of eukaryotic mRNA is aided by its distinctive structure (Fig. 2.9). Most of these species contain a "cap" at the 5′ end consisting of a 7-methylguanosine residue joined by a triphosphate linkage; this feature is added as transcription proceeds. At the 3′ end, most mRNAs have a run of adenine-containing nucleotides ("poly A tail") that is added at the completion of transcription and not coded for by the primary DNA sequence. The purpose of these features is unknown; presumably they add protection in the cell against degradation by cellular nucleases. The poly A tail allows isolation of the major mRNA fraction of a cell on columns containing complementary thymidine polymers (oligo-dT columns).

On the other hand, isolated RNA is sensitive to degradation by an extremely hardy ribonuclease ("RNase") that resists common sterilization procedures, thereby necessitating compulsively careful handling. Furthermore, single-stranded RNA tends to fold into double-stranded configurations called hairpins (Fig. 2.10) that stabilize the RNA but that confuse, for example, studies of RNA size, as mobility on agarose gels depends on conformation as well as size. Most single-stranded RNAs can form a number of hairpin structures, each of which differs sufficiently from the others to give several bands corresponding to one RNA species on a gel. (Double-stranded DNA molecules, of course, tend to assume one configuration so that each DNA fragment corresponds to one band on a gel.) Thus RNA gels must be run in the presence of strong (and poisonous) denaturing agents such as methyl mercury or formaldehyde, further complicating the manipulation of RNA.

With reference to oncogenes, two RNA methods are of major importance: (1) "Northern blotting" and variant RNA–DNA hybridization techniques; and (2) in vitro translation assays.

Filter-Bound RNA Assays

RNA hybridizes with DNA in solution; however, RNA does not have the same propensity to stick to filters as does single-strand DNA. Alwine and co-workers were the first to report that chemical (e.g., diazobenzylomethyl-) modification of nitrocellulose paper enabled RNA to become covalently bound to it. RNA thus immobilized can be reacted with DNA probes in solution and specific species detected by autoradiography as described above for DNA. Variations on the procedure include: (1) "dot blots" where RNA from particular cells or from in vitro transcription assays is fixed to filters; and (2) a gel separation and transfer procedure similar to that described by Southern for DNA (see above) that was

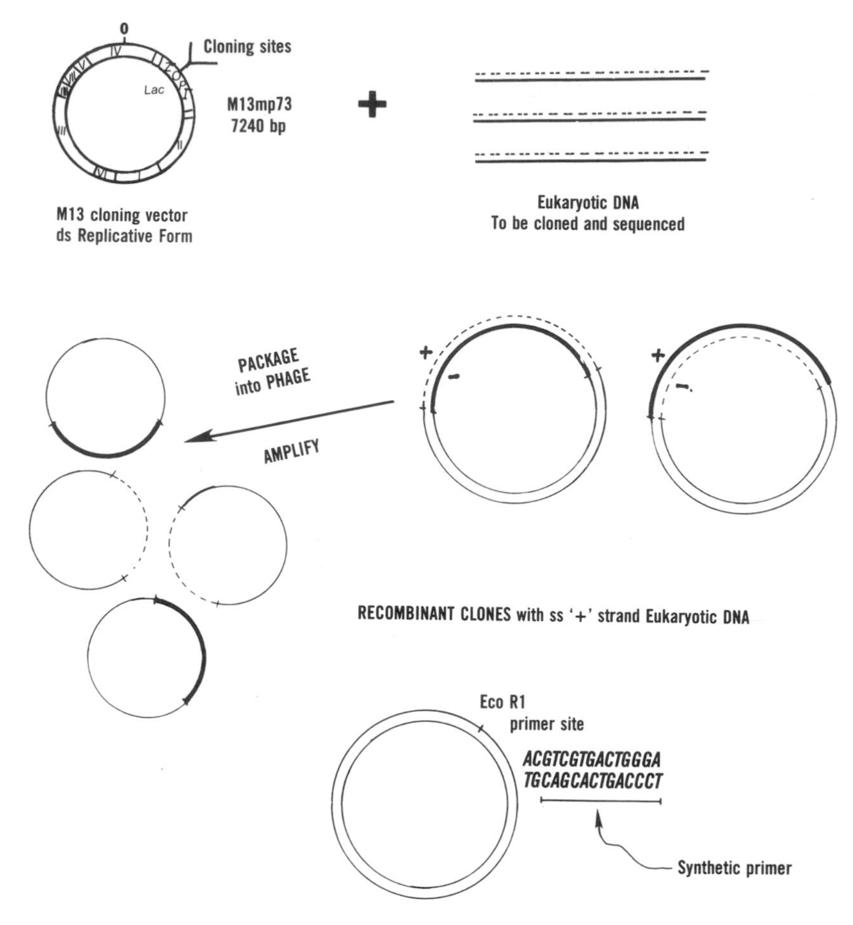

FIGURE 2.8. M13 bacteriophage cloning and dideoxy sequencing. (A) M13, an *E. coli* bacteriophage, has been modified to facilitate cloning and sequencing. One such modified M13 vector, M13mp73, takes advantage of an engineered segment of the *E. coli* lac operon inserted into the phage. This segment contains cloning sites for the restriction enzymes *Eco* R1, *Bam* H1, *Sal* 1, and *Pst* 1 (Z′, O, P, and I′ are genes of the Lac operon). Eukaryotic DNA is cut with one of these enzymes and mixed with the double-stranded replicative form (RF) of the phage. After amplification, single-stranded DNA is packaged into the phage particle; only the '+' strand is packaged. The amplified single strand is then sequenced using the Sanger dideoxy method. (B) Dideoxy sequencing (Sanger method) takes advantage of nucleotides missing both 2′ and 3′ OH groups. A primer is attached to the single strand at the 3′ end of the DNA to be sequenced; using a known cloning site, a single primer can be used to sequence any recombinant DNA (see text). Four reaction mixtures are set up, each containing the same DNA strand and attached primer, DNA polymerase 1, the four normal deoxynucleotides, and one each of the four dideoxynucleotides. The polymerase adds nucleotides to the primer until a dideoxynucleotide is incorporated, after which the chain cannot further elongate. After the reaction, each mixture is run on a separate lane of a polyacrylamide sequencing gel. The sequence of the primer is read directly from the gel. The original sequence is complementary to the primer.

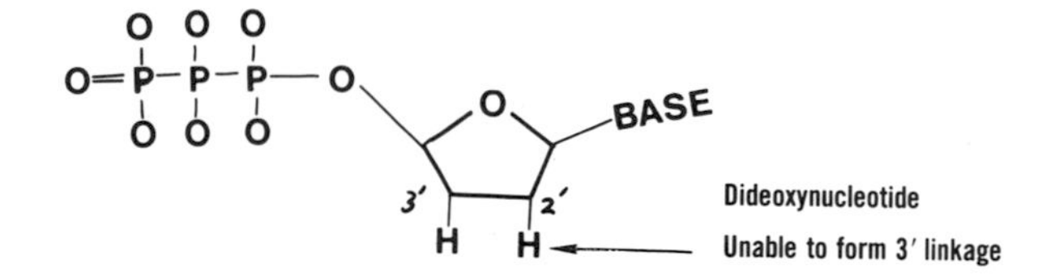

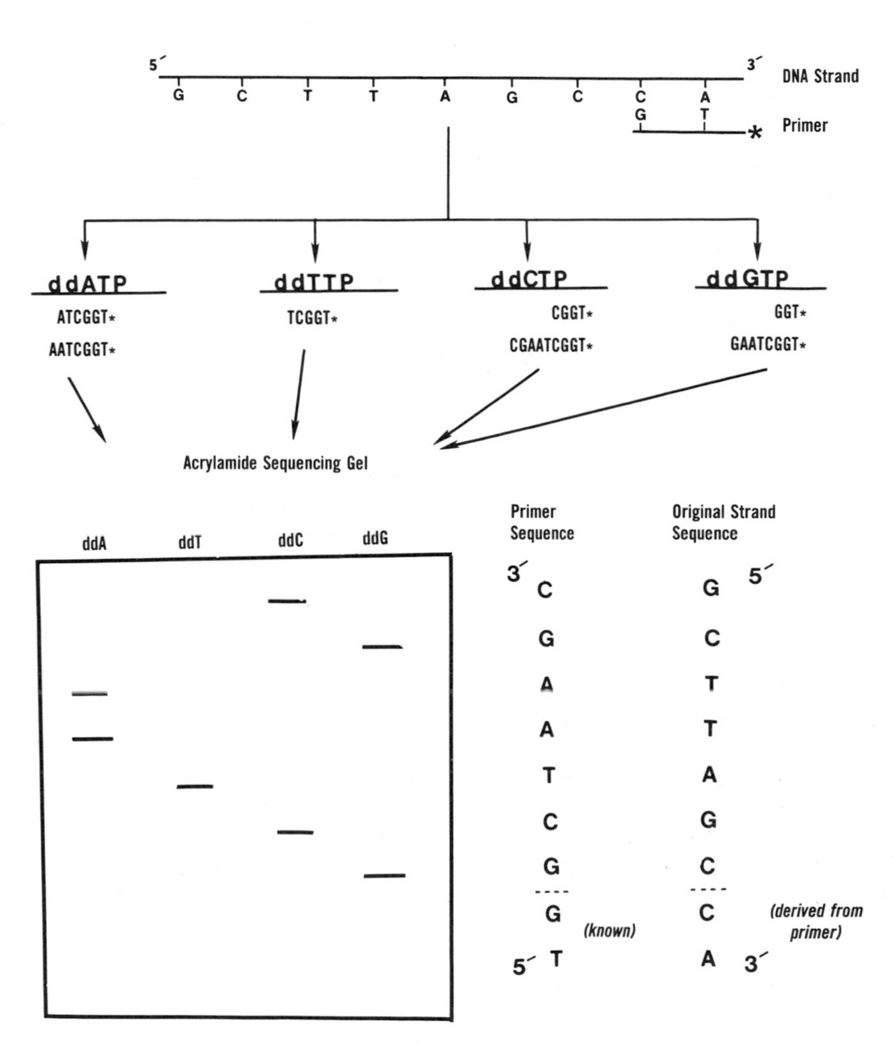

FIGURE 2.8. *Continued*

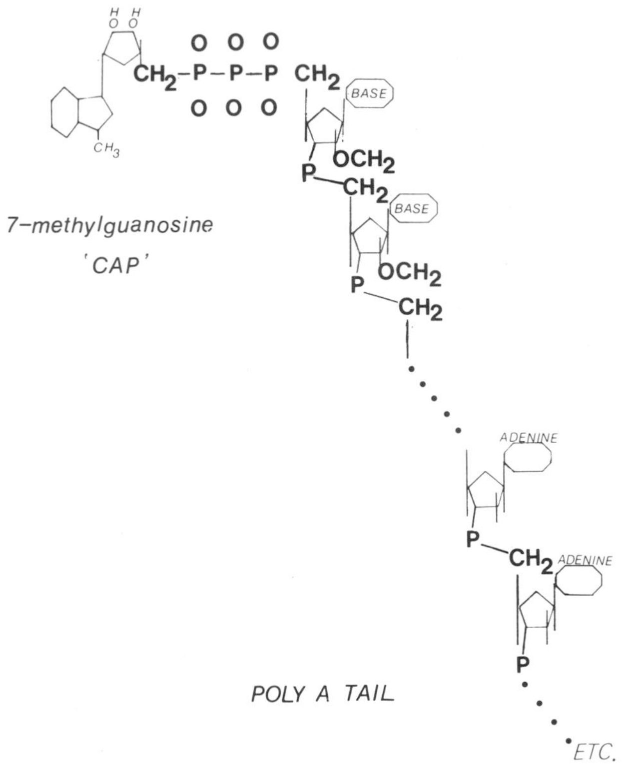

FIGURE 2.9. Eukaryotic mRNA showing a 5′ cap and 3′ poly A residues. This structure characterizes most eukaryotic cellular and viral mRNA.

dubbed "Northern blotting." With cloned DNA probes available, it is clear that these RNA–DNA hybridization procedures allows detection of transcripts corresponding to particular genes. The presence of particular transcripts and a crude estimation of abundance can be derived from the "dot blot" techniques. Comparison of sizes of natural genes, primary transcripts, processed transcripts, and cDNA is possible utilizing the Northern technique.

In Vitro Translation Assays

Analysis of the protein products of isolated mRNA is possible using an in vitro translation system. The protein-synthesizing machinery consisting of ribosomes, tRNAs, amino acids, and essential initiation and elongation cofactors can be isolated from lysates of reticulocytes (the immediate nucleated precursors of red blood cells). When mRNA molecules from virtually any source – gel purified, copied from DNA clones, or isolated from cells – are incubated at body temperature (37°C) with the buffer-stabilized reticulocyte lysates, polypeptides are assembled, programed by the added mRNA. These proteins can be analyzed by

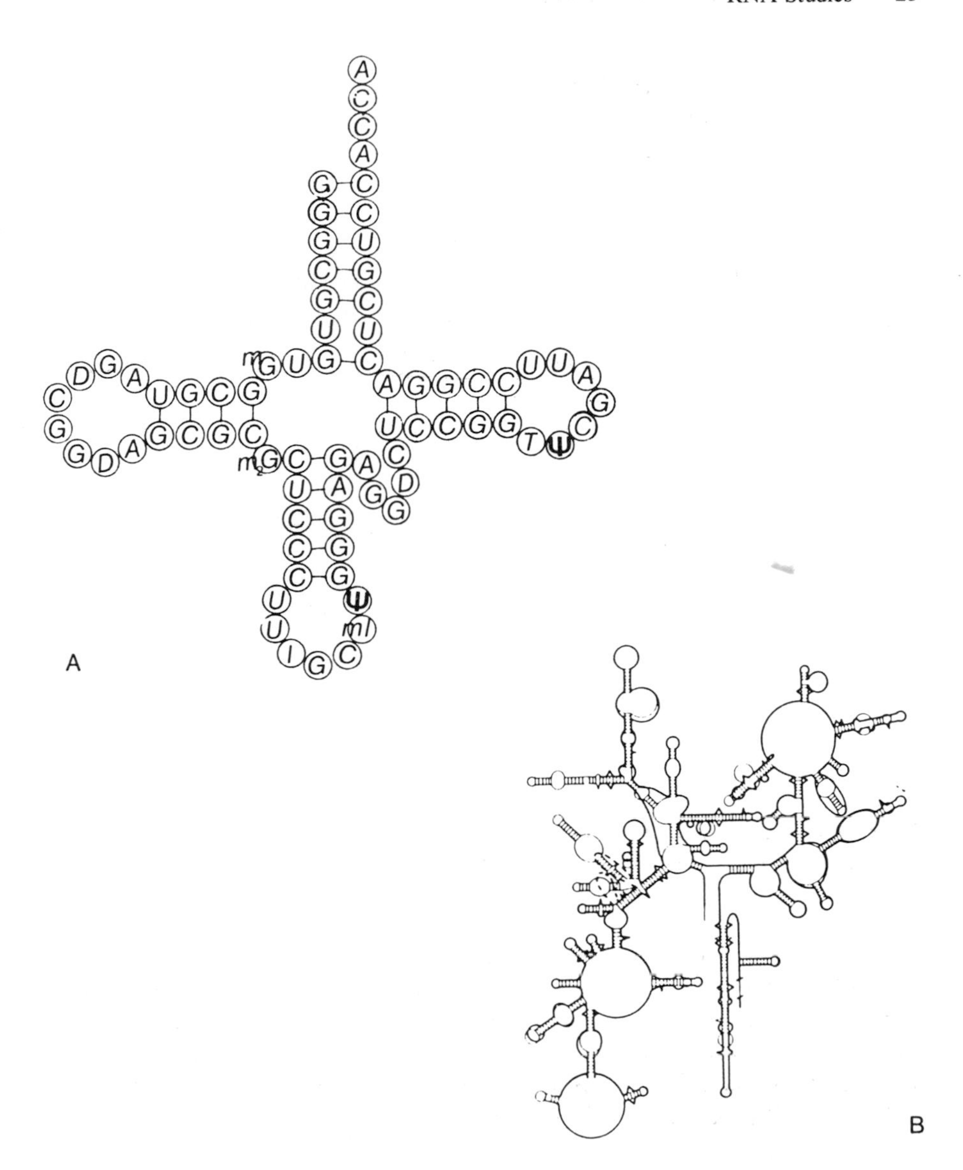

FIGURE 2.10. RNA species illustrating hairpin loops. (A) Yeast phenylalanine tRNA illustrating base-pairing. tRNAs contain both usual and modified nucleotide bases. A = adenine. C = cytosine. D = dihydrouridine. G = guanosine. I = inosine. m = methyl group. ψ = pseudouridine. T = thymidine. U = uracil. (B) 18S rRNA from *Xenopus laevis* illustrating complex hairpin configurations of larger RNA molecules. Messenger RNAs possess similar secondary structure. (Reproduced, with permission, from the Annual Review of Biochemistry, Vol. 53. © 1984 by Annual Reviews Inc.)

a variety of methods, e.g., on two-dimensional protein acrylamide gels or by chromatography. This technique has proved valuable for analysis of oncogene products: The isolated proteins have had molecular weights, isoelectric points (a measure of charge on the molecule), and behavior under various chromatographic conditions determined by this method. In vitro translation can also be used as a method of screening for clones if the final protein product is known: Pure mRNAs are isolated from the clones to be screened, translated in vitro, and the protein products compared. When an in vitro product matches characteristics of the known protein, the gene coding for it can be analyzed in more detail.

Protein Studies

The chemical characterization of proteins is as old as the science of biochemistry itself. In the context of oncogenes, it is necessary to establish what types of product these genes encode and what functions these products perform in normal cells and in their transformed counterparts. Numerous techniques are used to separate, isolate, and purify proteins to better understand their structure and function. These methods rely on size (gel filtration chromatography), charge (ion-exchange chromatography), isoelectric point (isoelectric focusing), hydrophobic nature (hydrophobic chromatography), as well as other physical properties of proteins. In addition, proteins can be purified by affinity for, or binding to, specific antibodies (see below) or various prosthetic groups. The latter method, affinity chromatography, has greatly increased the ability to purify proteins present in tiny quantities in cells or biological fluids.

Polyacrylamide Gel Electrophoresis

Although the above techniques provide a rough idea of protein size and charge characteristics, a more refined analytical technique for this purpose is polyacrylamide gel electrophoresis (PAGE). A polymerized gel of polyacrylamide is used to separate proteins on the basis of their size. Proteins are boiled with sodium dodecyl sulfate (SDS), a detergent, and the denatured proteins are applied to the gel origin. When an electric field is applied to the gel, the proteins migrate through the gel in rough proportion to their molecular size. SDS-PAGE allows determination of relative molecular weights (M_r) by comparison of unknown proteins to known size standards run in parallel on the same gel.

M_r values or molecular weights calculated by amino acid sequence derived from the cloned genes are given in the text in daltons. By combining all the above-mentioned isolation techniques, it is possible to prepare sufficient quantities of proteins to determine their primary amino acid sequence. Automated machines called microsequenators now are able to derive the primary protein sequence from a few micrograms of purified material.

ANTIBODIES

Because direct isolation and characterization of nonabundant proteins (including, certainly, those products of presumed regulatory genes) is exceedingly difficult if not impossible, an understanding of these molecules has been vastly facilitated by the knowledge that virtually all proteins are antigenic, i.e., can elicit an antibody response in an immunized animal. When a vertebrate, i.e., a rabbit, is injected with a foreign protein or antigen,[2] its immune system synthesizes antibodies against the foreign molecule. *Antibodies* are complex proteins that are "complementary" to the antigen in the sense that a lock is complementary to a key. The antibody is specific for the given antigen and binds only with it or closely related molecules. Thus isolation of the immunoglobulin (antibody-containing) fraction from the rabbit serum provides a reagent capable of reacting with the protein used to elicit the immune response. As can be imagined, however, rabbit serum contains many more antibodies than the particular one of interest, and these molecules may react nonspecifically in an assay to give spurious results. This background can be differentiated from the specific response by comparing immune and nonimmune serum from similar individuals (or from the same individual before and after immunization) (Fig. 2.11). A protein is a complex antigen and elicits the formation of several antibodies to various parts of itself (Fig. 2.12); these different antibody-generating regions on the same protein are called *epitopes*. When an immune serum contains more than one variety of antibody, it is called *polyclonal serum*.

It has long been recognized that individual antibody molecules are specific for one and only one epitope. A technique to generate individual monoclonal antibodies was developed by Koehler and Milstein in Cambridge, England, during the mid-1970s.

Natural antibody molecules are produced by plasma cells, terminal derivatives of B lymphocytes, that have been activated by exposure to particular antigens. Any one B cell and its progeny plasma cells can produce only one type of antibody; the plasma cell is a terminally differentiated cell that, in fact, contains a permanent rearrangement of its cellular DNA entailing actual loss of some DNA sequences not essential for production of the particular antibody.

If individual antigen-specific plasma cells could be isolated and grown, there would be an unlimited supply of any given antibody. Unfortunately, not only plasma cells but the B cells from which they are derived are like other terminally differentiated cells; they have a finite life-span and so can rarely undergo a sufficient number of divisions to generate the quantity of cells required to produce reagent quantities of antibody. Some plasma cells, however, are malignant: They

---

[2]An *antigen* is any substance that can elicit an immune response. Antigens include nucleic acids (nucleotide polymers) and polysaccharides (carbohydrate polymers) as well as polypeptides or proteins (amino acid polymers).

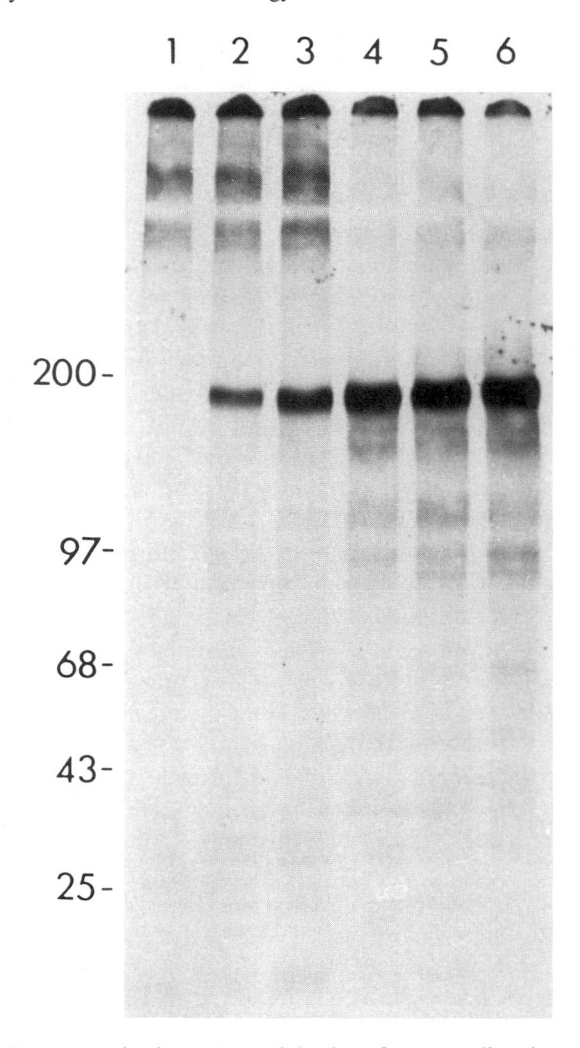

FIGURE 2.11. Representative immunoprecipitation of cancer cell antigens. In this experiment breast cancer cells were grown in tritiated galactose and glucosamine to label the sugars. Cells were lysed with detergent and immunoprecipitated with five monoclonal antibodies (columns 2–6) that recognize a 200-kilodalton cell surface glycoprotein. Column 1 is a nonbinding monoclonal antibody. The immunoprecipitate is boiled in SDS and applied to a polyacrylamide gel for electrophoresis. Labeled molecules are separated by molecular weight. The molecular weights of marker proteins are shown on the left. (Kindly provided by Dr. David Ring)

cause a cancer known as multiple myeloma. Myeloma cells are immortal and can be established in cell culture.

Koehler and Milstein found that fusion of an immortal myeloma cell with a mortal B cell resulted in production by the "hybridoma" cell of both myeloma and

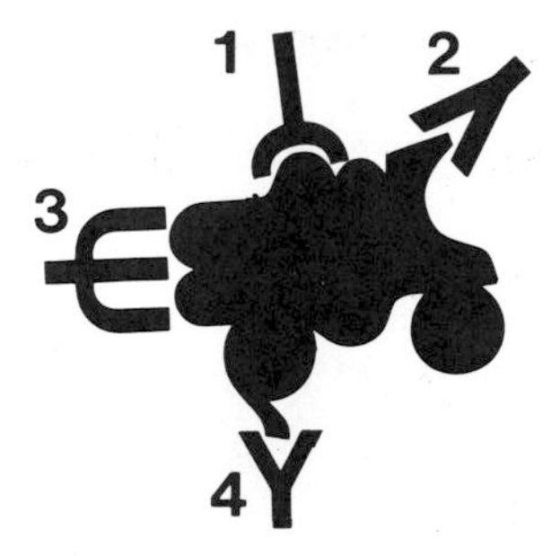

FIGURE 2.12. Protein with multiple antigenic determinants. A given protein may have several antigenic sites called epitopes. Each epitope has the capacity to elicit production of a unique antibody. The protein illustrated has four epitopes. Antibodies to these epitopes also are illustrated. Each antibody is produced by cells clonally derived from a single B cell. A mixture of these antibodies in serum constitutes a polyclonal antiserum. A single, monoclonal antibody can be derived by hybridoma techniques.

B cell antibodies. Subsequent work led to development of myeloma cell lines that no longer produce their own protein but that give immortality to antibody-generating B cells when fused to them. Thus a technique was developed to allow production of pure monoclonal antibodies to one epitope of any desired protein (Fig. 2.13): Mice are immunized with the protein of interest; they are later sacrificed, and lymphocytes are isolated from their spleens. Spleen lymphocytes and myeloma cells are fused using a polymer such as polyethylene glycol or a virus such as Sendai that facilitates fusing of the cell membranes. Antibody-producing hybrids then are selected using, for example, a radioimmunoassay (RIA), an assay using radioactive heterologous antibodies, e.g., goat anti-mouse antibodies[3] (Fig. 2.13).

More recently a colorimetric reaction has been developed: the enzyme-linked immunosorbent assay (ELISA) (Fig. 2.14). Here an enzyme such as peroxidase is attached to the heterologous "anti-antibody." The hybridoma antibodies are reacted against the antigen of interest, the mixture is washed to diminish nonspecific binding, and the enzyme-linked antibody is added (this molecule reacts with the specifically bound hybridoma antibodies). The assay plate is washed again to deplete nonspecific binding, and the enzyme substrate is added. In cases where the original hybridoma produces the antibody of interest, the antibody binds to its complementary epitope. The enzyme-linked heterologous antibody then binds to the antibody of interest. When the enzyme substrate is added, the enzyme reacts and changes the color of the solution. Alternatively, a fluorescent-dye-labeled antibody can be used that fluoresces under UV or laser light.

[3]Mouse immunoglobulins injected into goats elicit a polyclonal antibody response. The goat serum can be used to identify mouse monoclonal antibodies via an "anti-antibody" reaction.

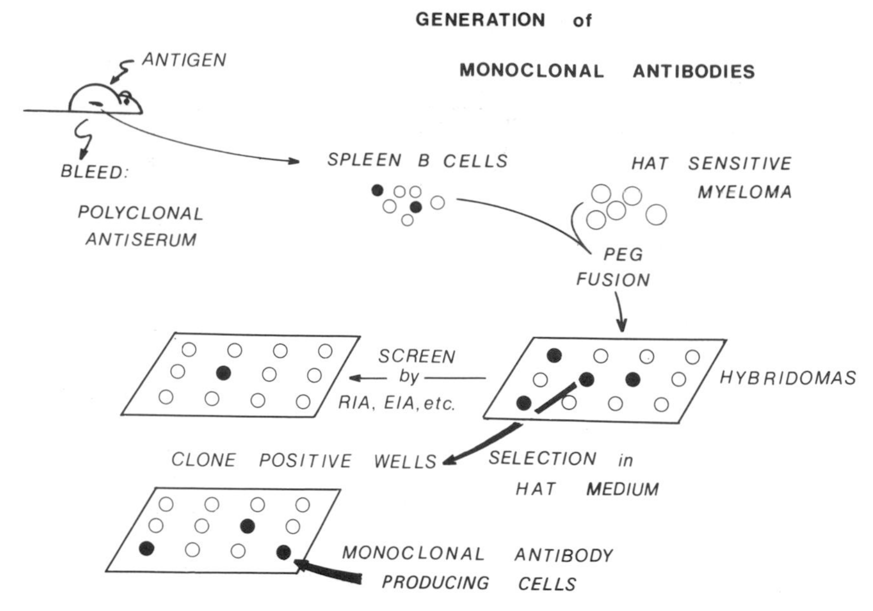

FIGURE 2.13. Generation of monoclonal antibodies. When animals are injected with an antigen, they produce a polyclonal response, i.e., many B cells are activated to make a variety of antibodies. Normal B cells have a definite life-span and are unable to grow indefinitely in culture. Myeloma cells are malignant B cells that can remain alive indefinitely in culture. Mutant myeloma lines lacking the enzyme hypoxanthine-guanine phosphoribosyl transferase (HGPRT) are unable to grow in a selective medium containing hypoxanthine, aminopterin, and thymidine (HAT). When spleen B cells from an immunized animal are fused with mutant myeloma cells using, for example, polyethylene glycol (PEG), only hybrid cells containing normal HGPRT genes (from the spleen B cells) and the immortalized phenotype (from the myeloma cell) can survive. Each hybridoma cell secretes a different (monoclonal) antibody. Hybridoma cells producing an antibody of interest are selected; e.g., by RIA or ELISA. Positive cells are probed by single-cell cloning, then expanded using large-scale cultures to produce reagent quantities of the monoclonal antibody.

Once a hybridoma cell producing an antibody of interest is isolated, the cell line derived from it can be grown indefinitely in culture, providing a hardy, reliable, and indefinite source of that antibody. Monoclonal antibodies can be routinely prepared against practically any antigen, limited only by the availability of selection procedures (i.e., if an antigen cannot be stuck to a plastic dish or used in the ELISA assay or immobilized by some other method, it would be difficult to test for an antibody against this antigen). Human monoclonal antibodies, derived from human B lymphocytes, have much potential use in the diagnosis and therapy of a variety of diseases including cancer (see Chapters 13 and 14) but have proved much more difficult to develop. (See review by Larrick and Bourla; the details are beyond the scope of this chapter.)

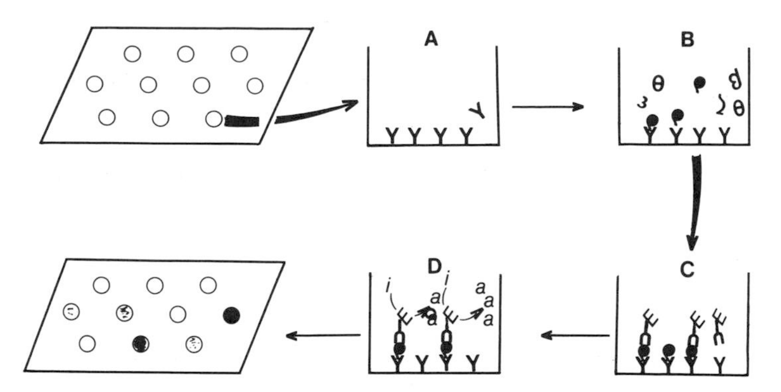

FIGURE 2.14. Principle of sandwich ELISA or enzyme immunoassay (EIA). Plastic microtiter plates (96 wells) or small beads are (A) coated with the "capture" antibody that recognizes one site on the antigen being assayed. The biological fluid (B) containing the antigen is added and the excess washed away. Next, the developing antibody (C) with an attached enzyme is added. Excess reagent is washed away. Finally, (D) the inactive (i) substrate for the enzyme is added and active (a) colored substrate is produced in proportion to the quantity of antigen captured by the first antibody. The plate is read visually or quantitatively in a microspectrophotometer plate reader.

Molecular techniques have progressed to the point where it is no longer necessary to isolate a protein in order to generate an antibody. From the DNA sequence of a gene, it is possible to predict the amino acid sequence of its protein product. It is now possible to synthesize an oligopeptide corresponding to the predicted amino acid sequence by techniques developed by Mike Hunkapiller and Leroy Hood at the California Institute of Technology. The oligopeptide conjugated to a larger protein carrier molecule is used as an immunogen and as an antigen for antibody screening.

IMMUNOASSAYS

Once polyclonal or monoclonal antibodies to particular proteins are available, it is possible to use them in a number of procedures to determine the presence and localization of those (and related) protein products within eukaryotic cells. Using immunoprecipitation, cellular proteins are labeled with a radioactive isotope, then isolated and reacted with an antibody attached to an insoluble matrix such as *Staphylococcus* protein A; this mixture is washed to eliminate nonspecific binding and then electrophoresed on polyacrylamide gel. Once the gel is stained, the immunoglobulin chains of the antibody are visible; usually the precipitated protein is not visible because as a nonabundant gene product it may be present in only minute quantities per cell. However, using x-ray film, an autoradiograph reveals a band corresponding to the immunoprecipitated protein. By this technique, not only the existence but the molecular weight and a rough estimate of the quantity of a nonabundant cellular protein can be established.

Cells and tissues can be isolated and fixed on microscope slides. Antibodies then can be used to localize proteins inside or outside cells, on the cell membrane, or in the nucleus or cytoplasm; in some cases it is even possible to achieve a detailed subcellular localization using an immunoperoxidase technique. The attached antibody is reacted with an anti-antibody linked to an enzyme (peroxidasc) as for thc ELISA assay; altcrnatively, the attached antibody can be localized by the immunofluorescence technique where the anti-antibody is linked to a fluorescent probe (Fig. 2.15). Using fluorescent probes, it is possible to sort cells of one population from those of another by flow cytometry using lasers to excite the fluorescence.

Immunoblots (Western Blots)

The immunoblot (Western blot) procedure is similar in concept to Southern (DNA) and Northern (RNA) blotting. Protein mixtures are separated by SDS-PAGE or isoelectric focusing and transferred to a filter support matrix, e.g., nitrocellulose, in a manner similar to that used for DNA and RNA (Fig. 2.6). Specific proteins are detected by poly- or monoclonal antibodies, which in turn are detected by anti-antibodies labeled radioactively, enzymatically, or by avidin–biotin complexes. The latter reaction allows visualization of the protein corresponding to the original antibody.

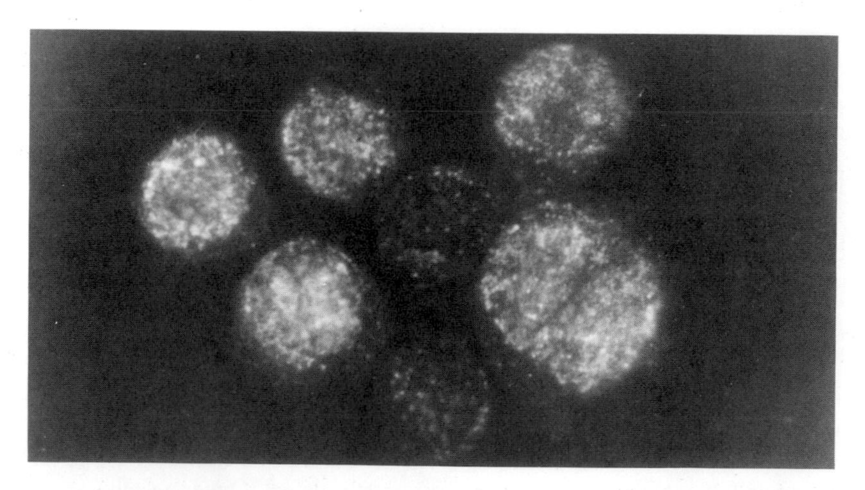

Figure 2.15. Localization of antigens by immunofluorescence. Antibodies are labeled with dyes such as fluorescein isothiocyanate that absorb light at one wavelength and emit light (as fluorescence) at another. These antibodies bind to their particular antigens when reacted with cells or tissues. The location of the antibodies (and thus the location of the antigen of interest) can be observed directly by viewing the cells or tissue under a special fluorescence-equipped microscope. The figure depicts lymphoblastoid cells reacted with an anti-Epstein-Barr virus antibody against nuclear antigens. Because the antigen is localized to the nucleus, only nuclei are visualized.

Cell Assays

Cultured Cells and Phenotypes

Attempts to maintain eukaryotic cells in culture have been made since early in this century. Two categories of culture can be distinguished: (1) *organ culture* in which whole organs or sections thereof are maintained apart from intact organisms in an artificial environment; these tissues remain close to their original form (i.e., the microscopic configuration of the cells themselves and their relation to each other are nearly identical to the original organ in the living animal); and (2) *cell culture* in which cells are isolated from an organ or tissue and dispersed to individual units, then maintained in this manner as separate cells without regard for their original microscopic anatomy or their configuration with respect to one another.

Organs generally contain more than one cell type and so do organ cultures. These cells tend to retain their original functional capabilities and interactions with one another; they do not differ substantially in their genetic composition from cells in the original organism. Dispersed cells of a cell culture tend to be more homogeneous; i.e., the culture generally contains only one cell type.

Cells obtained directly from the animal and cultured for a few cell divisions or generations are termed primary cultures. Such cells resemble the cell of origin anatomically and functionally but generally differ in some particulars, allowing them to live independently of the other cell types normally contained in an organ.

It is difficult to maintain differentiated eukaryotic cells in culture. Typically, they live, grow, and divide for several generations, but then they die. Even if provided with fresh nutrients, growth factors, or serum, most die. Rarely, however, a culture does not die, and the cells continue to divide generation after generation. Such immortal cell cultures are referred to as *cell lines*. Some change or changes occur in the cells to allow them to grow under less stringent conditions; for example, they may not require a certain growth factor, or they may no longer respond to signals of contact inhibition. They may become "dedifferentiated" and continue to grow and divide as more primitive cells do.

Nontransformed, primary culture cells look and act differently from their transformed counterparts. Phenotypically, the primary cells appear "normal" (Fig. 2.16A). They grow to a particular density and then stop; they do not resume growth until diluted. Fibroblasts, cells that are relatively easy to establish in culture, grow in a monolayer and stop growing when the cell edges reach each other; this process is termed *contact inhibition*. Single nontransformed cells rarely if ever are able to establish a colony in vitro. Such cells tend to retain their normal, diploid chromosome complement; they are homogeneous and do not change significantly until they die or undergo transformation.

Transformed cells, on the other hand, are frequently phenotypically abnormal (Fig. 2.16B). They do not necessarily resemble the cells from which they originally were derived, and they may be heterogeneous among themselves (*pleomorphism*). They grow in a nonordered way and tend to pile up; transformed fibroblasts appear to have lost their contact inhibition. Single transformed cells

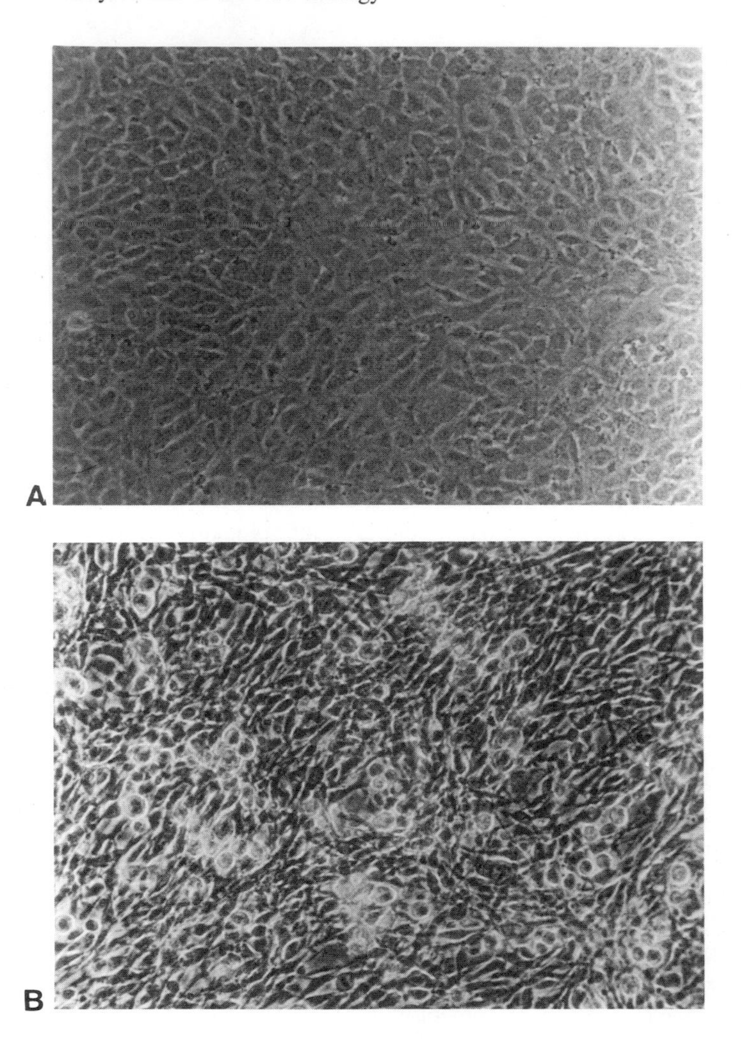

FIGURE 2.16. Normal and transformed cultured cells. (A) Normal cultured fibroblasts. (B) Transformed fibroblasts showing heaped-up, irregular cells.

can be cloned, e.g., in soft agar, where they grow and divide in isolation to form a discrete colony. Transformed cells may show altered chemical properties, e.g., the ability to be agglutinated by plant lectins. Transformed cells tend to be aneuploid; i.e., they may contain one copy of some chromosome(s) and three or more copies of another or others. Chromosomes can be rearranged or altered. No particular pattern appears to apply for all transformed cells; however, each transformed cell line tends to have a consistent chromosome complement. Chromosomes, biochemical characteristics, and phenotype of a transformed culture may evolve over time.

NIH 3T3 Cells and Transformation

Scientists working with early cell cultures were interested in studying the difference between normal cells and malignant cells; this interest, in fact, provided the incentive for much of the original cell culture work. To study neoplasia in cell culture, a major criterion was that transformed cells be able to induce tumors when injected into experimental animals. In fact most, though by no means all, cultured cells that fit the transformed phenotype are able to induce tumors when injected into experimental animals (see below). It soon became apparent that a major difficulty was the maintenance of normal cells in culture for sufficiently long periods so that they could be compared with their abnormal counterparts. Practically all cells became abnormal and transformed if retained in culture long enough.

One exception is a mouse fibroblast cell line, called NIH 3T3, that was developed by Green and co-workers during the early 1960s. Although not entirely "normal," this line nevertheless retained major normal characteristics: The cells retained their property of contact inhibition and grew and divided only to a monolayer. In addition, these cells were unable to induce tumors when injected into susceptible animals. It was recognized that 3T3 cells were not normal because they were immortal and capable of indefinite maintenance in culture; nevertheless, they became a standard of "normal" by which to compare transformed cells. NIH 3T3 cells are themselves quite easily transformed to cells that have lost their original morphology and contact inhibition properties and that are capable of forming tumors when injected into animals. Thus these cells are a sensitive assay for carcinogenesis: radiation, chemicals, and viruses are capable of inducing the transformed phenotype. Subsequently it was found that isolated DNA from cancer cells, when taken up by 3T3 cells, could result in their transformation. This observation forms the basis for the NIH 3T3 transformation "transfection" assay.

Transfection

The technique of introducing naked DNA into intact cells is called *transfection* (Fig. 2.17). DNA is mixed in a buffer containing calcium phosphate and applied to the cells. This solution is then removed and the cells placed in fresh medium where they can grow and divide. The exact mechanism by which cells take up the DNA is unknown; however, the transfection technique has wide applicability for prokaryotes as well as for eukaryotes. Transfection is used for the introduction of plasmid clones into bacteria and the introduction of cloned viral or cellular DNA into other eukaryotic cells. Examples of successful transfection include transformation as discussed above, as well as: (1) uptake of antibiotic-resistant plasmids permitting the growth of bacteria in the presence of antibiotics; (2) production of intact virus particles in transfected cells (viral genome expression after uptake); and (3) growth of eukaryotic cells in medium lethal to nontransfected cells (e.g., a "selection" medium such as hypoxanthine/aminopterin/thymidine medium inhibits cells without a functional gene for thymidine kinase).

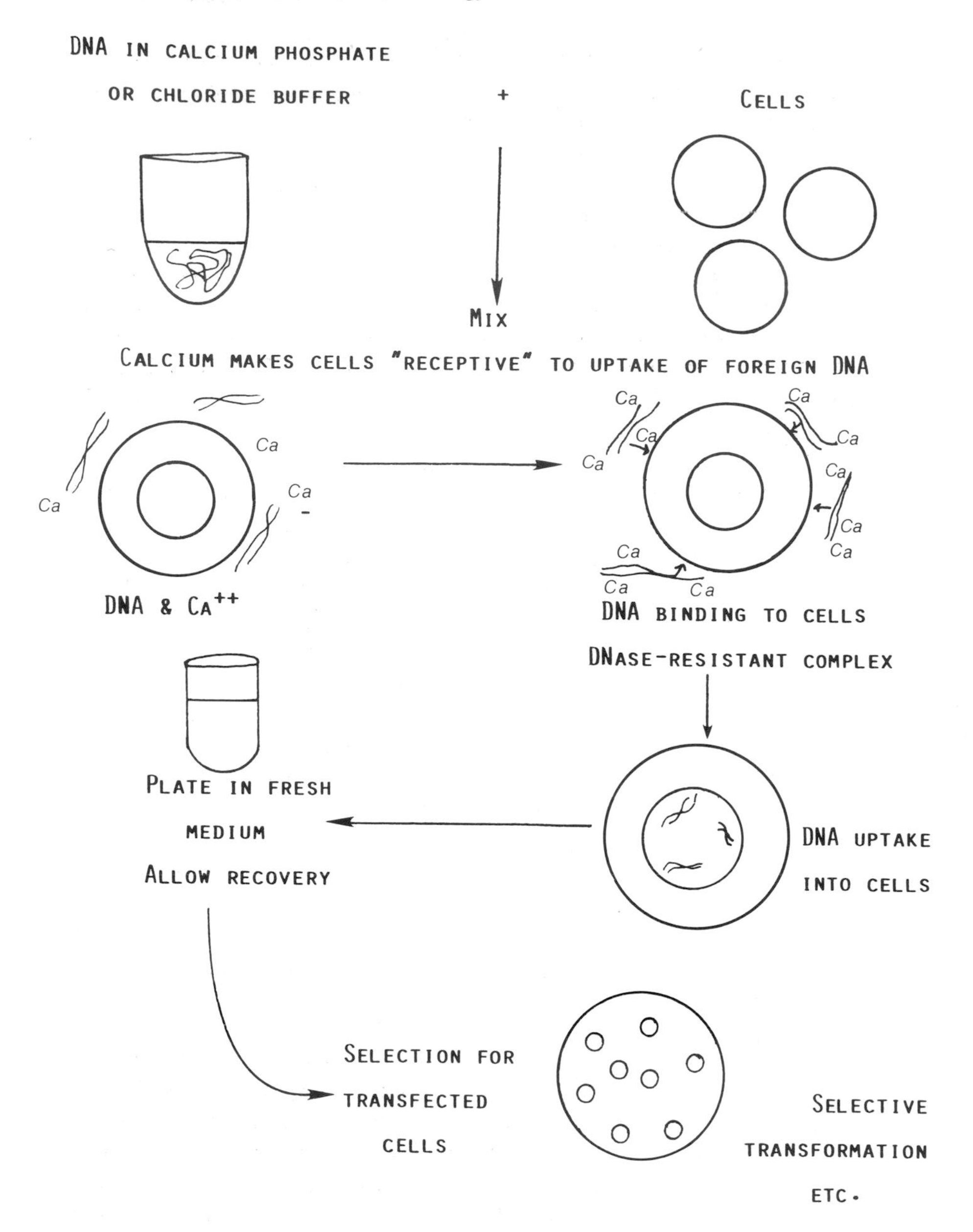

FIGURE 2.17. Calcium-mediated transfection. DNA in a calcium phosphate or chloride buffer is mixed with log phase (exponentially growing) cells. Cells are centrifuged and resuspended in cold medium (0–4°C). DNA in conjunction with calcium ions somehow forms complexes that bind to the cell membrane. DNA is taken up by the cell and arrives in the nucleus by an unknown mechanism. Cells then are resuspended in fresh medium and allowed to recover from exposure to the toxic calcium. Finally, recovered cells are plated under selective conditions to allow detection of successful transfectants.

The NIH 3T3 transfection assay has proved invaluable for isolation and characterization of both retroviral oncogenes and their cellular proto-oncogene counterparts, as is discussed in upcoming chapters.

Assays Using Intact Organisms

The ultimate test of whether a *presumed* carcinogen – a chemical, virus, gene – is in fact a *true* carcinogen is if it can produce tumors in animals. Early experiments attempted to determine if tumors were "transplantable," e.g., if human tumor tissue could induce tumors in mice or guinea pigs. Later the test to differentiate transformed from nontransformed cells in culture was induction of tumors in animals by a small inoculum (10^6 or fewer) of transformed cells when a similar number of nontransformed cells failed to do so.

In these early experiments tumor induction in animals was frequently thwarted by the host animal's immune system. In particular, if tumor cells from a different species or even strain of animal were injected into a host animal, they might fail to produce a tumor, not because they were incapable of doing so intrinsically but because they were recognized as foreign and were destroyed by the host's immune system. The discovery of "nude mice" (Fig. 2.18), which lack functional cellular immunity (because they lack a thymus, the organ responsible for maturation of T lymphocytes), has greatly facilitated animal assays of tumorigenicity. Using nude mice, a wide variety of entities including tumor cells, cultured cell lines, and cells treated with radiation, chemicals, or viruses can be tested for their ability to produce tumors in vivo without the complication of tumor rejection by the animal's immune system.

Animal assays are sensitive not only to the presence or absence of tumorigenicity but to the degree of malignancy. Tumors that grow rapidly, spread, and kill the host soon after inoculation are highly malignant. Less malignant cells grow slowly to a small tumor size, may not spread, and may or may not kill the host eventually. Cultured cells transformed by carcinogenic agents similarly have been observed to differ in their degrees of carcinogenicity.

The transformed morphology of tissue culture cells discussed above does not necessarily correlate with carcinogenicity in animal assays, nor do other indicators of transformation in culture such as aneuploidy (abnormal number of chromosomes) or the ability to be agglutinated by plant lectins. For example, "revertants" of transformed and carcinogenic cells frequently are both aneuploid and have a "transformed" morphology. Nevertheless, they are incapable of tumor induction in whole organisms, which is why they are termed *revertants*.

The most advanced assay for oncogenes currently available involves introducing these genes into the germ lines of mice or other species (see Chapter 12). These so-called *transgenic* animals are observed for the development of malignancy.

The study of oncogenes and carcinogenicity utilizes all the techniques described here. From manipulation of DNA molecules to detection of tumors

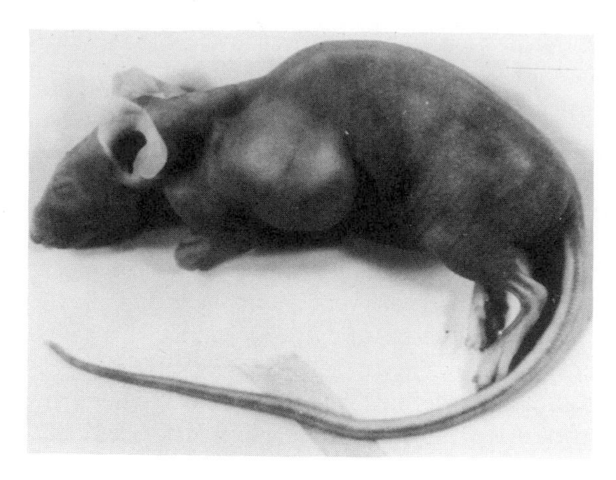

FIGURE 2.18. The immunodeficient hairless "nude" mouse can be used to grow xenografts of human tumors.

in animals, the search for the molecular mechanism of cancer continues. It is currently thought that oncogenes and their normal cellular counterparts provide an important clue in this regard.

Bibliography

Alwine JC, Kemp D, Stark G. Method for detections of specific RNAs in agarose gels by transfer to diazobenzyloxymethyl-paper and hybridization with DNA probes. Proc Natl Acad Sci USA 24:5350, 1977.

Crowe R, Ozer H, Rifkin D. Experiments with Normal and Transformed Cells. Cold Spring Harbor, NY: Cold Spring Harbor Laboratory, 1978.

Davis LG, Dibner MD, Batley JF. Basic Methods in Molecular Biology. New York: Elsevier, 1986.

Kohler G, Milstein C. Continuous cultures of fused cells secreting antibody of predefined specificity. Nature 256:495–497, 1975.

Krontiris TG, Cooper GM. Transforming activity of human tumor DNAs. Proc Natl Acad Sci USA 78:1181, 1981.

Larrick JW, Bourla JM. Prospects for the therapeutic use of human monoclonal antibodies. J Biol Response Modifiers 5:379, 1986.

Maniatis T, Fritsch E, Sambrook J. Molecular Cloning: A Laboratory Manual. Cold Spring Harbor, NY: Cold Spring Harbor Laboratory, 1982.

Milstein C. From antibody structure to immunological diversification of immune response. Science 231:1261, 1986.

Noller HF. Structure of ribosomal RNA. Annu Rev Biochem 53:119, 1984.

Pellicer A, Wigler M, Axel R, et al. The transfer and stable integration of the HSV thymidine kinase gene into mouse cells. Cell 14:133, 1978.

Rodriquez RL, Tait RE. Recombinant DNA Techniques: An Introduction. Reading, MA: Addison-Wesley, 1983.

Shih C, Padhy LC, Murray M, et al. Transforming genes of carcinomas and neuroblastomas introduced into mouse fibroblasts. Nature 290:261, 1981.

Shih C, Shilo B-Z, Goldfarb MP, et al. Passage of phenotypes of chemically transformed cells via transfection of DNA and chromatin. Proc Natl Acad Sci USA 76:5714, 1979.

Southern EM. Detection of specific sequences among DNA fragments separated by gel electrophoresis. J Mol Biol 98:503, 1975.

Watson JD, Crick FHC. Molecular Structure of Nucleic Acids: A structure for deoxyribose nucleic acid. Nature 171:737–738, 1953.

Watson JD, Tooze J, Kurtz DT. Recombinant DNA: A Short Course. Scientific American Books (distributed by W.H. Freeman, New York), 1983.

Wigler M, Pellicer A, Silverstein S, et al. Biochemical transfer of single-copy eukaryotic genes using total cellular DNA as donor. Cell 14:725, 1978.

3
Viruses and Oncogenes

Overview

Five groups of double-stranded DNA viruses and diploid RNA viruses of the retroviral group are associated epidemiologically, experimentally, or both with animal and human neoplasia. DNA tumor viruses associated with cancer on epidemiological grounds include papilloma viruses and urogenital cancer, herpes viruses and Burkitt's lymphoma and nasopharyngeal carcinoma, and hepatitis viruses and liver cancer. No biochemically defined oncogenes have yet been described for any of these viruses; however, preliminary evidence suggests the existence of transforming sequences in both papilloma and herpes viruses.

Although studied in great molecular detail with elucidation of clearly defined oncogenes, polyoma viruses and adenoviruses have not been associated with naturally occurring tumors in vivo. Their oncogenes are structurally dissimilar to retroviral oncogenes, yet these two gene classes appear to subserve similar functions.

The RNA tumor viruses were the first infectious agents unequivocally associated with neoplasia. A large number of such viruses have been isolated from chickens, rodents, cats, and monkeys. Many transduce one or two oncogene sequences necessary and sufficient for efficient cellular transformation. These oncogenes were derived originally from host sequences. Virally transduced oncogenes comprise some 20 independent species falling into a smaller number (some five to seven) of functional categories. The retroviral life cycle, employing a double-stranded DNA intermediate (the provirus) that integrates into host DNA, suggests a mechanism where by the virus could acquire cellular sequences.

Both RNA and DNA viruses are associated with cancer. There are five major categories of DNA tumor viruses, each of which has a double-stranded DNA genome. No known single-stranded DNA virus is associated with oncogenesis. Of the RNA viruses, only one group is associated with cancer: the single-stranded but diploid oncorna viruses.

DNA Tumor Viruses

Members from each of five double-stranded DNA virus families are associated with neoplasia (Table 3.1). They may be divided into two groups: (1) those associated on epidemiological grounds with certain human neoplasias; and (2) those that induce transformation of cultured cells in vitro and can cause tumors experimentally in animals but that appear to play no significant role in in vivo oncogenesis. As luck would have it, the viruses that have been most thoroughly characterized at the molecular and biochemical level are in the latter category because these viruses lend themselves to cell culture and in vitro model systems. The more intriguing viruses linked to in vivo tumorigenicity in humans unfortunately are larger and more complex, or they have no good model systems for their detailed biochemical study. In several cases it has not yet proved possible to propagate these viruses in vitro.

Viruses that are epidemiologically linked with human neoplasia are members of the papilloma virus family, the herpes virus family, and the hepatitis B virus. Each type has a unique life cycle in humans, and as yet no definite viral oncogenes have been identified for any of them.

Polyoma viruses and adenoviruses have been studied extensively and are rather similar in their life styles. They express genes with two basic functions: (1) early functions needed for viral DNA replication and RNA expression, and (2) late functions essential for production of viral capsid proteins and for packaging reactions to produce infective progeny virus. Each virus produces one or more well characterized proteins essential for transformation. These proteins are the products of viral oncogenes that are expressed among the early functions. These viruses, however, are not linked with any known human tumors.

Polyoma viruses, adenoviruses, and herpes viruses share a productive life cycle (in cells permissive for viral replication) that leads to the production of progeny virions and death of the host cell. Under these circumstances the *entire* viral genome is expressed, including the late functions coding for viral capside proteins and orchestrating packaging of viral DNA into capsids to produce infective progeny. Each virus also has a nonproductive life cycle in semi- or nonpermissive cells. In this case, only a *subset* of viral genes is expressed, including the region shown by deletion mapping to be essential for transformation. In contrast to cells that are productively infected, a small percentage of semiproductively infected cells do not die but become transformed to a neoplastic phenotype. In the case of adenoviruses and polyoma viruses, stable integration of portions of virus DNA into the host genome can be observed in the transformed cells.

Papilloma Viruses

Over 40 human papilloma viruses (HPVs) have been isolated to date, most of which have been discovered since 1978. Such viruses are also found naturally in cattle, rabbits, sheep, deer, and other species. Papilloma viruses are known to

TABLE 3.1. DNA tumor viruses.

Papilloma viruses
  Multiple species including human, bovine, rabbit (Shope), equine, canine, elk, deer, sheep

Herpes viruses
  Alpha subfamily: herpes simplex 1 and 2; equine herpes virus type 1
  Beta subfamily: cytomegalovirus (human)
  Gamma subfamily: Epstein-Barr virus (human); Marek's virus (avian); *Herpesvirus saimiri* (squirrel monkey); *Herpesvirus ateles* (spider monkey)
  Other: Lücké herpes virus (frog); *Herpesvirus sylvilagus* (rabbit)

Hepatitis B family
  Hepatitis B virus (human)
  Woodchuck (*Marmota monax*) hepatitis virus
  Ground squirrel (*Spermophilus beecheyi*) hepatitis virus
  Duck hepatitis virus (domestic duck, Pekin duck)

Polyoma virus
  Polyoma virus (murine)
  Simian virus 40 (monkey)
  BK virus (human)
  JC virus (human)

Adenovirus
  Human (at least 37 types), groups A, B, C, D, and E
  Simian, bovine, murine, canine, avian, porcine, ovine, equine, tree shrew

cause warts or papillomas, a benign epithelial neoplasm. In addition, nonhuman papilloma viruses have been causally linked to fibropapillomas (fibrous elements in epithelial tumors) and fibromas (pure fibroblastic tumors). Human papilloma viruses are associated with plantar warts, flat warts, and genital warts (condyloma acuminata) (Table 3.2). In addition, they are linked with a disease called epidermodysplasia verruciformis, a rare lifelong disease characterized by dis-

TABLE 3.2. Diseases associated with human papilloma viruses.

Plantar warts (verruca vulgaris): HPV-1, HPV-4
Hand warts (verruca vulgaris): HPV-2, HPV-4, HPV-7
Flat warts (verruca plana): HPV-3, HPV-5, HPV-9, HPV-10
Genital warts (condyloma acuminata): HPV-6, HPV-42
Epidermodysplasia verruciformis: HPV-3, HPV-9, HPV-12, HPV-14, HPV-15
Epidermodysplasia verruciformis with malignant conversion: HPV-5, HPV-8, HPV-10
Cervical cancer: HPV-6, HPV-10, HPV-11, HPV-16, HPV-18, HPV-31, HPV-33, HPV-39
Papillomas of meat handlers: HPV-7
Vulvar cancer: HPV-10
Laryngeal papillomas and condylomas: HPV-11
Oral focal hyperplasia: HPV-13
Bowenoid papulosis: HPV-16

seminating papillomas. Most importantly, there has been an epidemiological and biochemical link of certain papilloma virus isolates with human urogenital cancers, including cervical cancer, vulvar cancer, penile cancer, and malignant transformation of epidermodysplasia verruciformis lesions (Table 3.2).

Human urogenital cancers act epidemiologically like venereal diseases, suggesting transmission by an infectious venereally transmitted agent. Reid and coworkers in 1982 published histological evidence for subclinical papilloma virus infection in 91% of women with cervical cancer but in only 13% of matched controls. More recently, Mitchell et al. followed 846 women who in 1979 had cytological evidence of human papilloma virus infection. By 1985 carcinoma in situ was found in 30 women (compared with an expected value of two women for that cohort), giving a relative risk of 15.6 for development of cervical cancer after exposure to HPV. Moreover, women infected at a younger age appeared to be at even greater risk (relative risk 38.7 for those <25 years when HPV infection was documented). When male sexual partners of women with cervical carcinomas and precancerous lesions were examined, the men were found to have an increased incidence of genital papillomavirus infection supporting a role for venereal transmission of these viruses and suggesting the existence of a male viral reservoir.

Molecular and biochemical studies of oncogenesis by human papilloma viruses have been hampered by lack of a suitable cell culture system to prepare virus in large quantity and to study expression of viral genes. Nevertheless, the HPV genome exists as a closed, circular, unit-length DNA molecule that generally replicates as an episome, unassociated with host cell DNA. More recent evidence indicates that HPV-16 and HPV-18 genomes, at least, are integrated into cellular DNA in some cervical carcinomas and the cell lines derived from them. Current studies identify HPV DNA in more than 80% of cervical dysplasias and cancers and approximately 40% of vulvar and penile cancers.

More information is available on the molecular biology of certain bovine papilloma viruses (BPVs), as these viruses are readily tumorigenic in rodent cells in vitro. In vivo these viruses are associated with papillomas and fibropapillomas of bovine skin, teats, and udders. DNA sequence analysis of HPVs and BPVs have shown a similar genetic organization with analogous open reading frames (ORFs); these sequences have no stop codons and so represent potential protein coding sequences. The predicted amino acid sequences of analogous HPV and BPV proteins have significant homology; therefore it is thought that molecular details of cell transformation by BPVs are likely paralleled by those of HPVs.

Papilloma viruses (PVs) have double-stranded DNA genomes of approximately 8 kilobases (kb) (Fig. 3.1). Several PV genomes have been sequenced; and early and late regions have been defined by ORFs. The early region of both human and bovine papilloma viruses consists of 8 ORFs (E1 to E8), which are expressed as mRNA in papillomas as well as in murine cells transformed by BPVs. The late region consists of two ORFs (L1 and L2), expressed in papillomas (which produce mature virions), but are not required for cell transformation by BPVs. Thus papilloma viruses are analogous to other, better studied DNA tumor viruses

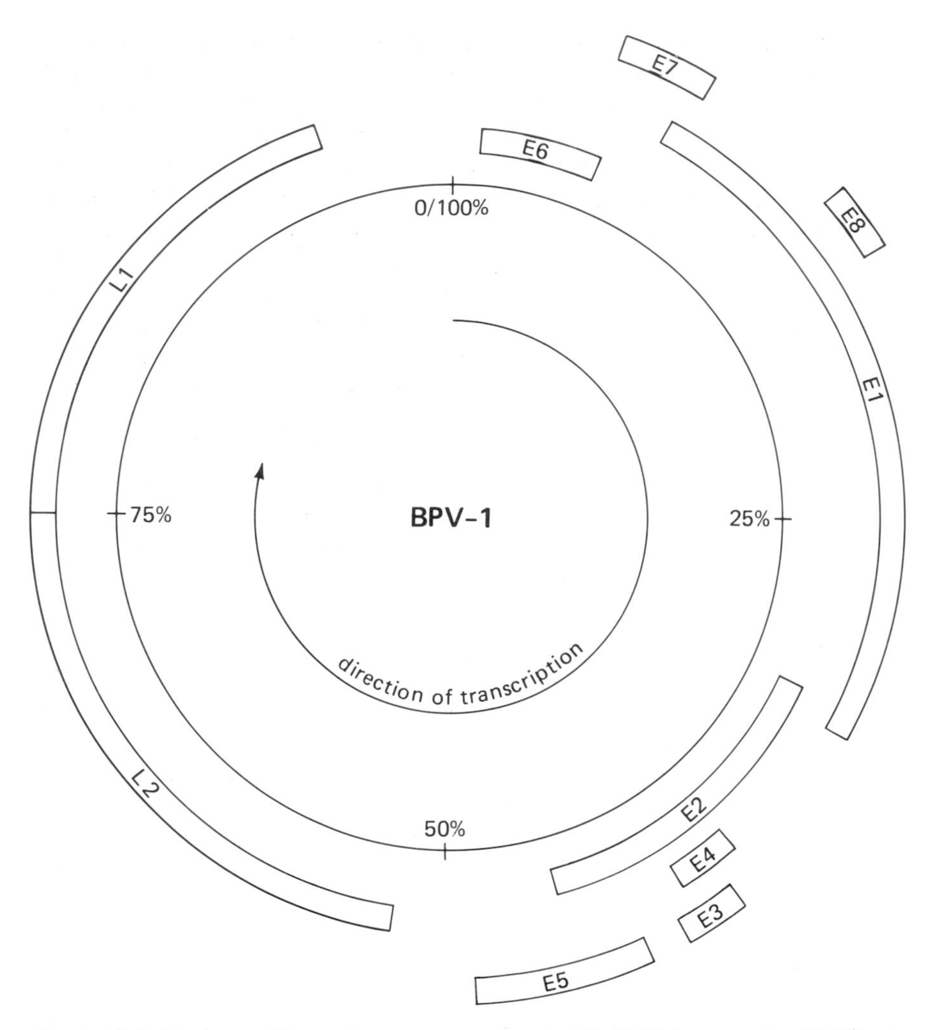

FIGURE 3.1. Bovine papilloma virus genome structure. The BPV-1 genome has 7945 base pairs. E1 through E8 represent early genes, and L1 and L2 are late genes. All genes appear to be transcribed from the same strand of DNA. Genes E5 and E6 have been associated with the transformed state.

including SV40, polyoma virus, and adenovirus (see below), all of which utilize early genes, but not late genes, during cell transformation.

Two genetically independent transforming genes, located in ORFs 5 and 6, have been defined for BPVs (Fig. 3.1). The analogous E6 region of HPV is retained and expressed as RNA in cervical carcinoma cells containing HPV DNA. Proteins in BPV-transformed cells corresponding to both E5 and E6 have been found using antisera obtained by immunization of rabbits with synthetic polypeptides derived using known nucleotide sequences of the E5 and E6 genes.

The E5 transforming regions have been defined by frame shift mutations: Insertion of a termination codon in the 3′ half of E5 ORF drastically reduces cell transformation, whereas correction of the frameshift by a second mutation restores transforming activity. The protein coded by the E5 region is the smallest transforming protein yet characterized with an apparent molecular weight of 6 or 7 kilodaltons (kD). It is a membrane-associated polypeptide with 68% hydrophobic amino acids and 34% leucine residues. The coding sequence for the 7-kD protein is present entirely within the 3′ half of the gene. The function of the protein has not been elucidated; however, it is not a protein kinase, nor is it associated with protein kinase activity. As stated, the major fraction of the protein is membrane-associated, but a small proportion is located in the nucleus. E5 ORF sequences are most strongly conserved among papilloma viruses, such as BPV-1 and deer papilloma virus, which induce both epidermal (epithelial) and dermal (fibroblastic) cell proliferation. Viruses such as HPV-1 that induce purely epidermal cell proliferation show little homology with BPV-1 E5 sequences. Studies to define the function of E5 protein are currently in progress.

Antisera to the E6 protein precipitates a 15.5-kD protein corresponding to the predicted molecular size of 127 amino acids of the E6 ORF. Cell fractionation localized approximately 50% of the protein to nuclear fractions and 50% to membrane fractions. The protein is rich in cysteine (~ 11% of its amino acids), and many of the cysteine moieties are arranged in repeats of cysteine–x–x–cysteine, a sequence proposed to be characteristic of nucleic acid binding proteins. E6 protein also contains a high proportion of basic amino acids: 17% arginine or lysine. The E6 region is conserved in many cells and tissues associated with the transformed or malignant phenotype including HeLa (a human cervical carcinoma cell line) and human cervical carcinoma tissue. The E6 gene can induce transformation of murine C127 cells but not NIH 3T3 cells (the E5 gene can transform both cell types); thus the E6 gene appears to be analogous to "immortalization-inducing," nucleus-associated oncogenes such as *myc* (see Chapter 5). Again, the function of E6 protein is not known, and studies are currently under way to elucidate it.

Cultured cells transformed by BPVs are able to induce tumors in nude (athymic) mice. Replication of BPVs is not required for transformation, as deletion mutants that are unable to replicate are still capable of provoking transformation. Independent expression of the E5 or E6 region is sufficient to induce transformation; concurrent expression of both regions does not appear to be necessary. Because late proteins are not produced during cell transformation, it is conceivable that the inability to package papilloma virus DNA may lead to overexpression of E5 or E6 (or both) and result in neoplastic transformation.

Human papilloma viruses are believed to complete their full productive cycle only in terminally differentiated squamous epithelial cells. Less differentiated proliferating epithelial cells do not support productive virus infection.

A potential beginning for molecular studies has been reported by Kreider et al. Papilloma virus from condylomata acuminata was shown to morphologically transform tissue grafts of human uterine cerivx grown in nude mice. Normal cervical grafts infected with this papilloma virus manifested cell alterations typical of condylomata acuminata, including enlargement of nuclei, hyperchromasia and binucleation, cytoplasmic clearing, and larger, thicker epithelium. Thus despite the current lack of success in propagating HPVs in cell culture, direct involvement of HPVs in dysplastic change of human cervical tissue has been demonstrated. In addition, this experimental system or a modification of it may prove useful for growing large quantities of HPVs for molecular investigations.

Although there is homology among the DNA sequences of various human and animal papilloma viruses, there is no significant homology between papilloma viruses and other DNA tumor viruses. Further molecular studies are needed to define the role of viral genes in the production of tumors in this, the only DNA tumor virus family proved explicitly to produce human neoplasms in vivo.

Herpes Viruses

Herpes viruses comprise several subfamilies of large, enveloped double-stranded DNA viruses. Members of the alpha, beta, and gamma subfamilies are associated with cell transformation in vivo, but only beta and gamma herpes viruses have been shown to induce experimental tumors in animals or to be associated with neoplasia in humans (Table 3.3). Some herpes viruses are known causes of cancer in their native host, i.e., Marek's disease of chickens and Lücké carcinoma of frogs. In addition, two primate herpes viruses, *Herpesvirus saimiri* of squirrel monkeys and *Herpesvirus ateles* of spider monkeys, are highly oncogenic in several nonhuman primates but do not cause tumors in their native species. Of the human herpes viruses, Epstein–Barr virus (EBV), a gamma virus, shows a strong epidemiological connection with human neoplasia in Burkitt's lymphoma (Africa) and nasopharyngeal carcinoma (Asia). A beta herpes virus, cytomegalovirus, has been strongly associated with Kaposi's sarcoma, a human skin tumor.

Table 3.3. Herpes virus families: associated tumors.

Alpha subfamily: transformation of cells in vitro
No known in vivo tumors
Beta subfamily: transformation of cells in vitro
CMV: Kaposi sarcoma in vivo (human)
Gamma subfamily
Epstein-Barr virus: Burkitt's lymphoma; nasopharyngeal carcinoma (humans), lymphomas in immunocompromised hosts, e.g., after organ transplantation
Marek's virus: T cell lymphoma (chickens)
H. saimiri: malignant lymphomas (species other than that of origin, the squirrel monkey)
H. ateles: malignant lymphomas (species other than that of origin, the spider monkey)
Other
Lücké virus: renal adenocarcinomas (*Rana pipiens*)
Herpes B lymphotropic virus: lymphoproliferative disorders

Herpes viruses are among the largest animal viruses, with a genome size on the order of 140 to 170 kb. Because of their genome complexity and the lack of good cell culture systems to study transformation (herpes viruses are cytotoxic to so many cell types that it is difficult to define their transforming activity in vitro), molecular and biochemical characterization of herpes virus genes and gene products is in its infancy. It has never been definitively shown that herpes viruses encode conventional oncogenes; however, studies of cloned portions of several herpes virus genomes are suggestive.

Three alpha-type herpes viruses have been studied in some detail: herpes simplex virus (HSV) types 1 and 2 and equine herpes virus (EHV). Specific DNA restriction fragments from each of these viruses have been shown to transform cultured cells. HSV-1 possesses one transforming region encoding a glycoprotein, a DNA binding protein, and an origin of viral DNA replication. It has not yet proved possible to detect viral DNA sequences in cells transformed by the cloned viral DNA. HSV-2 has two transforming regions judging by transfection studies with cloned DNA fragments. Somewhat surprisingly, neither of these regions is related by sequence homology to the HSV-1 transforming region. One 22-kb transforming fragment from HSV-2 [the "morphological transforming region" (mtr) III] appears to encode two sequential steps in transformation: "immortalization" of cells and "conversion to tumorigenicity." Again, no specific HSV-2 DNA sequences are detectable in cells transformed by HSV-2 DNA fragments, suggesting that the herpes simplex viruses may transform by a "hit and run" mechanism analogous to a chemical carcinogen. Specific EHV sequences, on the other hand, appear to be integrated into host DNA in EHV-transformed cells and may be associated with "maintenance" of the transformed phenotype (by analogy with other oncogene-containing DNA tumor viruses and retroviruses).

Early studies were thought to demonstrate a correlation between HSV-2, which causes genital herpes infections in humans, and human urogenital cancer. However, further investigations, including prospective clinical studies, have shown that previous infection by HSV-2 imposes no significant risk to the patient of developing cervical cancer. Cervical cancer development is much more strongly correlated with infection by certain human papilloma viruses (see above). Thus no human neoplasm is well correlated with any of the alpha herpes viruses.

Human cytomegalovirus (CMV) (Fig. 3.2), a beta herpes virus, has two transforming regions demonstrated by transfection studies with DNA restriction fragments. One of the regions is partially homologous to the *mtr*III region of HSV-2. In addition, sequences homologous to the 5′ end of the cellular/retroviral oncogene *myc* are detectable in the CMV genome. It is not explicitly known if the *myc*-like sequences correspond to one of the viral transforming regions; these *myc*-like sequences were mapped to several regions of the CMV genome. Again, transformed cells do not appear to retain specific CMV sequences.

Cytomegalovirus is ubiquitous in the human population and is associated commonly with disease. Several human cancers, including those of the prostate,

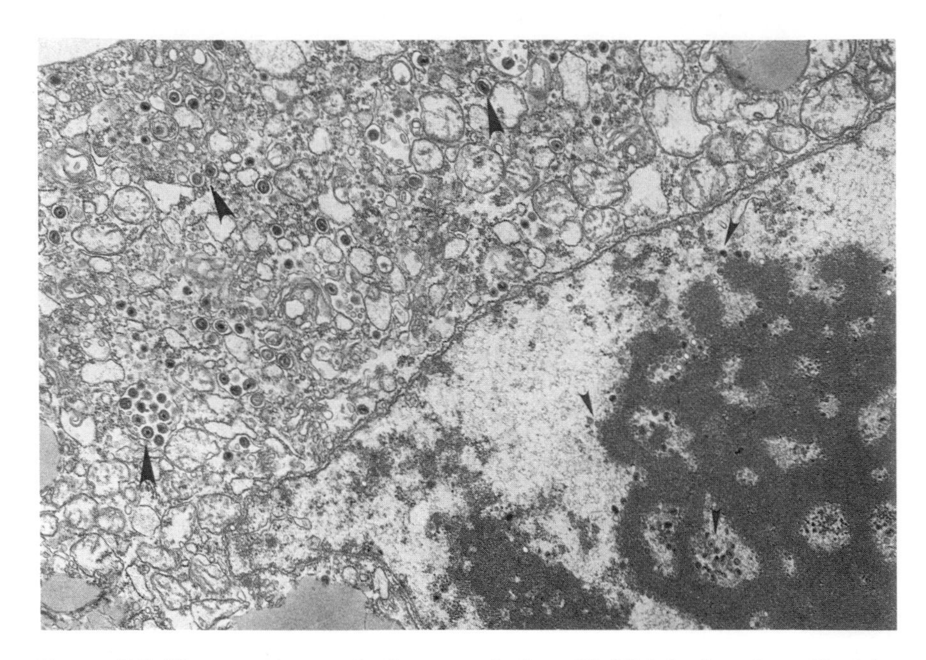

FIGURE 3.2. Electron micrograph of cytomegalovirus (CMV) infecting a lung. Particles are visible within the cytoplasm (large arrowheads) and intranuclearly (small arrowheads) ×16,100. (Kindly provided by S. Kunkel, R. Kunkel, and Ted Beals.)

colon, and cervix, have detectable CMV sequences. However, normal human tissue is found as frequently to be associated with CMV sequences, suggesting merely that CMV is ubiquitous and not necessarily associated with neoplastic transformation in these cases. A stronger case can be made for association of CMV with Kaposi's sarcoma. Not only are viral sequences found integrated in tumor cell DNA, but epidemiological studies have linked the two. Thus at least one human cancer is associated with a beta herpes virus.

Several gamma herpes viruses are associated with cancer in vivo (Table 3.3), epidemiologically in the case of Epstein–Barr virus in humans, and causally in the case of Marek's disease of chickens and *H. saimini* and *H. ateles* of monkeys.

Marek's herpes virus causes a T cell lymphoma in chickens that can be prevented by vaccination. T cell lymphomas also are induced by *H. saimiri* and *H. ateles* in marmosets and other New World monkeys, although the natural host species (squirrel monkey for *H. saimiri* and spider monkey for *H. ateles*) are spared neoplastic transformation by their respective herpes viruses.

In 1964 Epstein and co-workers discovered EBV in Burkitt's lymphoma cells, a B cell lymphoma. The virus subsequently was shown to be oncogenic in vitro and in vivo in marmosets. Primary EBV infection is the cause of infectious mononucleosis in humans, and organ transplant patients appear to be predisposed to B cell lymphomas induced by this virus. Burkitt's lymphoma occurs predominantly among African children (see Fig. 6.7). Children in Uganda who developed

Burkitt's lymphoma were shown to have high titers of antibodies to the viral capsid antigen. In fact, patients with such high antibody titers were shown to have a 30-fold increased risk of developing Burkitt's lymphoma compared to individuals with low or average titers. Multiple copies of circular, episomal full-length EBV DNA also are found in Burkitt's lymphoma tumor tissue.

Epstein-Barr virus is additionally associated with nasopharyngeal carcinoma, an epithelial cell tumor occurring predominantly among Cantonese Chinese adults in Southeast Asia. These patients also exhibit high antibody titers to viral capsid antigens; and as with Burkitt's lymphoma, multiple copies of EBV DNA have been found in virtually all nasopharyngeal tumors.

Currently, there are limited data on the putative EBV transforming genes or proteins. Studies of Burkitt's lymphoma have shown translocation of a portion of chromosome 8 to chromosomes 14, 2, or 22. The translocated region of chromosome 8 has been shown to contain the cellular *myc* gene. Chromosomes 14, 2, and 22 contain gene loci for immunoglobulin heavy chain and light κ and λ chain genes, respectively, that are normally expressed in B lymphyocytes.

Regions endemic for Burkitt's lymphoma also are noted for the presence of holoendemic malaria. It has been suggested that the massive B lymphocyte proliferation provoked by EBV or malarial infestation may promote translocation of the c-*myc* gene, resulting in development of a neoplastic clone of B cells. Although EBV can replicate in epithelial cells of the nasopharnyx a "co-carcinogen" has not been identified in cases of nasopharyngeal carcinoma (see Chapter 10 for details of the *myc* gene).

One other herpes virus, the currently unclassified Lücké virus of frogs is associated with in vivo oncogenesis, i.e., renal adenocarcinomas in the American leopard frog *Rana pipiens*. Renal tumors can be induced in frogs by inoculating tadpoles with the virus. Further details of the mechanism of transformation or putative transforming genes are not available as the virus has not yet been grown successfully in cultured cells.

Gallo and co-workers at the National Cancer Institute isolated a novel human B lymphotropic virus (HBLV) from the blood of several individuals with lymphoproliferative disorders. The virus is morphologically similar to viruses of the herpes virus family but can be distinguished by host range, antigenicity, and other characteristics.

Hepatitis Viruses

Four hepatitis viruses have so far been described (Table 3.1). Of these viruses, three have been associated with primary liver cancers (hepatocellular carcinoma = hepatoma) in their native species: human hepatitis B virus (HBV); woodchuck hepatitis virus (WHV); and duck hepatitis B virus (DHBV). Lack of suitable in vitro culture systems for these viruses has hampered molecular and biochemical studies of oncogenesis; however, there is exceptionally strong epidemiological evidence linking HBV with human hepatoma.

Hepatocellular carcinoma is not common in the United States and other western nations; however, worldwide it ranks among the most common malignant neoplasms. In Taiwan, for example, it accounts for 20% of all cancers and is the second leading cause of all deaths in the population. In an extensive 1981 study, Beasley and co-workers showed the relative risk of hepatoma in carriers of HBV (shown by positive hepatitis B surface antigen, HBsAg) to be more than 200 times that of non-HBV carriers. Factors other than chronic infection with HBV were not necessary to explain the epidemiological studies of hepatoma. This result is in contrast to those with EBV-induced lymphomas and nasopharyngeal carcinomas where environmental and genetic factors seem to be required in addition to primary virus infection.

Other studies have elucidated a unique mechanism of hepatitis virus replication (Fig. 3.3). The genome consists of a partially double-stranded DNA molecule 3000 to 3500 base pairs in length (depending on the species) consisting of a complete "−" strand and an incomplete "+" strand. The "+" strand is completed to various degrees in different virions; completion of this strand is not necessary for packaging of the genome within the viral capsid. After infection of a susceptible cell, the viral DNA matures to a complete double-stranded molecule that then is transcribed to yield a full-length "+" strand RNA molecule termed the *pregenome*. The pregenome is packaged with the DNA polymerase to yield an immature core particle. The RNA is transcribed by a reverse transcriptase activity of the hepatitis virus DNA polymerase to yield a full-length "−" DNA strand. The pregenome RNA is destroyed in the process. The "−" strand serves as a template for synthesis of the "+" strand, which is partially completed by the time the immature core is coated with envelope protein to become a finished virion. Subsequently, the infectious coated virion is exported from the cell.

In acutely infected cells, HBV DNA exists free in the host cytoplasm. In chronically infected cells and in hepatomas, viral sequences also are found integrated into the host genome. Chronically infected cells may contain only free DNA without detectable integrated sequences; however, tumor cells always contain some detectable integrated viral sequences. Complete genomes may be found, but just as frequently subgenomic fragments, often rearranged, are the only detectable viral sequences. Human HBV DNA appears to integrate at a specific viral sequence in the single-stranded part of the viral DNA, but not into any specific region of the liver cell genome. No particular region of the hepatitis viral DNA was found to be conserved in all clones of hepatomas examined. In addition, virus replication and viral gene expression are generally absent in poorly differentiated hepatocytes of advanced tumors. These findings have not suggested an unequivocal mechanism of oncogenesis by the hepatitis virus.

►

FIGURE 3.3. Proposed mechanism of hepatitis B virus replication.

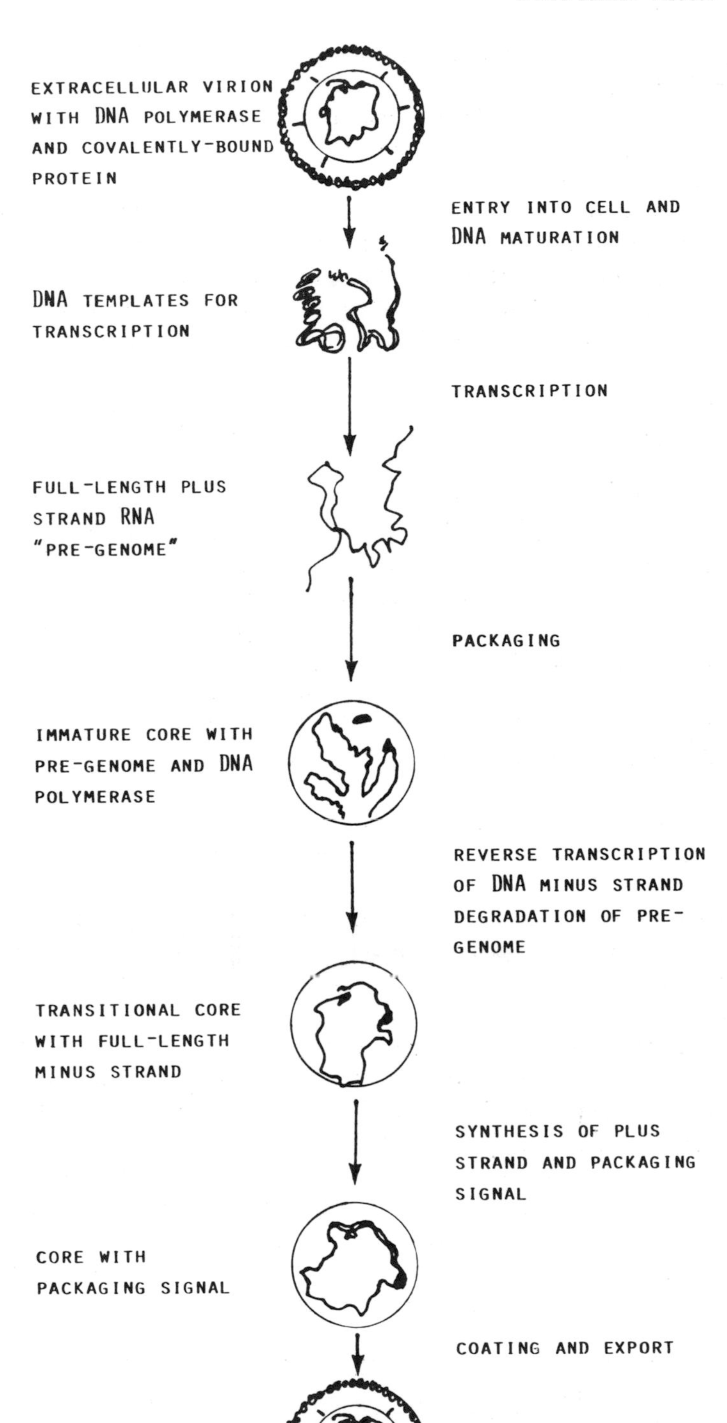
EXTRACELLULAR VIRION WITH DNA POLYMERASE AND COVALENTLY-BOUND PROTEIN
ENTRY INTO CELL AND DNA MATURATION
DNA TEMPLATES FOR TRANSCRIPTION
TRANSCRIPTION
FULL-LENGTH PLUS STRAND RNA "PRE-GENOME"
PACKAGING
IMMATURE CORE WITH PRE-GENOME AND DNA POLYMERASE
REVERSE TRANSCRIPTION OF DNA MINUS STRAND DEGRADATION OF PRE-GENOME
TRANSITIONAL CORE WITH FULL-LENGTH MINUS STRAND
SYNTHESIS OF PLUS STRAND AND PACKAGING SIGNAL
CORE WITH PACKAGING SIGNAL
COATING AND EXPORT
EXTRACELLULAR VIRION

Some further knowledge of the protein products of hepatitis virus genes has been gained (Fig. 3.4). Nucleotide sequence analysis of HBV DNA has determined that there are four consistent and conserved open reading frames on the "−" DNA strand and none on the "+" strand. The four presumed genes are termed S, C, P, and X. S codes for the surface antigen protein of the viral envelope. C encodes the viral core protein. The P region partially overlaps regions C and X and completely overlaps region S. The translation product is the hepatitis DNA polymerase, a 90,000-dalton basic protein resembling retrovirus reverse transcriptase. The X region is more mysterious. In several subtypes examined, the protein product was predicted to be 145 to 154 amino acids in length. Although its function is unknown, a polypeptide reacting with antibodies generated against a synthetic peptide corresponding to the X region of the HBV

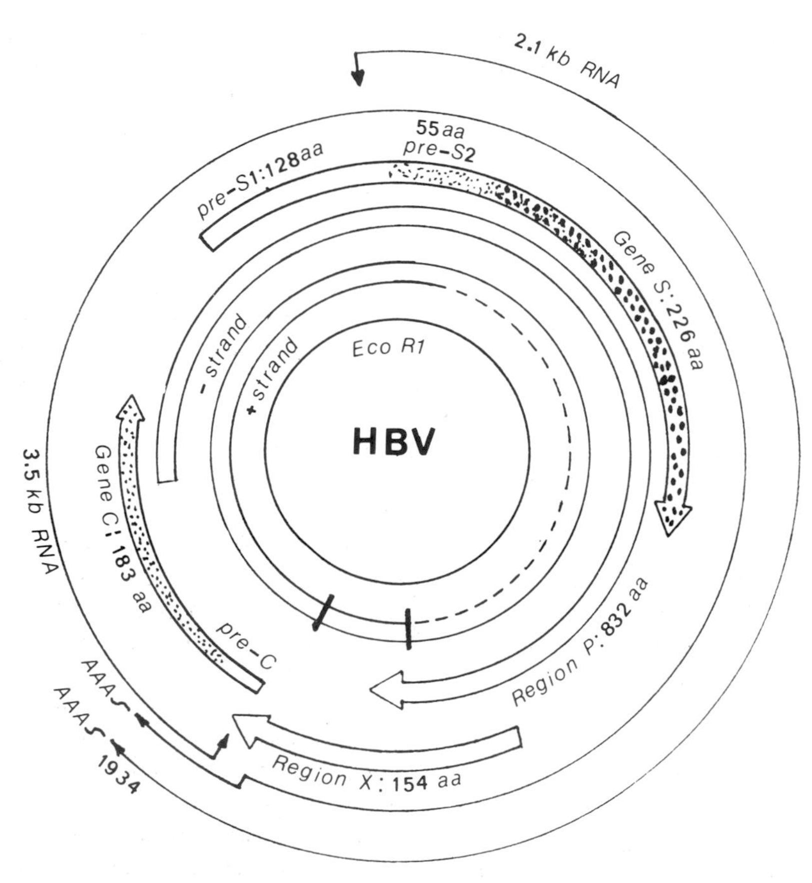

FIGURE 3.4. Hepatitis B (HBV) genome and protein products. Genes S, C, and region P code for viral structural proteins. Region X may encode a transforming region, although this theory has not been proved.

genome was found in HBV-infected liver cells. Patients with hepatoma appear to have serum antibodies reactive against X region synthetic peptides.

Of all the DNA viruses, hepatitis virus is the most closely related to retroviruses (see below) in its capacity to code for three structural proteins and potentially a fourth transforming protein, and its replicative cycle involving reverse transcription. However, hepatitis viruses have not been shown to transform cultured cells or to induce tumors in experimental animals. As yet there is no definitive evidence for a hepatitis virus oncogene or for integration of hepatitis virus adjacent to any known cellular oncogene to account for its epidemiologically established oncogenic potential.

Polyoma Viruses

Viruses in the polyoma family have been described for mice (polyoma), monkeys (simian virus "SV"40), and humans (BK and JC viruses) (Table 3.1). In 1960 polyoma virus was the first DNA virus shown to transform cultured cells. Each of the viruses in the family readily transforms cells in culture and is highly oncogenic for newborn rodents. However, none is associated with in vivo tumorigenesis in its native host. With the exception of the human JC virus (JCV), none of these viruses has been associated with a known disease in the native host. JCV appears to cause progressive multifocal leukoencephalopathy, a rare and often fatal demyelinating disease associated with decreased immunocompetence in humans. Generally, however, this virus does not cause recognized disease: JCV as well as BKV, SV40, and polyoma virus are ubiquitous in their native species, but none appears to play a significant role in the etiology of any cancer. Nevertheless, despite their lack of in vivo oncogenicity, polyoma virus and SV40 are among the most intensively studied DNA tumor viruses because they grow readily in tissue culture cells.

SV40 and polyoma virus genomes are small, covalently closed circular DNA molecules (Fig. 3.5). Polyoma virus has 5292 base pairs, and SV40 has 5243 base pairs. These viruses have similar but nonidentical protein products. Gene products can be divided into two general categories during productive viral infection: early and late. Early genes are expressed before viral DNA replication begins. The transforming functions of these viruses are encoded within the early region. Late genes are transcribed only during productive viral infection of permissive host cells. In non- or semipermissive host cells, only the early genes (including the transforming functions) are expressed. In most cases, cell transformation is transient and is termed "abortive"; however, a small percentage of cells are found to stably integrate the viral DNA and continue to express viral early genes. These cells become permanently transformed.

The transforming genes (oncogenes) of polyoma virus and SV40 have been characterized in detail. SV40 encodes a large T antigen of 94 kD and a small t antigen of 17 kD (Fig. 3.5). Large T antigen is a multifunctional protein and appears to mediate initiation of viral replication, autoregulation of viral early transcription, induction of viral late transcription, activation of rDNA

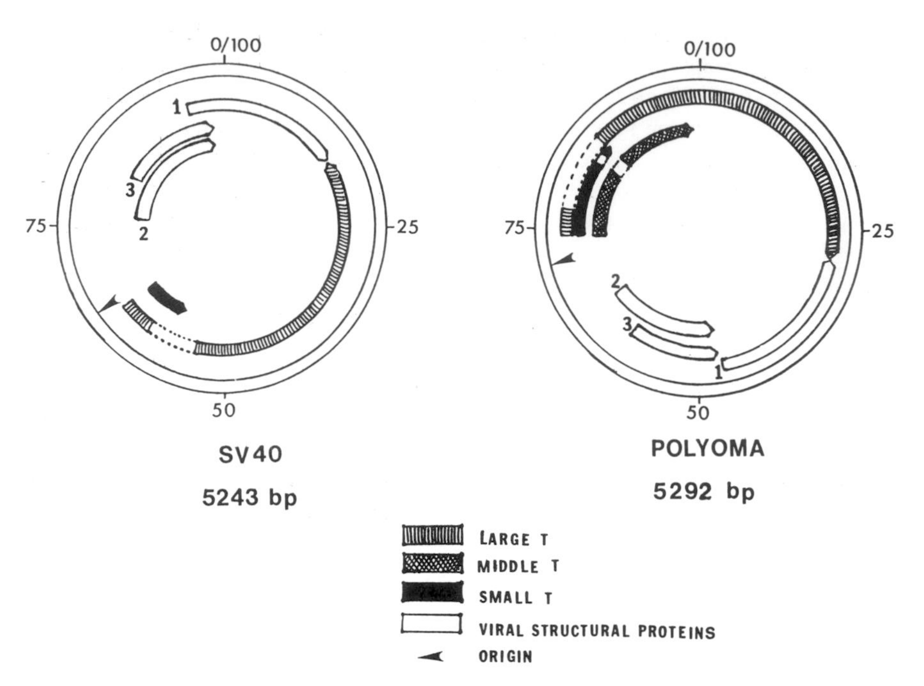

FIGURE 3.5. SV40 and polyoma viruses. The small, middle, and large T antigens are early proteins. Viral structural proteins (1, 2, and 3) are late proteins.

transcription, induction of cellular enzyme synthesis, and activation of RNA polymerase and transcription of cellular genes. It shows sequence-specific binding to viral DNA. It also binds cellular DNA and a cellular protein associated with the transformed state called p53 (see Chapter 10). In addition, it possesses ATPase activity. The SV40 large T protein is associated with DNA in the host cell nucleus for the most part, but a subfraction appears to be associated with the plasma membrane. The large T protein is essential for initiation of cell transformation, and continued expression is required for maintenance of the transformed phenotype. The *N*-terminal portion of large T protein appears essential to immortalize cultured cells. The SV40 small t antigen is not as well characterized. It is not essential for establishment or maintenance of the transformed phenotype but under some circumstances plays an enhancing role.

Polyoma virus encodes three T antigens: large T, middle T, and small t of 100, 55, and 22 kD, respectively (Fig. 3.5). Middle T antigen is essential for cell transformation and appears to encode the major transforming functions of the virus. Middle T antigen alone is sufficient to transform established cell lines, but it cannot transform primary cultures without large T antigen function (see below). Functional middle T antigen has been localized to both cytoplasmic (predominant) and nuclear membrane structures. Middle T protein also appears to be associated with the product of the cellular *src* gene. The large T antigen of polyoma virus has been localized to the nucleus. It is not as multifunctional as the SV40 large T protein (some of its functions have been taken by middle T protein). Similarly to SV40 large T protein, polyoma virus large T protein binds to the origin of viral DNA replication and possesses ATPase activity. It is required in conjunction with middle T protein to transform primary cells. A function needed for cultured cell "immortalization" (the function that complements middle T protein to transform primary cells) has been localized to the *N*-terminal domain of large T protein. Polyoma small t protein is required for the simultaneous transformation of primary rodent cells by middle T and large T proteins and for tumor induction by middle T protein. It is not absolutely required for cell transformation, as once primary cells have been immortalized by large T protein they are subsequently transformable by middle T protein without the requirement for small t protein. No further characterization of polyoma small t protein has yet been carried out.

The conclusions of studies on transforming genes of polyoma viruses are that they are of viral origin and are required in a multistep process to effect cell transformation.

Adenoviruses

There are five groups (A through E) and some 37 isolates of human adenoviruses (Table 3.4). In addition, adenoviruses have been detected in many animal species including monkeys, cattle, mice, dogs, sheep, pigs, birds, and horses. Ad12 and Ad18 were shown to induce tumors in newborn rodents in 1962, making them the first human viruses with demonstrated oncogenic potential. All human adenoviruses thus far examined can transform rodent fibroblast cell cultures; however, they differ in their ability to cause tumors in rodents. Group A viruses cause tumors with high efficiency, whereas group B viruses cause low efficiency, low incidence tumors. Group D viruses specifically induce mammary fibroadenomas in female rats. Groups C and E have no known oncogenic potential in vivo. Despite their demonstrated oncogenic potential for rodents, extensive

Table 3.4. Properties of human adenoviruses.

Group	Type	Cell transformation	Animal tumors
A	12, 18, 31	+	High incidence of tumors in newborn hamsters
B	3, 7, 11, 14, 16, 21	+	Low incidence of tumors in newborn hamsters
C	1, 2, 5, 6	+	None
D	8, 9, 10, 13, 15, 17, 19, 10, 22–30	+	Mammary fibroadenomas in rats
E	4	+	None

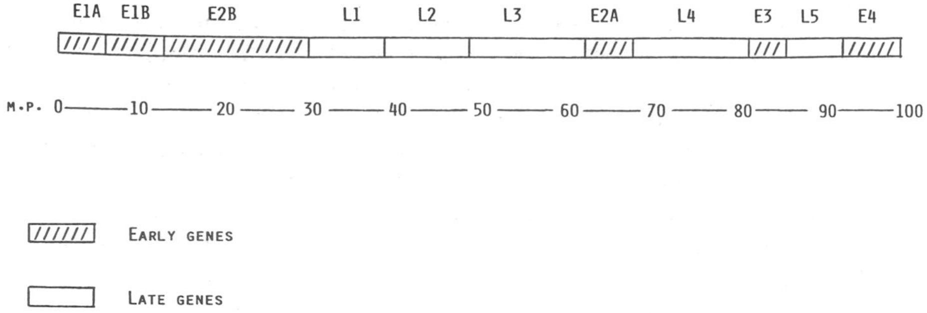

E1A E1B E2B L1 L2 L3 E2A L4 E3 L5 E4

M.P. 0—10—20—30—40—50—60—70—80—90—100

EARLY GENES

LATE GENES

FIGURE 3.6. Adenovirus genome map showing approximate locations of adenovirus Ad2 early and late genes. Distances on the approximately 36-kb genome are given in map units (mp) of 0 to 100. Genes E1 through E4 are transcribed during the early phases of adenovirus productive infection. The adenovirus-transforming region is located in the E1 region. Late proteins L1 through L5 are mainly viral structure proteins. A specific class of mRNAs also is transcribed from the L1 region early in adenovirus infection. Adenovirus gene expression is complex. E2B, E2A, and E4 are transcribed from the opposite DNA strand as the other genes. Each gene has multiple transcripts and proteins generated by differential splicing. Specific leader sequences begin each transcript. Three late leaders in the E2B region serve for each of the late genes. For further details see the Green and Green et al. references.

investigation has revealed no sign of adenovirus sequences in human cancers representing approximately 90% of the cancer occurring in the United States.

Adenoviruses are larger than polyoma viruses, averaging genome sizes of 35 to 40 kb. Nevertheless, they have been extensively studied, and a representative virus, Ad2, has had its genome fully mapped with respect to translation units and protein products. As with polyoma viruses, the adenoviruses possess early and late genes; the adenovirus oncogenes are among the early genes (Fig. 3.6). This discussion focuses primarily on adenovirus transforming genes. (For further information on the adenovirus life cycle, see the Green reference.)

Two regions of the adenovirus genome have been implicated in transformation: E1A and E1B. The E1 region is located within the left 11 to 12% of the viral genome. The E1A gene is transcribed to two early mRNAs of 12S and 13S. These RNAs have a common precursor, and their size difference is generated by splicing. The 13S mRNA codes for a polypeptide of 289 amino acids (32 kD), whereas the 12S codes for a 243 amino acid peptide (26 kD). The only difference between the proteins is that the 32-kD protein contains 46 additional internal amino acids. The *N* and *C* termini are identical. When isolated from infected cells, the E1A antigens have proved to be unexpectedly complex. Multiple species of several molecular weights have been identified, presumably reflecting unknown post-translational modifications. The E1B gene also produces two major mRNAs of 22S and 13S. Three T antigens have been identified of 53 kD (from the 22S mRNA), 19 kD (from both 22S and 13S mRNA), and 20 kD (from unidentified mRNA).

The two major E1A T antigens and the E1B 19 kD T antigen are sufficient to induce transformation. E1A gene products function in the initiation and maintenance of cell transformation as well as activation of viral early genes. Cloned E1A genes can immortalize cells and cause aneuploidy, but the cells retain morphological and growth characteristics of untransformed cells. Either of the E1A products can establish indefinite growth of cultured cells, but both are required for full neoplastic transformation. E1A proteins alone can "partially transform" established cell lines; both immortalization and "partial transformation" functions are localized in the *N*-terminal portions of the proteins. E1A functions also are required, directly or indirectly, for maintenance of cell transformation.

E1A antigens appear to be localized to the host cell nucleus (immunofluorescence). They may function to activate transcription of cellular genes, a role dissociable from the transformation functions.

Of the E1B T antigens, only the 19-kD protein has been found to be essential for cell transformation. It is found in both the cytoplasm and the nucleus by immunofluorescence. Immunoelectronmicroscopy has localized E1B 19-kD antigen to distinct cytoplasmic membrane structures at the nuclear periphery. An additional function of the 19-kD antigens may be to protect newly replicated viral and also host cellular DNA.

The E1B 53-kD protein appears to be required for high tumorigenicity of transformed cells in nude mice. It is also present in the nucleus and cytoplasm and has been reported to possess a protein kinase activity. (It is not clear whether the latter is a property of the 53-kd antigen itself or of an associated host cell protein.)

The precise roles played by the E1B proteins in transformation and oncogenesis have been difficult to elucidate because the genes overlap in different reading frames, and it is difficult to isolate one by cloning without interference by others. It has been suggested that the E1B 53-kD antigen may influence cell growth by binding to the cellular p53 fraction; however, this action has not been firmly established.

An interesting facet of the adenovirus story was the finding by Schrier and coworkers in 1983 that products of E1A (especially the 289 aa protein) can inhibit transcription of MHC class 1 antigens in transformed rat cells. This effect is seen only with group A adenoviruses, in particular Ad12, the group with the highest potential for oncogenicity (see Chapter 10).

DNA Tumor Virus Oncogenes: Summary

Transforming genes have been well characterized for only two classes of DNA tumor viruses: polyoma viruses and adenoviruses. Unlike retroviral oncogenes, which appear to be originally of cellular origin and are not required for the retrovirus life cycle (see below), the DNA tumor virus oncogenes are purely of virus origin and are intimately involved in the life cycles of these viruses. Nevertheless, there are similarities between the two groups of genes extending to their ability to "complement" each other in tumorigenic assays (see Chapter 5). The ability of organisms as divergent as viruses and vertebrates to encode such similar functions implies a fundamental necessity for these functions in life forms on earth.

RNA Tumor Viruses

Retroviruses provide a link between neoplasia due to viral infection and the oncogenic potential contained within normal cellular genes, genes that are essential for eukaryotic development.

All RNA tumor viruses are retroviruses, but not all retroviruses are associated with tumors. The first infectious agent found to be oncogenic was a filterable extract of a chicken leukemia described in 1908 by the Danish workers Ellermann and Bang. This agent subsequently proved to be retrovirus. In 1911 Peyton Rous, of the Rockefeller Institute, demonstrated that cell-free filtrates of chicken sarcomas could induce sarcomas in other chickens. Because the filters used were known to retain bacteria, a filterable agent, i.e., a virus, was established as the infectious cause of chicken sarcoma. This virus, the Rous sarcoma virus (RSV) became the prototypic RNA tumor virus. By 1938 it was established that a single RSV particle was capable of infecting a normal cell leading to transformation of the cell, as well as producing infectious viral progeny.

Mice, as well as chickens, were suspected to have virus-induced cancers. During the early 1930s it was noted that several inbred strains of mice had unusually high incidences of spontaneous leukemia. At the same time a mouse strain called C3H was bred for a high incidence of mammary cancer.[1]

Murine leukemias were transmissible by injection of leukemic cells from one animal to another. However, proof that viruses were the causal agents was not forthcoming until the early 1950s, when Ludvik Gross demonstrated that 1-day-old mice, inoculated with a cell-free extract of a murine leukemia, developed leukemias later in life with a frequency of approximately 50%. Later studies established that the earlier efforts to transmit leukemia via cell-free extracts were frustrated by the use of older mice for recipients. Older mice proved capable of developing an immune response against the virus, and so no tumor would develop. In addition, murine leukemia viruses were subsequently shown to have a complex system of host range and restriction such that some mice resist infection by some strains of virus but are susceptible to infection by other strains. (These issues are beyond the scope of this book; for more detailed discussion, see Weiss, et al.)

During the 1930s studies of mouse mammary tumors showed that these cancers were maternally transmitted via milk. In 1942 John Bittner passed murine milk through filters that trap bacteria; by subsequent transmission of mammary tumors to susceptible mice, he demonstrated a viral cause. These

[1]Coincidentally, C3H mice were seen to have a low incidence of spontaneous leukemia, establishing that the propensity to develop one type of cancer was not correlated with an increased tendency to develop a different malignancy.

viruses became known as the mouse mammary tumor viruses (MMTVs). Both mouse leukemia viruses and the MMTVs were shown to be retroviruses.

During the 1950s and 1960s a plethora of RNA tumor viruses were isolated, not only from chickens and mice but also from rats, cats, woolly monkeys, and gibbon apes. After much intensive searching, the first retrovirus associated with a human malignancy, HTLV-I, was isolated by Robert Gallo's group in 1978 (see Chapter 4). Studies were facilitated by the development of cell culture, techniques to grow and purify the viruses, and assay systems. Electron microscopy permitted direct visualization of the viral particles, and a classification based on particle morphology was established.

A list of the more commonly studied viruses, including species of origin and associated oncogene, is given in Table 3.5. Retroviruses as a group also have been shown to induce lymphomas, erythroblastomas, and several nonmalignant disorders in addition to leukemias, sarcomas, and mammary carcinomas. Certain

TABLE 3.5. Retroviral oncogenes.

Oncogene	Prototype virus(es)	Species of origin
abl	Abelson murine leukemia virus (Ab-MLV)	Mouse
erb A, *erb* B	Avian erythroblastosis virus (AEV)	Chicken
ets	E26	Chicken
fes, *fps*	Fuginami sarcoma virus (FuSV)	Chicken
	Snyder-Theilen feline sarcoma virus (St-FeSV)	Cat
	Feline sarcoma virus (GA-FeSV)	Cat
fgr, *yes*	Gardner-Rasheed feline sarcoma virus (GR-FeSV)	Cat
	Yamaguchi 73 sarcoma virus (Y73)	Chicken
	Esh sarcoma virus (ESV)	Chicken
fms	Susan McDonough feline sarcoma virus (SM-FeSV)	Cat
fos	FBJ-murine sarcoma virus	Mouse
	FBR-murine sarcoma virus	Mouse
mos	Moloney murine sarcoma virus (Mo-MSV)	Mouse
myb	Avian myeloblastosis virus (AMV)	Chicken
	E26	Chicken
myc	Myelocytomastosis-29 virus (MC29)	Chicken
	Mill-Hill-2 virus (MH2)	Chicken
	CMII/OK10 viruses	Chicken
raf	Mill-Hill-2 virus (MH2)	Chicken
	Murine sarcoma virus strain 3611 (MSV-3611)	Mouse
ras Ha	Harvey murine sarcome virus (Ha-MSV)	Rat
	Rasheed rat sarcoma virus (RaSV)	Rat
(*bas*)	BALB murine sarcoma virus	Mouse
ras Ki	Kirsten murine sarcoma virus (Ki-MSV)	Mouse
rel	Reticuloendothelial virus strain T	Turkey
ros	UR2	Chicken
sis	Simian sarcoma virus (SSV)	Woolly monkey
	Parodi-Irgens feline sarcoma virus (PI-FeSV)	Cat
ski	SKV770	Chicken
src	Rous sarcoma virus (RSV)	Chicken

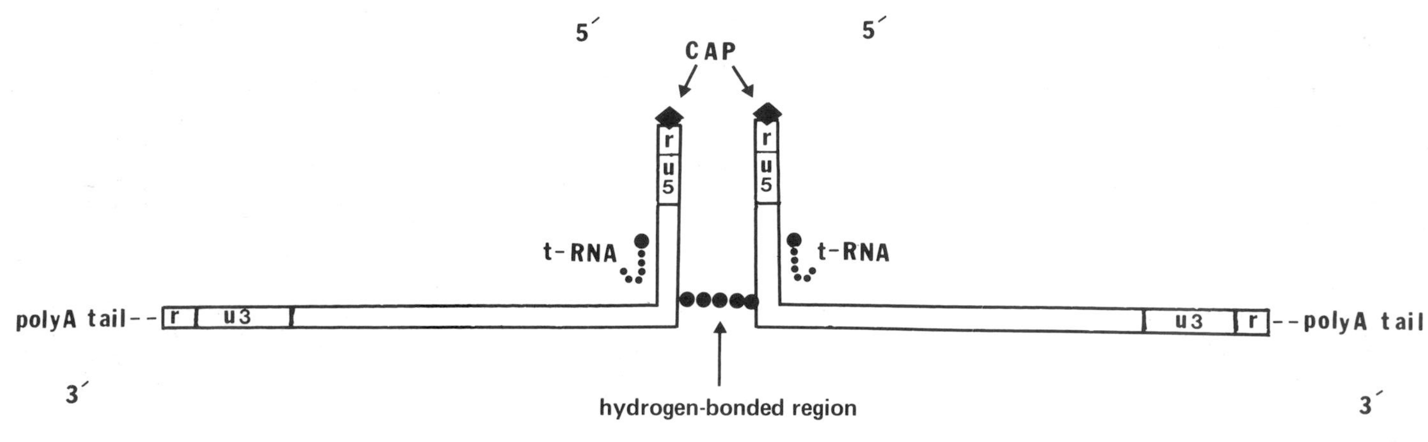

FIGURE 3.7. Retrovirus genome structure. Retroviruses, including RNA tumor viruses, have diploid RNA genomes. Each strand has a redundant region (r) at the extreme 5′ and 3′ ends and unique sequences (U5 and U3) at the 5′ and 3′ ends, respectively. The sequence U3-r-U5 makes up the LTR of the DNA provirus (see text). Each of the RNA strands has a "cap" structure at the 5′ end and a poly A tail at the 3′ end similar to a typical eukaryotic mRNA. In addition, each strand has an associated tRNA molecule near the 5′ end that acts as a primer for DNA synthesis during reverse transcription. The two RNA strands are hydrogen-bonded near the 5′ end, a configuration thought to aid reverse transcription of the single-stranded RNA to a double-stranded DNA provirus.

viruses have been found to induce more than one tumor type, depending on conditions and the type of cell infected.

Retroviruses are characterized by single-stranded RNA genomes (Fig. 3.7); the size varies from 3.5 to 9.0 kb. Each virion contains two copies of the RNA bound noncovalently at the 5′ end and are considered diploid. The reason for this unique genome structure is not clear. Conceivably it could be required for virus replication (see below).

The RNA tumor viruses can be classified into three general categories: (1) complete-transforming viruses: viruses that are replication-competent and that induce tumors rapidly and at high frequency; (2) slow-transforming viruses: viruses that are replication-competent but that induce tumors with low frequency (if at all) and only after long latent periods (most transforming viruses in nature belong to this category); and (3) defective-transforming viruses: viruses that are not replication-competent but that induce tumors rapidly and at high frequency. Viruses from the latter class require replication-competent "helper viruses" for infectious transmission from cell to cell. Infectious isolates of defective-transforming viruses invariably contain the replication-defective virus with its helper virus. The helper virus most frequently is a slow-transforming or leukemia virus (category 2).

The prototype, and (ironically) the only example of a complete transforming virus, is the Rous sarcoma virus (RSV). Its genome structure, which can serve as a generalized structure for all retroviruses, is shown in Figure 3.8. Each of the single-stranded RNA subunits has a 5′ cap and a 3′ poly A tail, thus resembling a typical eukaryotic messenger RNA. RSV and other replication-competent retroviruses contain three viral genes encoding (1) the protein capsid (*gag* = group-specific antigen); (2) RNA-dependent DNA polymerase, "reverse trancriptase" (*pol*); and (3) the envelope antigens (*env*), which are embedded in the membrane surrounding the viral capsid and that determine host range and mediate virus binding to host cells during infection. In addition, RSV contains the transforming gene *src*, for "sarcoma."

Every retrovirus contains a unique tRNA hydrogen bonded near the 5′ end of each genomic RNA strand. In addition, the two genomic strands are linked noncovalently in a poorly understood manner, also near the 5′ end (Fig. 3.7).

An example of a replication-competent slow-transforming retrovirus is the avian leukosis virus (Fig. 3.8). Its genome is similar to that of RSV, but it does not contain the *src* or any other transforming gene. Each of the viral structural genes is intact, and they occur in the same order as for RSV: 5 *gag-pol-env*-3′.

Defective-transforming viruses are exemplified by the various murine sarcoma viruses (MuSVs) as well as some feline sarcoma viruses (Fig. 3.8). Each of these viruses carries an intact transforming gene similar to *src* (collectively called *onc* genes for their oncogenic potential) but is missing one or more of the viral structural genes. Most frequently, the missing gene is all or part of *env*, but deletions including parts of *gag* and *pol* have been seen as well.

The fact that one transforming virus is sufficient to provoke a tumor in animals or a focus of transformation in cultured cells suggested that a gene(s) carried by

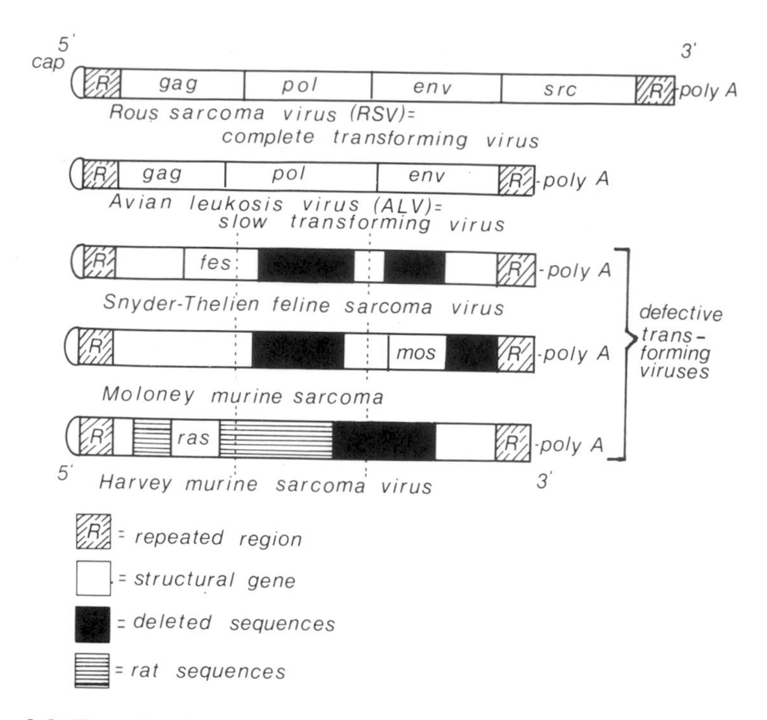

FIGURE 3.8. Examples of RNA tumor viruses. Comparison of genomes structures: Rous sarcoma virus is a replication-competent, rapidly transforming virus. Avian leukosis virus is a replication-competent but slowing transforming virus. There are also several replication-defective but rapidly transforming viruses.

the virus is directly responsible for the effect. However, only during the last one or two decades have biochemical and genetic techniques become sufficiently sophisticated to provide support for that idea. A major breakthrough by independent workers Howard Temin and David Baltimore in 1970 was elucidation of the retroviral life cycle spearheaded by the discovery of the unique viral polymerase, reverse transcriptase.

Once a retrovirus has infected a cell, RNA-dependent DNA polymerase (reverse transcriptase) synthesizes double-stranded DNA copies of the single-stranded RNA genome (Table 3.6; Fig. 3.9). This synthesis occurs in the cytoplasm utilizing the hydrogen-bonded tRNA as primer (Table 3.6; see Lowy for details). Several sequences present at the 5′ and 3′ ends of the viral RNA are of interest (Fig. 3.7). The viral RNA contains a small sequence of 30 to 60 bases called "r" (for "redundant") at the extreme 3′ and 5′ ends. A unique sequence U5 (80–120 nucleotides) occurs next at the 5′ end, followed by the structural genes, and finally a unique sequence U3 (200–1200 nucleotides) at the 3′ end lying just upstream of the 3′ r sequence. During synthesis of the double-stranded DNA "provirus," the U3-r-U5 sequence is repeated at both the 5′ and 3′ ends of the viral

TABLE 3.6. Proposed sequence in retroviral replication and expression.

Phase I: early events
- Adsorption to cell surface.
- Virion entry via interaction between viral envelope glycoprotein and host cell receptors.
- "Uncoating": preparation for virion expression.

Phase II: synthesis of free (unintegrated) viral DNA
- Synthesis of DNA " – " strand.
- Synthesis of DNA " + " strand.
- Degradation of viral RNA.

Phase III: integration of viral DNA
- Steps of integration not known in detail; however, termini of LTRs required. Circular form with two LTRs now appears to be necessary.
- Integration in same orientation as free linear viral DNA, i.e., LTR-*gag-pol-env*-LTR.
- Integration in many potential (perhaps random) host sites.
- Deletion of two terminal nucleotides from each end of free viral DNA and duplication of four to six host integration site nucleotides at each end of the viral DNA.
- Circular free viral DNA found in infected cells; may contain one or two copies of LTR sequences. The form with two LTRs appears to be the immediate precursor for proviral integration.

Phase IV: expression of viral DNA
- Transcription occurs for integrated viral DNA.
- Host-encoded enzymes used for viral transcription.
- LTR contains signals for promotion, initiation, enhancement, and polyadenylation of RNA synthesis.
- Fifty percent of viral RNAs synthesized serve as mRNA.
- Fifty percent of viral RNAs synthesized are incorporated into complete virions.
- V-*onc* genes expressed similarly to other viral genes: most are synthesized as *gag-onc* fusion transcripts, encode *gag-onc* fusion proteins.

Phase V: viral protein synthesis and virion assembly
- The *gag* products are core proteins.
- The *pol* product is reverse transcriptase.
- The *env* products are glycosylated envelope proteins.
- Complete virion consists of diploid RNA genome with tRNAs, reverse transcriptase, and core proteins.
- Envelope of virion is formed at the cellular plasma membrane as virus is released from the cell by budding.

DNA. This sequence, ranging from 280 to 1300 bases (with most variation occurring in the U3 sequence), is called the *long terminal repeat* (LTR).

Once synthesized, proviral DNA somehow becomes integrated into the host genome. Each successfully infected cell contains at least one integrated provirus, and most cells ultimately contain several (usually 4–10 copies). The LTR sequences are required for integration and a 5-base-pair sequence of host DNA is directly repeated at each end of the provirus. Analysis of host sequences at integration sites has revealed nothing unique, suggesting that proviruses may integrate at multiple cellular locations, probably at random. The proviral orientation in the host DNA is exactly the same as for free linear viral DNA; that is, there is no permutation of sequences such as occurs for bacteriophage λ, for example (see Lowy).

The integrated proviral DNA is expressed using host enzymes for RNA transcription and protein synthesis. The LTRs act as promoter and enhancer regions. In addition, they encode a polyadenylation signal. Viral structural proteins are synthesized from *gag*, *pol*, and *env* genes. In replication-competent viruses, these proteins are combined in some manner with single-stranded diploid RNA genomes to yield complete virions. Fully assembled progeny virus buds from the cell surface without killing the cell; that is, there is no lytic phase in the retroviral life cycle (Figs. 3.9 and 3.10).

What of the transforming retrovirus oncogenes? Experiments with temperature-sensitive and deletion mutants have explicitly shown that these genes are *not* necessary for virus replication but *are* required specifically for transformation of the host cell. Viruses without oncogenes transform cells slowly, if at all, over a period of several weeks to months and induce animal tumors at a similar slow frequency with low yield. A small number of virally transduced *onc* genes are known (approximately 20 isolates comprising five to seven categories; see Table 3.5). These categories include growth factors, growth factor receptors, cytoplasmic second messengers and protein kinases, and nuclear proteins. Several independently isolated retroviruses have been found to carry the same oncogene. A few retrovirus isolates carry more than one (generally two) oncogenes, e.g., avian erythroblastosis virus carrying the *erb* A and *erb* B genes. Each gene is associated with a particular tumor type (e.g., sarcoma, erythroblastoma), although a few have been associated with two or more neoplastic diseases (Table 3.5).

Evidence indicating the cellular origin of retroviral oncogenes was provided during the late 1970s by Bishop, Varmus and co-workers, who showed that radioactive *onc* gene probes hybridized specifically to host DNA in *uninfected* cells. Originally done with *src* and normal uninfected chicken DNA, the studies were expanded to include all the known oncogenes and a vast array of vertebrate and invertebrate DNAs. The results were surprising and exciting: All normal DNAs tested appeared to contain sequences homologous to the viral *onc* genes. The host cellular sequences have been termed *proto-oncogenes*, or c-*onc* genes, and the homologous viral genes are called v-*onc* genes. The implication is that nontransforming retroviruses are able to pick up or *transduce* these host cellular genes, thereby endowing the viruses with transforming potential. In most cases the gain of host transforming genes is at the expense of viral replication genes (i.e., the viruses become replication-defective transforming viruses). The transduced proto-oncogene also appears to accumulate alterations that convert it into a more potent transforming gene. Such alterations may occur during the transduction process itself, as well as during propagation of the resultant virus. The mutant virus capable of cell transformation has a selective advantage. An understanding of the transduction process may well provide insight into how normal cellular genes are converted to transforming genes (see Chapter 5).

Another aspect of retroviral replication appears important in the generation of transforming v-*onc* genes. Reverse transcription, the process whereby retroviral RNA is converted into DNA prior to integration, is highly error-prone. At each

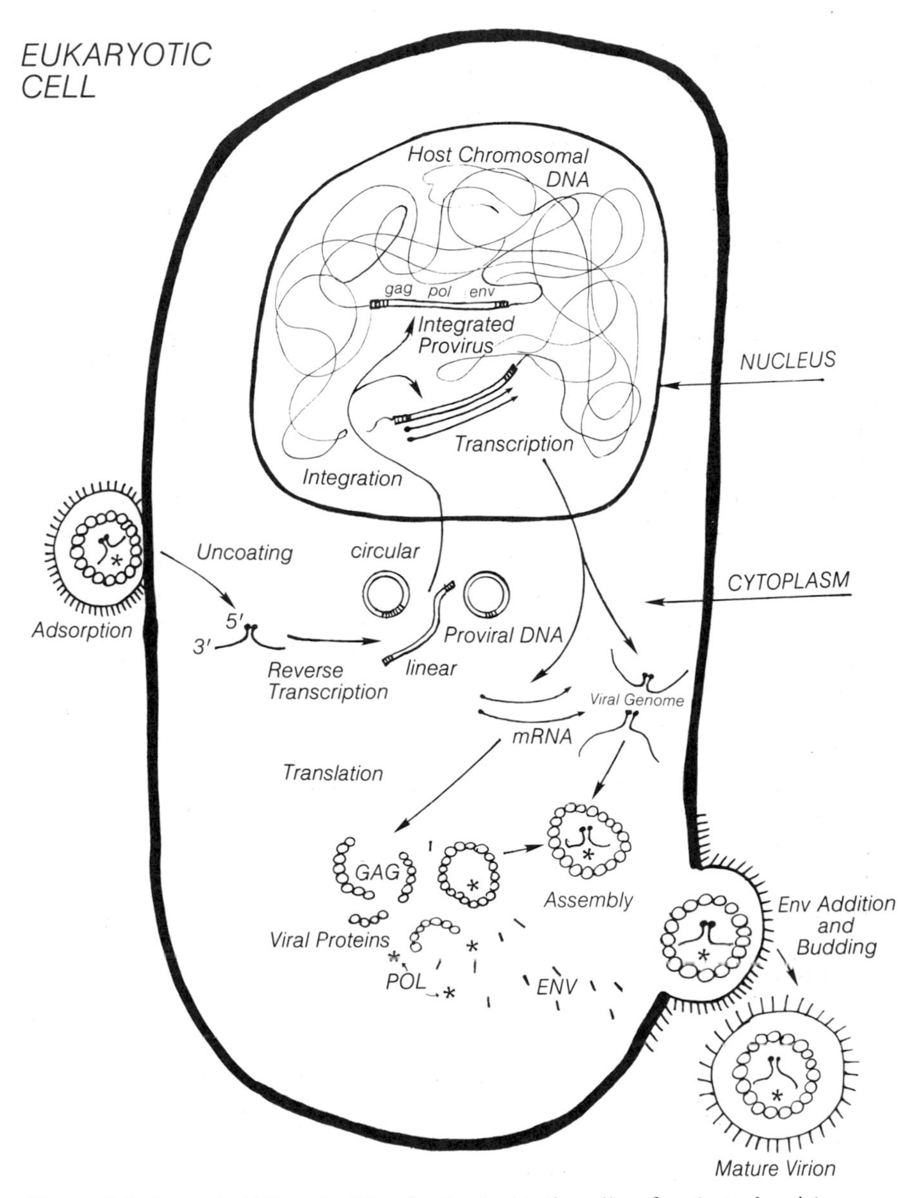

FIGURE 3.9. Retroviral life cycle. Virus is adsorbed to the cell surface (cytoplasmic) membrane; it is then "uncoated," and the diploid RNA genome acts as a template for reverse transcription. The linear provirus is thought to circularize; circles containing two copies of the LTR are thought to be the immediate precursors for proviral integration. Integrated proviral DNA is transcribed to RNA using host RNA polymerase. Viral RNA serves both as template for protein synthesis and as viral genome. Once viral proteins are synthesized, an assembly reaction incorporates diploid genomes into capsids. The virus particle then buds from the cell surface; viral *env* proteins are embedded in the cellular membrane, and each virion is coated with an envelope derived from the host cell membrane and containing virus-specific *env* antigens.

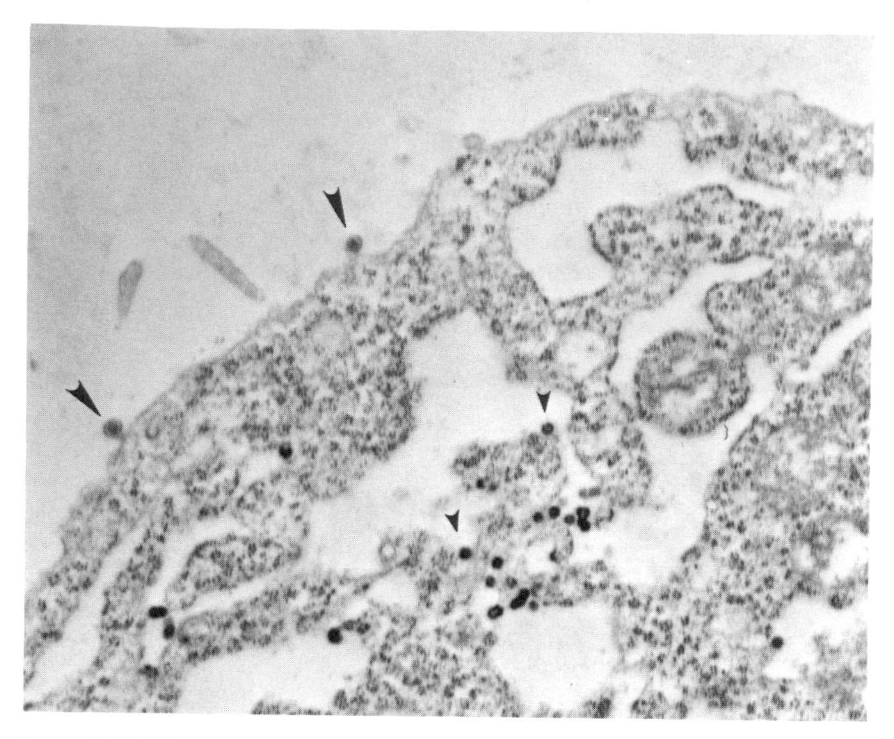

FIGURE 3.10. Electron micrograph of retrovirus budding from an infected cell. This hybridoma cell was infected with a type c retrovirus. Viral particles can be seen budding from the cell membrane (large arrowheads) and within the cytoplasm (small arrowheads). ×21,000. (Kindly provided by S. Weiss and N. Benson)

replication cycle, as many as 0.5% of viral RNA sequences reverse-transcribed to DNA are copied incorrectly. Though many of these errors produce nonviable viruses or inconsequential changes, the rare base conversions that result in a transforming oncogene give the viral mutant and its host cell a clear selection advantage. Experimentally, Rous sarcoma virus variants carrying the c-*src* proto-oncogene are not transforming; but on propagation, transforming mutants are derived. Thus retroviral transduction of cellular proto-oncogenes and their subsequent propagation during viral replication conspire to select mutant cellular genes capable of transformation.

The functions of retroviral oncogenes and their potential role in human neoplasia are discussed in subsequent chapters.

BIBLIOGRAPHY

General

Bishop JM. Cellular oncogenes and retroviruses. Annu Rev Biochem 52:301, 1983.
Bishop JM. Viral oncogenes. Cell 42:23, 1985.

Varmus HE. Form and function of retroviral proviruses. Science 216:812, 1982.
Varmus H, Levine AJ (eds). Readings in Tumor Virology. Cold Spring Harbor, NY: Cold Spring Harbor Laboratory, 1983.
Viral oncogenes. Cold Spring Harbor Symp Quant Biol vol. 44, 1980.

DNA Viruses

Barrasso, R, De Brux J, Croissant O, et al. High prevalence of papillomavirus-associated penile intraepithelial neoplasia in sexual partners of women with cervical intraepithelial neoplasia. New Eng J Med 317:916, 1987.
Beasley R, Lin C, Hwang L, et al. Hepatocellular carcinoma and hepatitis B virus: a prospective study of 22,707 men in Taiwan. Lancet 2:1129, 1981.
Bikel I, Montano X, Agha M, et al. SV40 small t antigen enhances the transformation activity of limiting concentrations of SV40 large T antigen. Cell 48:321, 1987.
Clertant P, Seif I. A common function for polyoma virus large T and papilloma virus E1 proteins? Nature 311:276, 1984.
Epstein M, Achong B, Barr Y. Virus particles in cultured lymphoblasts from Burkitt's lymphoma. Lancet 1:702, 1964.
Green M. Transformation and oncogenes: DNA viruses. In: Fields BN (ed), Virology. New York: Raven Press, 1985.
Green M, Brackman K, Lucher L, et al. Antibodies to synthetic peptides targeted to the transforming genes of human adenoviruses: an approach to understanding early viral gene function. Curr Top Microbiol Immunol 109:167, 1983.
Henderson A, Ripley S, Heller M, et al. Chromosome site for Epstein-Barr virus DNA in a Burkitt tumor cell line and in lymphocytes growth-transformed in vitro. Proc Natl Acad Sci USA 80:1987, 1983.
Kreider JW, Howeh M, Wolfe S, et al. Morphological transformation in vivo of human uterine cervix with papilloma virus from condylomata acuminata. Nature 317:639, 1985.
Macnab JCM, Walkinshow SA, Cordiner JW, et al. Human papillomavirus in clinically and histologically normal tissue of patients with genital cancer. N Engl J Med 315:1052, 1986.
Mitchell H, Drake M, Medley G. Prospective evaluation of risk of cervical cancer after cytological evidence of human papilloma virus infection. Lancet 1:573, 1986.
Rassoulzadegan MZ, Naghashfar A, Cowie A, et al. Expression of the large T protein of polyoma virus promotes the establishment in culture of "normal" rodent fibroblast cell lines. Proc Natl Acad Sci USA 80:4354, 1983.
Reid R, Stanhope C, Herschman B, et al. Genital warts and cervical cancer. I. Evidence of an association between subclinical papilloma virus infection and cervical malignancy. Cancer 50:377, 1982.
Ruley HE. Adenovirus early region 1A enables viral and cellular transforming genes to transform primary cells in culture. Nature 304:602, 1983.
Ruley HE, Moomaw JF, Maruyama K. Avian myelocytomatosis virus myc and adenovirus early region 1A promote the in vitro establishment of cultured primary cells. Cancer Cells 2:481, 1984.
Salahuddin SZ, Ablashi DV, Markham DD, et al. Isolation of a new virus, HBLV, in patients with lymphoproliferative disorders. Science 234:596, 1986.
Schrier P, Bernards R, Vaessch R, et al. Expression of class I major histocompatibility antigens switched off by highly oncogenic adenovirus 12 in transformed rat cells. Nature 305:771, 1983.

Tanaka K, Isselbacher KJ, Khoung G, et al. Reversal of oncogenesis by the expression of a major histocompatibility complex class I gene. Science 228:26, 1985.

Tiollais P, Pourcel C, Dejean A. The hepatitis B virus. Nature 317:489, 1985.

Viruses. 2nd ed, revised. Cold Spring Harbor, NY: Cold Spring Harbor Laboratory, 1985.

RNA Viruses

Gallo RC, Wong-Staal F. Retroviruses as etiologic agents of some animal and human leukemias and lymphomas and as tools for elucidating the molecular mechanism of leukomogenesis. Blood 60:545, 1982.

Hayward WS, Neel BG, Astrin SM. Activation of a cellular onc gene by promoter insertion in ALV-induced lymphoid leukosis. Nature 290:475, 1981.

Iba H, Takeya T, Cross FR, et al. Rous sarcoma virus variants that carry the cellular src gene instead of the viral src gene cannot transform chicken embryo fibroblasts. Proc Natl Acad Sci USA 82:4424, 1984.

Lowy DR. Transformation and oncogenes: retroviruses. In: Fields BN (ed), Virology. New York: Raven Press, 1985.

Neel BG, Hayward WS, Robinson HL, et al. Avian leukosis virus-induced tumors have common proviral integrations sites and synthesize discrete new RNAs: oncogenesis by promoter insertion. Cell 23:323, 1981.

Payne GS, Bishop JM, Varmus HE. Multiple arrangements of viral DNA and an activated hos oncogene in bursal lymphomas. Nature 295:209, 1982.

Shih CK, Linial M, Goodenow MM, et al. Nucleotide sequence t' of the chicken c-myc coding region: localization of a noncoding exon that is absent from c-myc transcription in most avian leukosis virus-induced glycoproteins. Proc Natl Acad Sci USA 81:4657, 1984.

Spector DH, Varmus HE, Bishop JM. Nucleotide sequences related to the transforming gene of avian sarcoma virus are present in DNA of uninfected vertebrates. Proc Natl Acad Sci USA 75:4102, 1978.

Takeya T, Hanafusa H. Structure and sequence of the cellular gene homologous to the RSV src gene and the mechanism for generating the transforming virus. Cell 32:881, 1983.

Weiss R, Teich N, Varmus H, Coffin J (eds). RNA Tumor Viruses: Molecular Biology of Tumor Viruses. 2nd ed. Cold Spring Harbor, NY: Cold Spring Harbor Laboratory, 1982.

4
Human T Cell Lymphotropic/Leukemia Viruses

Overview

Several human retroviruses have been identified from hematopoietic cells. Viruses related to animal oncorna viruses cause T cell malignancies. Viruses of the lentivirus group cause the acquired immunodeficiency syndrome. Whether other retroviruses cause other human diseases or malignancy is a subject of intense investigation.

Human retroviruses were not identified until 1978 when isolates were obtained from two American Black patients with mature T cell malignancies originally diagnosed as mycosis fungoides and Sézary syndrome. Since then a family of human retroviruses has been identified consisting in at least three diverse subgroups (Fig. 4.1). These viruses, the human T lymphotropic/leukemia viruses (HTLVs), share a number of properties (Table 4.1). HTLV-I and HTLV-II appear to be human oncorna viruses, whereas HTLV-III (now referred to as human immunodeficiency virus, HIV) belongs to the lentivirus group. They infect T cells primarily of the helper/inducer (CD4) subset. Such T cells play a central role in regulation of the immune response by promoting the differentiation of antibody-producing B cells, activating macrophages, and promoting the generation and expansion of cytotoxic T cells.

The HTLVs cause a wide spectrum of morphological and functional changes in infected T cells ranging from formation of multinucleated giant cells, stimulation of T cell mitogenesis, and secretion of lymphokines to outright cell death. These viruses appear to have a common origin and a genomic organization and life cycle similar to those of animal retroviruses. They share a related Mg^{2+}-dependent reverse transcriptase, a similar major capsid protein (p24), and immunologically cross-reactive core and envelope antigens.

Despite many shared features, the viruses of this family cause at least two diseases that are of opposite natures. HTLV-I, the first of these viruses to be identified, is the cause of adult T cell leukemia (ATL), an often fatal malignant *proliferation* of CD4 lymphocytes (Tables 4.2 and 4.3). Just as lethal is the illness associated with HIV, the acquired immunodeficiency syndrome (AIDS), characterized by CD4 lymphocyte *depletion* and *immunosuppression*. Not surprisingly,

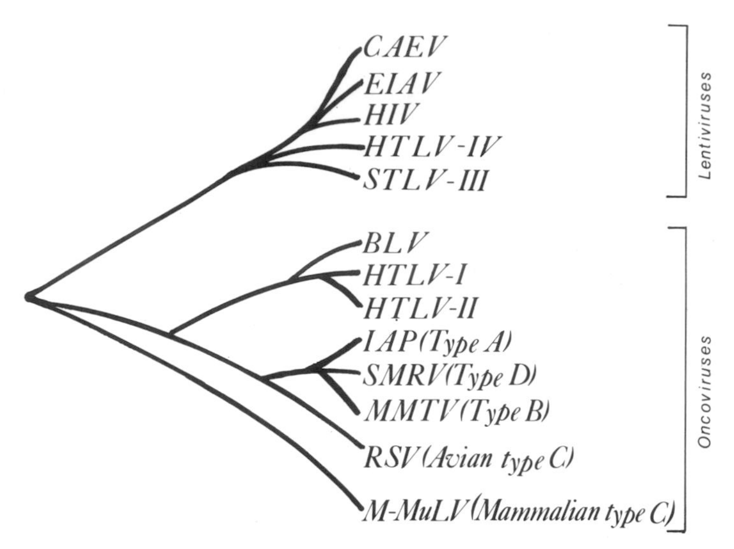

FIGURE 4.1. Classification of human T lymphotropic viruses. Note the relation of HIV and HTLV-I to other retroviruses. This depiction of genetic distance is based on a Fitch-Margoliash phylogenetic tree of the reverse transcriptase genes. CAEV = canine arthritis and encephalitis virus. EIAV = equine infectious anemia virus. HIV = I and II human immunodeficiency virus I and II. HIV-I was originally called HTLV-III, human T lymphotropic virus III or lymphodenopathy-associated virus (LAV). STLV-III = simian T lymphotropic virus III. BLV = bovine leukemia virus. HTLV-I, II, and IV = human T lymphotropic leukemia virus I, II, and IV. IAP = intracisternal A particle. SMRV = squirrel monkey retrovirus. MMTV = murine mammary tumor virus. RSV = Rous sarcoma virus. M-MuLV = Moloney murine leukemia virus. (Types A, B, C, and D refer to electron microscopic morphology)

HTLV-I causes transformation and immortalization of normal human T cells in vitro, whereas HIV infects and kills T cells.

HTLV-II initially was isolated from a T cell malignancy called hairy cell leukemia. This uncommon virus subgroup is related to HTLV-I in form and function.

Origin of the HTLVs

The vertebrate retrovirus family is divided into three subfamilies that differ in their biological properties: oncornaviruses, lentiviruses, and spumaviruses. The oncornaviruses are notable for their oncogenic potential. In contrast, the lentiviruses cause slow, progressive inflammatory diseases in vivo and often kill infected cells in vitro. The spumaviruses cause no apparent disease in animals but do cause vacuolation of tissue culture cells.

HTLV-I and HTLV-II belong to a family of lymphotropic oncornaviruses that

TABLE 4.1. Characteristics of the HTLVs.

Probable African origin
Exogenous, primate retroviruses
Lymphotropic for CD4, helper-inducer T cells
Transmission: blood, congenital, sexual (semen)
Promote syncytial formation in vitro
Promote T cell transformation or death in vitro
Severe immune system dysregulation in vivo
Major core protein: p24/25
Virion size: 100 to 120 nm diameter
Magnesium-dependent reverse transcriptase of 100 kD
HTLV-I: transforming virus associated with adult T cell leukemia
HIV: cytopathic lentivirus associated with AIDS; also infects macrophages, B lymphocytes, and certain cells of the brain

includes simian T cell virus I (STLV-I) and bovine leukemia virus (BLV) (Fig. 4.1). STLV-I is found in Old World monkeys in Africa and Asia, but to date it has not been found in primates native to the Western Hemisphere. HTLV-I is endemic in Africa, Central and South America, the Caribbean, and coastal areas of southwestern Japan.

Seroepidemiology of HIV suggests that it is endemic in Haiti and sub-Saharan Africa. Essex and co-workers have obtained results suggesting that the simian viruses (STLV-IIIs) related to HIV are common in Old World monkeys; however, these viruses do *not* produce an AIDS-like illness in the infected animals. It has been postulated that the human population acquired the virus from this endemic source. The epidemiology of HIV infections among western populations is different than among Africans, where the virus has predominantly heterosexual spread. One hypothesis suggests that the virus has undergone genomic changes leading to the now lethal varieties of HIV found in the United States and Western Europe.

Gallo and co-workers believe that HTLV-I may have spread to the Americas via slave trade and to Japan by European seamen (Wong-Staal & Gallo, 1985). HIV may have entered Haiti via immigrant Blacks from Africa and spread to North America and Europe through intimate contact of homosexuals with infected individuals. The current AIDS epidemic has been intensified by promiscuous sexual behavior and contaminated blood products.

TABLE 4.2. Characteristics of HTLV-I, the primary cause of adult T cell leukemia.

1. *All* leukemia cells contain at least one copy of the HTLV-I provirus.
2. Nontransformed lymphocytes do not contain HTLV-I.
3. Each HTLV-I virus induced malignancy is clonal.
4. Each cell in the transformed lymphocyte clone contains virus integrated into the same site; however, this site may vary from patient to patient.
5. Transformation by HTLV-I is probably mediated by *trans*-acting viral genes that activate the transcription of cellular growth genes.

TABLE 4.3. Characteristics of adult T cell leukemia.

- Malignant proliferation of CD4 T lymphocytes
- Distinct T cell morphology: vacuolated lymphoid cells with lobulated nuclei
- High frequency of interleukin 2 receptor-positive T cells
- Circulating antibodies to HTLV-I in >90% of patients and spouses
- Onset during adulthood
- Frequent cutaneous infiltration, lymphadenopathy, and hepatosplenomegaly
- Short survival time
- Hypercalcemia, often with lytic bone lesions
- Distributed worldwide, but clustered in certain locations
    - Islands of Kyushu and Shikoku in southwestern Japan
    - Caribbean basin
    - Blacks in southeastern United States

Transformation by HTLV-I and HTLV-II and Adult T Cell Leukemia

As described in Chapter 3, retroviruses can be divided into two general categories. The "chronic" retroviruses lack a true oncogene and are replication-competent (e.g., the slowly transforming viruses). They require a long latency for disease induction because they produce transformation by a *cis*-acting mechanism: The integration of viral long terminal repeat (LTR) promoter or enhancer sequences or both is thought to activate nearby cellular genes (see Chapter 5 for further details). This process is inefficient, and only a rare cell gives rise to a tumor.

The "acute" retroviruses are often replication-defective, but because they carry a cell-derived gene (oncogene) that codes for a product that can initiate or maintain transformation (or both) they rapidly produce disease in vivo and can directly transform cells in vitro. They do not require a *specific* integration site (although they must be integrated somewhere in the host genome) and can be said to induce transformation by a *trans*-acting mechanism.

HTLV-I and HTLV-II share properties with both the acute and chronic animal retroviruses. Adult T cell leukemia appears after a long latent infection with HTLV-I, yet HTLV-I can directly transform T cells in vitro. This group of viruses does not carry known cell-derived oncogenes, nor do they have a conserved site of provirus integration. Instead, they have a conserved gene called *tat*-I (transactivator of transformation) that correlates with transformation but is unrelated to known retroviral genes or proto-oncogenes. This gene, like the transforming genes of DNA tumor viruses (see Chapter 3), functions to activate host cell transcription. Transformation is postulated to result from selective transcription of host cell genes that affect lymphocyte proliferation. Experiment with transgenic mice have shown tat incorporation in these animals results in multiple soft tissue tumors suggesting tat may function as an oncogene in this circumstance (see Chapter 12). Alternatively HTLV-I may cause expansion of the proliferating T

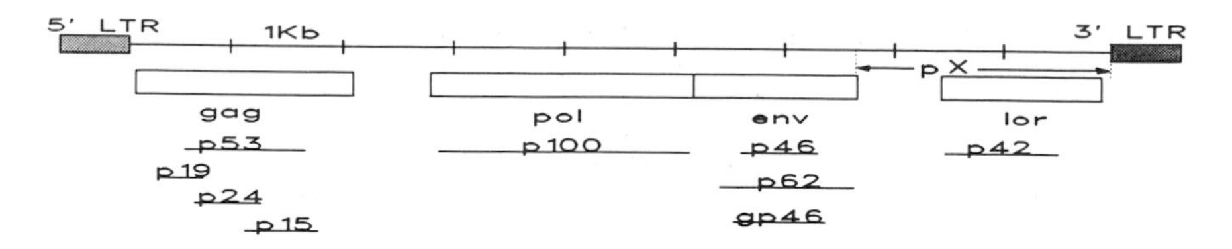

FIGURE 4.2. Proviral structures of HTLV-I and HTLV-II viruses. Genes and the derivative proteins are shown.

cell pool allowing an increased frequency of chromosomal translocation or rearrangements that may result in activation of endogenous protooncogenes such as c-myc (see Chapters 5 and 6).

Genome Structure of HTLVs

Several HTLV isolates have been completely sequenced. The proviruses are about 9 kilobases (kb) long, and the genome has essentially the same structural organization as the animal retroviruses outlined in Chapter 3 (Fig. 4.2), i.e., 5′ *gag-pol-env* 3′ flanked on both ends by regulatory LTRs.

When sequences of the LTRs of HTLV-I and HTLV-II are compared, homologies are limited to regions at the 5′ and 3′ ends; these conserved regions are necessary for viral integration and transcription. Tandem 21 base pair repeats also are conserved. These regions appear to function as transcriptional enhancer elements. Retroviral enhancer sequences are known to influence tissue tropism and the leukemogenicity of leukemia viruses.

The HIV viruses are actually a group of distinct but related viruses (Fig. 4.1). Any two isolates may differ by as much as 10% in their coding sequence. The greatest variability has been found in the extracellular portion of the viral envelope gene, whereas the transmembrane portion is highly conserved. In addition to structural *gag* and *env* genes and a gene that encodes the reverse transcriptase (*pol*), the genome of all HIV strains encodes at least three additional polypeptides (Fig. 4.3). The *tat*-III gene includes two coding exons, one located just 5′ to

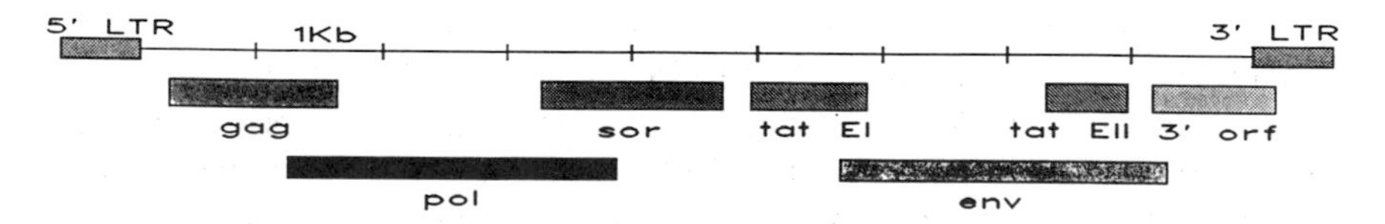

FIGURE 4.3. Proviral structure of human immunodeficiency virus. Note that several of the genes overlap one another.

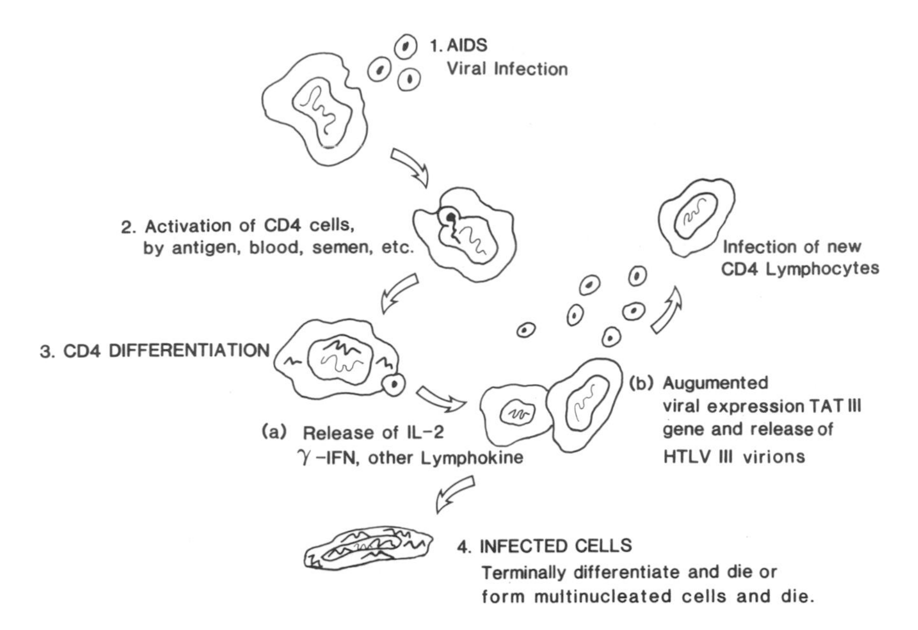

FIGURE 4.4. Biological events leading to HIV (HTLV-III) infection, viral expression, and death of CD4 cells. At the present time it is unclear whether T helper cells are depleted by formation of multinucleated giant cells, terminal differentiation, or a more direct cytotoxic mechanism.

env and a second located within an alternative reading frame of the *env* gene. It encodes the 24-kD transactivator protein, which acts to augment expression of genes under control of the HIV LTR. Expression of *tat*-III in infected cells greatly stimulates production of viral products by increasing translation of mRNA. (Similar genes *tat*-I and *tat*-II in HTLV-I and HTLV-II function at the transcriptional level and are responsible for the lymphocyte transformation caused by these viruses.) HIV also contains two genes with no equivalents within the HTLV-I and HTLV-II families. The *sor* gene, whose 5′ end overlaps the *pol* gene, encodes a protein of 23 kD. Although antibodies to this protein are found in sera from AIDS patients, its function is not known. Finally, a 3′ open reading frame (ORF) is located four base pairs downstream from the stop codon of the *env* protein. This gene extends into the U3 element of the 3′ LTR. It codes for a 27-kD protein, which also elicits antibodies. Its function also is unknown at present. Figure 4.4 depicts the possible events surrounding HIV infection of CD4 lymphocytes. These lymphocytes are thought to form multinucleated giant cells by cell fusion or to undergo terminal differentiation and cell death.

TABLE 4.4. Acquired immunodeficiency syndrome.

1. HIV infection causes death of CD4 (helper/inducer) T cells, which increases susceptibility to various opportunistic infections.
2. High risk groups include the following.
   a. Homosexual and bisexual men
   b. Intravenous drug abusers
   c. Recipients of blood products, particularly hemophiliacs
   d. Sexual partners and infants of high-risk populations
   e. Prostitutes and their contacts
3. There is a high prevalence of Kaposi's sarcoma in gay AIDS patients.
4. Symptoms and typical pathogens are as follows.
   a. Pulmonary symptoms (dyspnea, fever, cough, chest pain)
      Pneumocystis carinii
      Mycobacterium avium intracellulare
      Mycobacterium tuberulosis
      Legionella spp.
      Cytomeglovirus spp.
      Cryptococcus neoformans
   b. Gastrointestinal symptoms (diarrhea, cramping, fever)
      Cryptosporidium
      Isospora spp.
      Cytomegalovirus
      Campylobacter
   c. Central nervous system (dementia, encephalopathy, meningitis)
      Toxoplasma gondii
      Cytomegalovirus
      Cryptococcus neoformans
      Papovavirus (J-C virus) (progressive multifocal leukoencephalopathy)
   d. Fever of unknown origin, weight loss
      Mycobacterium avium-intercellulare
      Mycobacterium tuberculosis
      Toxoplasma gondii
      Other atypical opportunistic infections
      Malignancy: lymphoma, Kaposi's sarcoma
5. Chronic infection: Virus may establish latent infection in CNS tissues as well as lymphoid tissue. Incubation of several years is possible.
6. Diagnosis: Made on clinical grounds. Serum HIV antibody and antigen tests confirm infection; isolation of virus confirms diagnosis.
7. Treatment: Supportive and specific for each pathogen. New antiviral chemotherapies are under investigation. A drug that inhibits viral replication, 3′-azido-3′-deoxythymidine, has been released.
8. Vaccine: Defective live virus, killed virus, and recombinant subunit vaccines are being developed for clinical testing.

HIV and AIDS

The acquired immunodeficiency syndrome (Table 4.4) is an epidemic disease characterized by severe opportunistic infections. These infections are the consequence of the immunological chaos caused by CD4 helper/inducer T cell depletion following infection with HIV. The sera from more than 95% of patients with AIDS and pre-AIDS as well as a significant number of high-risk homosexuals, intravenous drug abusers, and hemophiliacs have specific anti-HIV antibodies; and as many as 20% to 40% of the entire population of some Central African countries are now known to be anti-HIV antibody positive. Because HIV only transiently infects human CD4 lymphocytes before they are killed, isolation and cloning of the virus had to await the discovery of a cell line permissive for viral replication but resistant to its cytopathic effect. This cell line, H9, was developed by Mikulas Popovic working at the National Cancer Institute under Robert Gallo. Prior to this discovery, workers at the Pasteur Institute, under Luc Montagnier, also had isolated a virus called lymphadenopathy-associated virus (LAV) from a patient with the AIDS-related complex (ARC). Subsequently, these viruses were shown to be closely related members of the human lentivirus family.

Although HIV is distantly related to oncornaviruses HTLV-I and HTLV-II by DNA homology criteria, it shares many properties with lentiviruses, e.g., cytopathic effects in vitro, neurotropism, morphology, and DNA sequence homology (Table 4.5). Studies of more than 100 HIV isolates by Southern hybridization have shown as much as 20% divergence in the amino acid sequences of the envelope proteins. In fact, another human immunodeficiency virus, HIV 2, has been associated with AIDS in West Africa. This result does not bode well for development of successful recombinant vaccines to envelope proteins. However, there are several highly conserved membrane receptor attachment sequences that might be exploited for this purpose.

TABLE 4.5. Lentiviruses.

1. Cause exogenous nononcogenic chronic latent infections
2. The *tat* gene(s): transactivating transcriptional regulation
3. Large (90–135 kD) glycosylated envelope protein
4. A 40-kD transmembrane protein
5. Major core protein p24
6. Caprine arthritis encephalitis virus (CAEV): crippling arthritis and central nervous system infection
7. Equine infectious anemia virus (EIAV): intermittent anemia, fever, and immune complex glomerulonephritis
8. Bovine visna-like virus (BUV): lymphadenopathy and persistent lymphocytosis
9. Visna: chronic pneumonitis, encephalitis, and wasting disease of sheep
10. Zwoegerziekte: pneumonia and meningoencephalitis in sheep
11. Progressive pneumonia virus (PPV): pneumonia in sheep and goats

Rous described the first oncogenic retroviruses nearly 70 years ago. Why did it take so long to demonstrate conclusively the existence of human retroviruses? Part of the answer is the fact that none of the HTLVs appears to carry oncogenes that can directly transform T cells without special growth-promoting lymphokines such as interleukin 2 (T cell growth factor). Part of the answer is that until the emergence of ATL and AIDS, no consistent horizontal transmission or clustering of leukemia or human cancer could be demonstrated. Another part of the answer lies in the advances in molecular biology that have permitted the identification of viral "footprints" in the absence of transmissible virus. Finally, the more common forms of human leukemias and solid cancers still are not associated with viruses. Their genesis perhaps is related to other mechanisms that activate proto-oncogenes (see Chapter 5). However, it is conceivable that by using sensitive DNA probe techniques human retroviral genes may be found to be associated with other malignancies.

Bibliography

HTLVs: Reviews

Haase AT. Pathogenesis of lentivirus infections. Nature 322:130, 1986.

Weiss A, Hollander H, Stobo J. Acquired immunodeficiency syndrome: epidemiology, virology and immunology. Annu Rev Med 36:545, 1985.

Wong-Staal F, Gallo RC. Human T-lymphotropic retroviruses. Nature 317:395, 1985.

HTLV-I and HTLV-II

Barré-Sinoussi F, Chermann J, Rey F, et al. Isolation of a T-lymphotropic retrovirus from a patient at risk for acquired immune deficiency syndrome (AIDS). Science 220:868, 1983.

Greene WC, Leonard WJ, Wano Y, et al. Transactivator gene of HTLV-II induces IL-2 receptor and IL-2 cellular gene expression. Science 232:877, 1986.

Kalyanaraman V, Sarngadharan M, Miyoshi I, et al. A new subtype of human T-cell leukemia virus (HTLV-II) associated with a T-cell variant of hairy cell leukemia. Science 218:571, 1982.

Poiesz B, Ruscetti F, Gazdar A, et al. Detection and isolation of type C retrovirus particles from fresh and cultured lymphocytes of a patient with cutaneous T-cell lymphoma. Proc Natl Acad Sci USA 77:7415, 1980.

HIV and AIDS

Centers for Disease Control. Kaposi's sarcoma and Pneumocystis pneumonia among homosexual men in New York and California. MMWR 30:305, 1981.

Clumeck N, Mascart-Lemone F, De Mauberge J, et al. Acquired immune deficiency syndrome in black Africans. Lancet 1:478, 1983.

Curran JW, Lawrence DN, Jaffe H, et al. Acquired immunodeficiency syndrome (AIDS) associated with transfusion. N Engl J Med 310:69, 1984.

Dalgleish AG, Beverley L, Clapham PR, et al. The CD4 (T4) antigen is an essential component of the receptor for the AIDS retrovirus. Nature 312:20, 1984.

Gallo R, Salahuddin S, Popovic M, et al. Frequent detection and isolation of cytopathic retroviruses (HIV) from patients with AIDS and at risk for AIDS. Science 224:500, 1984.

Guyader M, Emerman M, Soniga P, et al. Genome organization and transactivation of the human immunodeficiency virus type 2. Nature 326:662, 1987.

Hahn BH, Shaw GM, Arya SK, et al. Molecular cloning and characterization of the HIV virus associated with AIDS. Nature 312:166, 1984.

Kanki PJ, Bakin F, Boup SM, et al. New human T-lymphotropic retrovirus related to simian T-lymphotropic virus type III (STLV-III-AGM). Science 232:238, 1986.

Klatzmann D, Barré-Sinoussi F, Nugeyre MT, et al. Selective tropism of lymphadenopathy associated virus for helper-inducer T-lymphocytes. Science 225:59, 1984.

Levy JA, Hoffman AD, Kramer SM, et al. Isolation of lymphocytopathic retroviruses from San Francisco patient with AIDS. Science 225:840, 1984.

McDougal M, Kennedy S, Sligh JM, et al. Binding of HIV/LAV to T4$^+$ T cells by a complex of the 110 K viral protein and the T4 molecules. Science 231:382, 1986.

Montagnier L, Chermann J-C, Barré F, et al. A new human T-lymphotropic retrovirus: characterization and possible role in lymphadenopathy and acquired immune deficiency syndrome. In: Gallo RC, Essex M, Gross L (eds), *Human T-Cell Leukemia/Lymphoma Virus*. Cold Spring Harbor, NY: Cold Spring Laboratory, 1984, p. 363.

Popovic M, Sarngadharan M, Read E, et al. Detection, isolation, and continuous production of cytopathic retroviruses (HIV) from patients with AIDS and pre-AIDS. Science 224:497, 1984.

Robert-Guroff M, Kalyanaraman V, Blattner W, et al. Evidence for human T-cell lymphoma/leukemia virus infection of family members of human T-cell lymphoma/leukemia virus positive T-cell leukemia/lymphoma patients. J Exp Med 157:248, 1983.

Shaw GM, Hahn BH, Arya SK, et al. Molecular characterization of human T-cell leukemia (lymphotropic) virus type III in the acquired immune deficiency syndrome. Science 226:1165, 1984.

Sodroski J, Goh WC, Rosen C, et al. Role of the HTLV-III/LAV envelope in syncytium formation and cytopathicity. Nature 322:470, 1986.

Sodroski J, Rosen C, Haseltine W. Trans-acting transcriptional activation of the long terminal repeat of human T-lymphotropic viruses in infected cells. Science 225:381, 1984.

Starcich B, Ratner L, Josephs SF, et al. Characterization of the long terminal repeat sequences of HIV. Science 227:538, 1985.

Vieira J, Frank E, Spira TJ, et al. Acquired immune deficiency in Haitians: opportunistic infections in previously healthy Haitian immigrants. N Engl J Med 308:125, 1983.

Zagury D, Bernard J, Leonard R, et al. Long-term cultures of HIV infected T cells: a model of cytopathology of T cell depletion in AIDS. Science 231:850, 1986.

AIDS Vaccine

Alter HJ, Eichberg JW, Masur H, et al. Transmission of HIV infection from human plasma to chimpanzees; an animal model for AIDS. Science 226:549, 1984.

Chakrabarti S, Robert-Guroff M, Wong-Staal F, et al. Expression of the HTLV-III envelope gene by a recombinant vaccinia virus. Nature 320:535, 1986.
Fisher AG, Ratner L, Mitsuya H, et al. Infectious mutants of HTLV-III with changes in the 3′ region and markedly reduced cytopathic effects. Science 233:655, 1986.
Hu S-L, Kosowski SG, McDalrymple J. Expression of AIDS virus envelope gene in recombinant vaccinia viruses. Nature 320:537, 1986.
Lasky LA, Groopman JE, Feenie CW, et al. Neutralization of the AIDS retrovirus by antibodies to a recombinant envelope glycoprotein. Science 233:209, 1986.
Sarin PS, Sun DK, Thornton AH, et al. Neutralization of HTLV-III/LAV replication by antiserum to thymosin α_1. Science 232:1135, 1986.

5
Cellular Proto-oncogenes

Overview

Retroviral oncogenes originally were derived from genes in eukaryotic cells. The seminal discovery that DNA sequences within normal, uninfected, nonmalignant cells were homologous to retroviral oncogenes was made in 1976. Cellular proto-oncogenes have exon and intron structures typical of eukaryotic genes. Some exon sequences are well conserved among vertebrate and invertebrate species. The retroviral life cycle suggests a mechanism whereby cellular genes may be transduced by the viruses. The conservation of cellular proto-oncogenes among species suggests a fundamental role for them. These proto-oncogenes can be grouped according to their function or location in the cell: growth factors, growth factor receptors, nuclear proteins, and membrane proteins. The subcellular locations and functions of these proteins suggest that in normal cells they play a role in growth, development, and differentiation. Aberrant growth and development are characteristic of cancer, and the hypothesis that abnormal "activated" oncogenes contribute to the neoplastic state is therefore attractive. Changes activating normal proto-oncogenes to transforming oncogenes could occur at the DNA, RNA, or protein level. In vivo carcinogenesis is clearly a multistep process. The fact that single activated oncogenes are unable to transform normal primary cells, whereas co-introduction of at least two activated oncogenes does lead to transformation supports the genetic basis for this multistep process.

Retroviral Oncogenes Are of Cellular Origin

RNA tumor viruses frequently contain genes that render the viruses rapidly transforming. These transforming genes are not necessary for viral growth and development. In 1976, using DNA restriction analysis and Southern blotting, Bishop, Varmus, and their co-workers discovered that normal, nonmalignant avian cells contain sequences homologous to retroviral oncogenes. These sequences were designated *proto-oncogenes*. Subsequently, a wide variety of vertebrate and nonvertebrate cell types were tested. All vertebrate and many nonvertebrate cells, including *Drosophila* (fruit fly) and yeast, were shown to contain sequences related to retroviral oncogenes. Each cell contains many proto-oncogenes. Because divergent life forms have growth and development in common, it is hypothesized that proto-oncogenes play some central role in these processes. On the other hand, sequences closely related to DNA tumor virus transforming genes have not been found in noninfected eukaryotic cells.

Structural analysis of eukaryotic proto-oncogenes has shown them to be typical of average genes in the organisms in which they are found. They contain exon coding sequences divided by intervening (intron) sequences. Various species of organisms show relative conservation of exon sequences, whereas introns are more divergent, a property common to many eukaryotic genes. A given proto-oncogene might be well conserved among species, whereas different proto-oncogenes are not particularly similar within a given species (Fig. 5.1). For example, mammalian and avian *myc* genes share a common exon–intron structure, whereas mammalian *myc* and *ras* genes are dissimilar. Yeast *ras* genes are typical of other yeast genes in their exon–intron structure but share sequence homology with mammalian and other vertebrate *ras* genes.

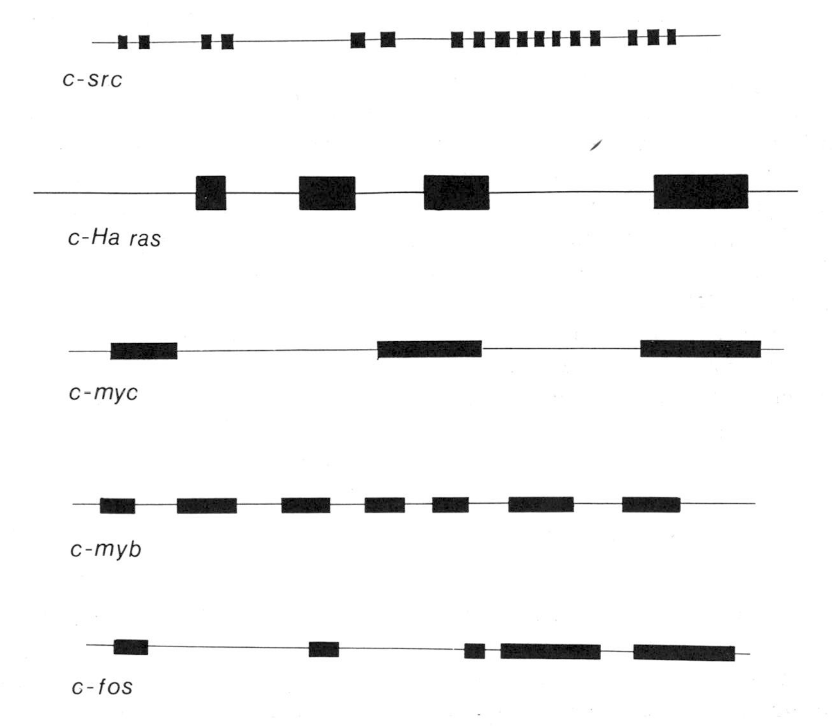

FIGURE 5.1. Examples of cellular proto-oncogenes. Different proto-oncogenes have different exon–intron structures. The general structures of these genes are shared among many species.

TABLE 5.1. Retroviral oncogenes.

Oncogene	Virus	Species	Protein product	Action	Subcellular localization
SRC family					
src	Rous sarc.v.[a]	Chicken	pp60	Tyrosine kinase	Plasma membrane[b]
yes	Y73 avian sarc.v.	Chicken		Tyrosine kinase	
ros	UR2 avian sarc.v.	Chicken		Tyrosine kinase	Cytoplasm, plasma membrane
rel	Reticuloendotheliosis v.	Chicken			
fes(fps)	Snyder-Theilen feline sarc.v. (Fujinami sarc.)	Cat (chicken)	p92(p98)	Tyrosine kinase	Cytoplasm, plasma membrane
fms	McDonough feline sarc.v.	Cat	gp140	Tyrosine kinase	Plasma membrane
fgr	Gardner-Rasheed feline sarc.v.	Cat		Tyrosine kinase	
abl	Abelson murine leuk.v.[c]	Mouse	p150	Tyrosine kinase	Plasma membrane
RAS family					
H-*ras*	Harvey murine sarc.v.	Mouse	p21	Binds GDP or GTP	Plasma membrane
K-*ras*	Kirsten murine sarc.v.	Mouse	p21	Binds GDP or GTP	Plasma membrane
Other					
myc	Avian myelocytomatosis v. MC29	Chicken	p58	Binds DNA	Nucleus
myb	Avian myeloblastosis v. AMV	Chicken	p75	Binds DNA	Nucleus
ets	Avian erythroblastosis v. E26	Chicken			
erb B	Avian erythroblastosis v.	Chicken	Truncated EGF[d] receptor	Growth factor receptor analog	Plasma membrane
fos	FBJ murine leuk.v.	Mouse	p55	Transcription factor	Nucleus
raf(mil)	3611 Murine sarc.v. (avian MH2 v.)	Mouse (chicken)		Serine, threonine kinase	Cytoplasm
mos	Moloney murine sarc.v.	Mouse		Serine, threonine kinase	Cytoplasm
sis	Simian sarc.v.	Monkey	PDGF[e] B-chain	Growth factor analog	Cytoplasm, extracellular

[a] Sarc.v. = sarcoma virus.
[b] Located on inner surface of plasma membrane.
[c] Leuk.v. = leukemia virus.
[d] EGF = epidermol growth factor.
[e] PDGF = platelet-derived growth factor.

TABLE 5.2. Oncogene products.

Oncogene products
Growth factor
Growth factor receptor
Tyrosine protein kinase
Serine/threonine kinase
GTP binding protein
DNA-associated protein
Transcription factor
Other membrane-associated proteins

Approximately 20 retroviral oncogenes are known, comprising five to seven groups defined by subcellular location or known function (Tables 5.1 and 5.2). The groups include protein kinases (tyrosine and serine/threonine), growth factors and growth factor receptors, GTP binding proteins, other membrane proteins, and nuclear proteins. Proto-oncogene sequences corresponding to each of the known oncogenes have been found in humans, mice, and chickens, the species most intensely studied. Chromosome locations of most proto-oncogenes have been mapped for humans; some proto-oncogenes also have been mapped for several other species. The human chromosome locations of the mapped proto-oncogenes are shown in Figure 5.2. These genes are dispersed among nearly every chromosome and have no special position relative to centromeres or chromosome termini.

How Did Retroviruses Acquire Cellular Genes?

The retroviral life cycle itself suggests a mechanism whereby cellular genes could have become incorporated into viral genomes (Fig. 5.3). As described in Chapter 3, retroviral DNA sequences are integrated in (apparently) random regions of the eukaryotic cellular genome. It is conceivable that molecular recombination sometimes puts cellular sequences under control of retroviral promoters. There is evidence that at times fusion occurs between viral and cellular DNA sequences such that the transcribed RNA is also a fusion product. The cellular RNA portion contains its introns and exons, typical of eukaryotic precursor mRNAs immediately after transcription. These RNAs are processed in the usual manner retaining exons side by side and excluding introns. The result is an RNA containing 5′ viral sequences, possibly viral 3′ sequences (e.g., the precursor for Rous sarcoma virus) and cellular exon sequences. Integrated proviruses generally exceed one copy per cell. Therefore in a given cell, normal viral RNA without recombined cellular sequences would also be expected. As transcription proceeds, multiple copies of both fusion and normal viral RNA are produced. Because 5′ viral sequences are retained in the recombinants, the fusion RNAs should be able to form dimer structures with other fusion RNAs or with normal viral RNAs and, provided they are the appropriate size, be packaged into viral capsids. During the

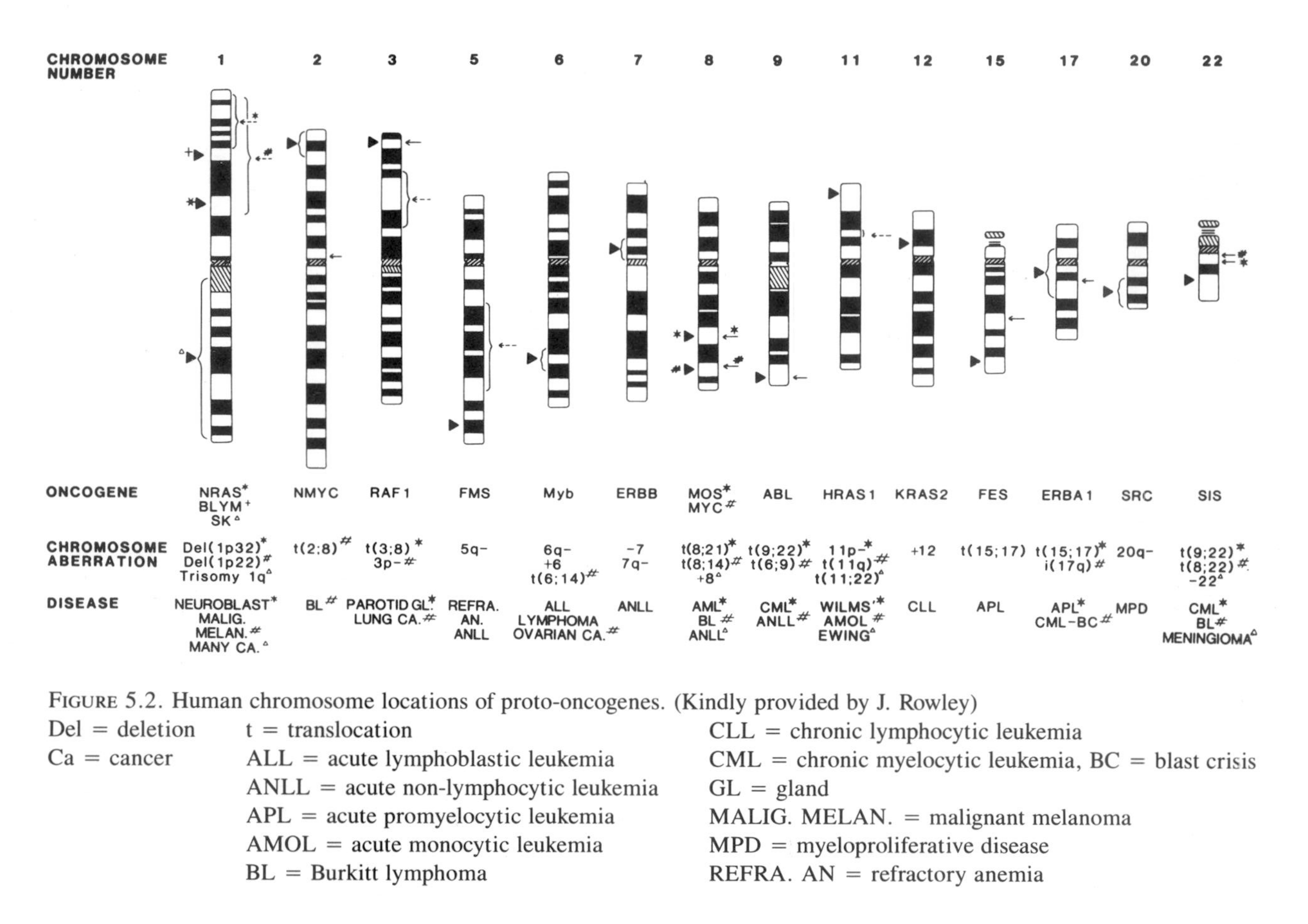

FIGURE 5.2. Human chromosome locations of proto-oncogenes. (Kindly provided by J. Rowley)

Del = deletion

t = translocation

Ca = cancer

ALL = acute lymphoblastic leukemia

ANLL = acute non-lymphocytic leukemia

APL = acute promyelocytic leukemia

AMOL = acute monocytic leukemia

BL = Burkitt lymphoma

CLL = chronic lymphocytic leukemia

CML = chronic myelocytic leukemia, BC = blast crisis

GL = gland

MALIG. MELAN. = malignant melanoma

MPD = myeloproliferative disease

REFRA. AN = refractory anemia

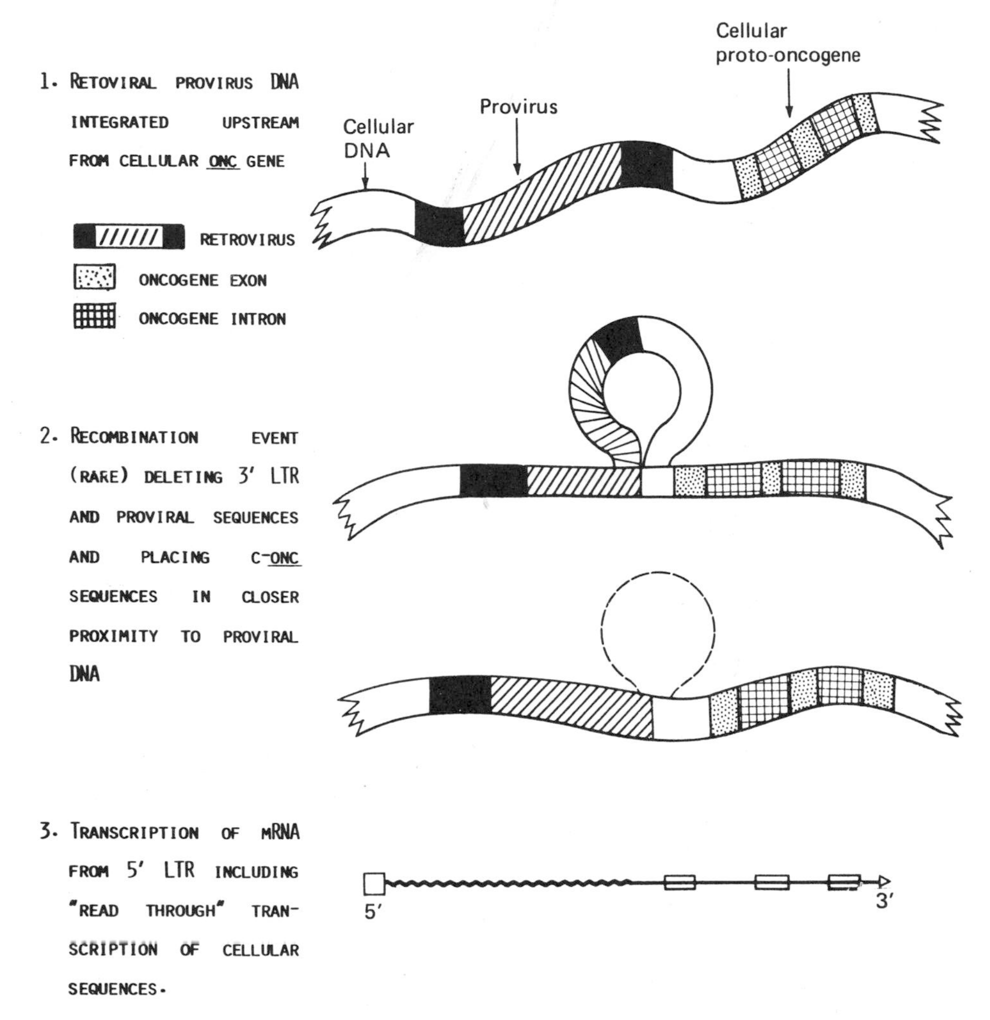

FIGURE 5.3 Hypothetical mechanism whereby a cellular proto-oncogene could be incorporated into a retroviral genome. See text for additional details. (Continued following page)

next round of viral infection, fusion RNAs containing processed proto-oncogene sequences linked to viral promoters and structural genes should be expressed as regular retroviral RNA. In particular, proviral DNA copies are made and incorporated into host genomes. Recombinant viruses would be preferentially selected in the natural state if neoplastic transformation and tumor production occurred as a result of viral infection.

The basic requirement for viral transduction is that viral and host genomes join or recombine into a unit that can be packaged as an infectious virus. The minimal unit requires 5′ and 3′ sequences that comprise the proviral long terminal repeats

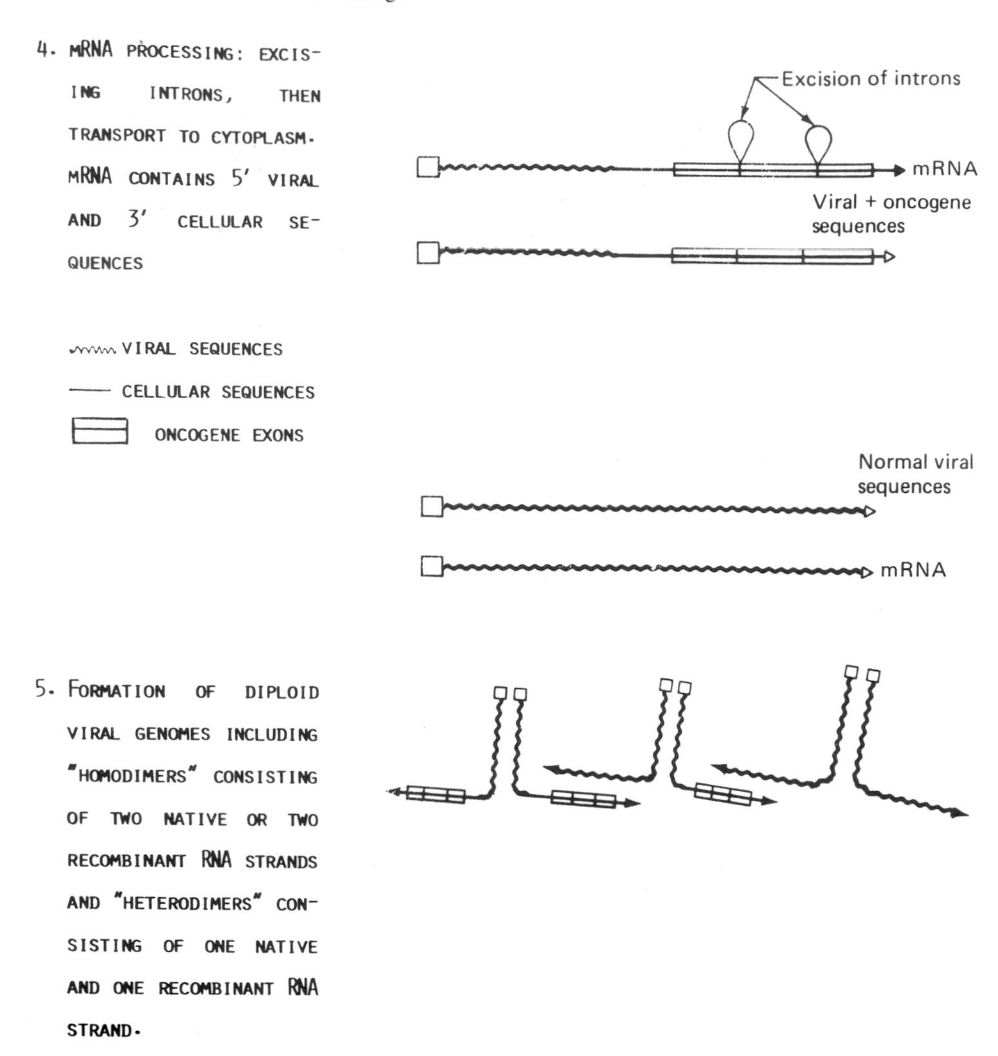

FIGURE 5.3. *Continued.*

(LTRs). Certain 5′ retroviral sequences are needed to provide a site for initiation of proviral DNA synthesis (primer binding site); and a packaging signal (Psi) is necessary for incorporation of viral mRNA into mature virions. All other functions (*gag*, *pol*, and *env*) may be provided by co-infecting helper viruses. Thus during the transduction process, a host proto-oncogene must somehow recombine at the 5′ and 3′ ends with viral sequences. Though the exact process is not clear, there are reasonable data suggesting that the 5′ recombination occurs between DNA molecules and the 3′ recombination occurs during reverse transcription, i.e., between RNA molecules after packaging and before the next round of proviral integration into newly infected cells. Evidence for DNA recom-

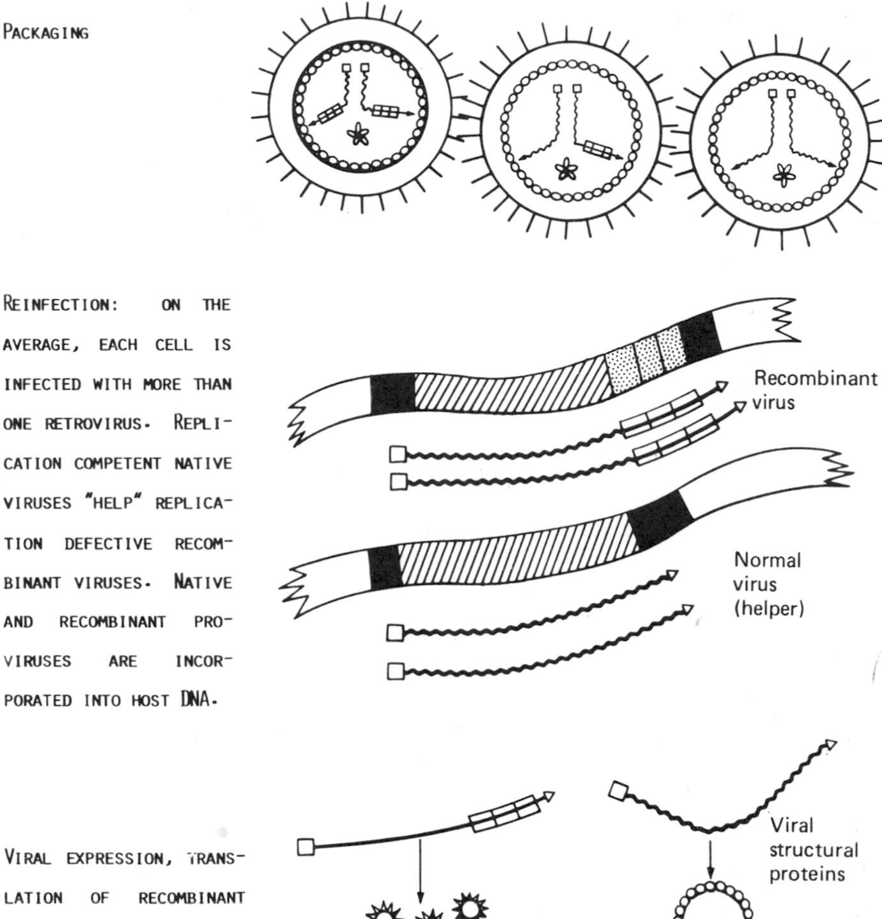

8. VIRAL EXPRESSION, TRANSLATION OF RECOMBINANT VIRAL mRNA PRODUCES ONC PRODUCT.

FIGURE 5.3. *Continued.*

bination is seen in the 5′ portion of certain v-*oncs* (*src*, *myc*) where remnant intron sequences are present. This finding can be explained by integration of a virus 5′ to a cellular proto-oncogene, deletion of 3′ viral sequences, and readthrough from the 5′ proviral LTR. Evidence for 3′ recombination via an RNA intermediate is seen in the v-*fps* gene of PRC II virus, where the virus contains the poly (A) tract derived from c-*fps* mRNA in addition to the poly (A) sequences in the 3′ LTR. Thus during either process truncation of 5′ and 3′ cellular sequences may occur and may contribute to the transforming potential of the v-*onc* gene. [Other examples include *erb* B (see Chapter 8) and *fos* (see Chapter 10).]

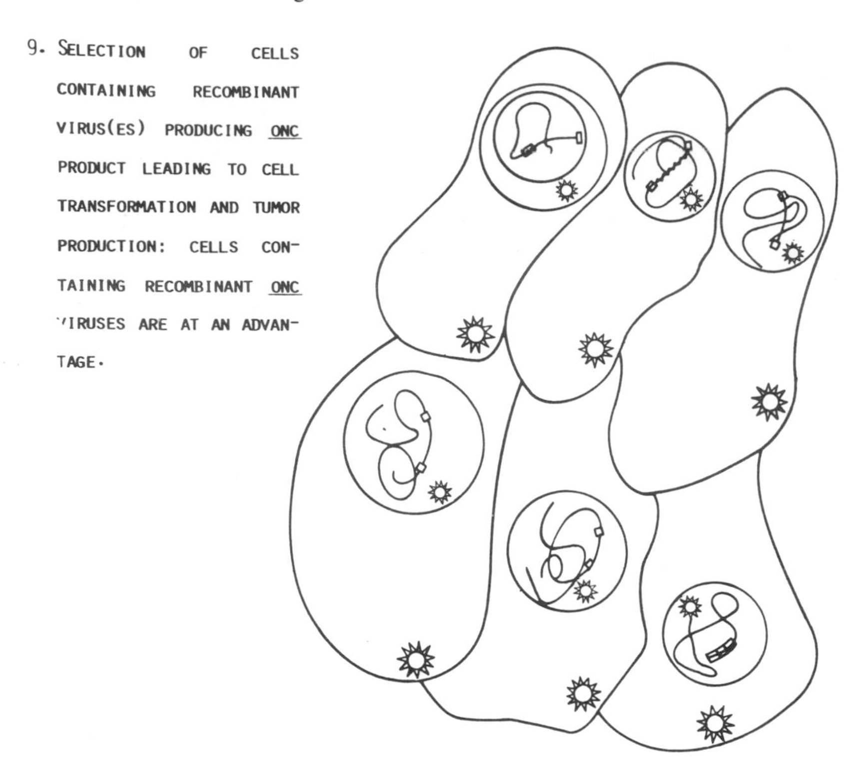

FIGURE 5.3. *Continued.*

It might be anticipated that this mechanism would result in fusion RNAs between viral sequences and random cellular genes. However, nontransforming cellular sequences alone would not be naturally selected because they would not result in tumor production. Even if the viruses were viable and could infect cells productively, the nontransforming genes would be lost by dilution with normal viral RNAs in subsequent rounds of infection. On the other hand, the avian erythroblastosis virus transduces two genes, *erb* A and *erb* B. The *erb* A sequences are homologous to the thyroid hormone receptor, whereas *erb* B is related to the gene for epidermal growth factor receptor (see Chapter 8). This virus has been selected for its transforming ability and might have resulted from a double recombination event. Although plausible, the actual details of the transduction mechanism have not been entirely elucidated.

The sites of oncogene incorporation within retroviral genomes are highly variable. For example, the *src* oncogene is located at the extreme 3′ end of the Rous sarcoma virus, whereas other viral oncogenes are fused with *gag* sequences (*myc* and *erb* B) or *env* sequences (*sis*). Murine and avian retroviruses have been the primary sources of retroviral oncogenes. As discussed in Chapter 4, few human retroviruses have been described to date, and none of these has yet been shown to carry a typical oncogene.

In addition to the oncogenes found in retroviruses, oncogenes have been detected by DNA transfection studies, e.g., genes belonging to the *ras* family. These genes are related to, but not identical with, v-*ras*-associated proto-oncogenes. Furthermore, proto-oncogenes related to v-*myc* have been detected in a variety of human tumors by hybridization of the DNA with *myc* probes. These genes include N-, R-, and L-*myc* (see Chapter 10). The locations of these genes also have been mapped to discrete regions of the human genome (Fig. 5.2).

Insertional Activation of Cellular Proto-oncogenes

Many retroviruses that lack cellularly derived oncogenes can transform cells by insertion near cellular proto-oncogenes. The juxtaposition of promoters and enhancers in the LTR portions of these viruses can augment expression of adjacent or nearby cellular genes. Common sites of viral integration have been identified using retroviral sequences as probes. A number of proto-oncogenes have been localized to these sites. Some of the common cellular proto-oncogenes identified in this fashion are listed in Table 5.3.

Most commonly, the cellular proto-oncogenes are activated by augmented transcription (e.g., *int*-1, *int*-2, *pim*). In other cases the cellular genes are interrupted by insertion of the provirus (e.g., *myc*, *erb* B). In some cases the proviral insertion is associated with secondary mutations in the proto-oncogene.

TABLE 5.3. Insertional activation of proto-oncogenes.

Gene	Insertional agent[a]	Tumor	Species
c-*myc*	ALV, CSV, RPV	B cell lymphoma	Chicken
	Mo-MLV, MCF-MLV	T cell lymphoma	Rat, mouse
	FeLV	T cell lymphoma	Cat
c-*erb* B	ALV	Erythroblastosis	Chicken
c-*Ha-ras*	MAV	Nephroblastoma	Chicken
c-*mos*	IAP	Plasmacytoma	Mouse
c-*myb*	Mo-MLV	Plasmacytoid lymphosarcoma	Mouse
int-1	MMTV	Mammary carcinoma	Mouse
int-2	MMTV	Mammary carcinoma	Mouse
Mlvi-1	Mo-MLV	T cell lymphoma	Rat
Mlvi-2	Mo-MLV	T cell lymphoma	Rat
Mlvi-3	Mo-MLV	T cell lymphoma	Rat
pim-1	MCF-MLV, Mo-MLV	T cell lymphoma	Mouse
pvt/mis	Mo-MLV	T cell lymphoma	Mouse

[a] ALV = avian leukosis virus. CSV = chicken synctial virus. FeLV = feline leukemia virus. IAP = intracisternal A particle. MAV = myeloblastosis-associated virus. MCF-MLV = mink cell focus-forming murine leukemia virus. MMTV = mouse mammary tumor virus. Mo-MLV = Moloney murine leukemia virus. RPV = ring-necked pheasant virus.

Cellular Proto-oncogenes: Normal Growth and Differentiation Genes

There are almost 40 identified proto-oncogenes including those transduced by retroviruses and those revealed by studies of tumor cells and DNA, e.g., translocation, insertional mutagenesis, and transfection of NIH-3T3 cells (see Chapter 11).

The prevalence and conservation of proto-oncogenes among divergent eukaryotes suggests a fundamental role for such genes. An obvious suggestion, especially in light of their transforming potential, is a role in cell growth, development, and differentiation. What has so far been learned about oncogene products?

Most characterization of oncogene products has been accomplished with viral oncogene proteins. However, the close correlation between the structures of proto-oncogenes and viral oncogenes makes it reasonable to presume that the "normal" eukaryotic proteins are highly similar to their viral counterparts. In fact, as we shall discuss, analysis of gene sequences has revealed subtle differences between some proto-oncogenes and viral oncogenes that may account for differences in normal versus transforming functions: Without exception, viral oncogenes contain mutations, deletions, or insertions relative to the corresponding proto-oncogene.

Proto-oncogene Function in Normal Cells

There appear to be a limited number of oncogenes and an even smaller number of functional oncogene groups. As mentioned previously, products of viral oncogenes include protein kinases, GTP binding proteins, chromosome/DNA binding proteins, growth factors, and growth factor receptors (Table 5.2). Cellular proto-oncogene products are expected to parallel these functions. Various *onc* proteins have been localized to discrete subcellular locations: nucleus, cytoplasm, membranes (nuclear or cytoplasmic), and the cell surface (Fig. 5.4). The ubiquitous distribution of proto-oncogenes throughout the vertebrate phylum, and in nonvertebrates as well, suggests a fundamental role for these genes. RNA transcription studies have revealed the expression of proto-oncogenes in a wide variety of normal (noninfected and nonmalignant) cells and tissue types. Some of these genes are expressed in many tissue types, whereas others are restricted to one or a few tissues. Mouse embryo studies have demonstrated that differential expression of proto-oncogenes is detectable during normal development. What roles could these gene products be playing?

In the case of growth factors and grow factor receptors, the role seems obvious. Many differentiated cells possess growth factor receptors on their cell surfaces and respond specifically to one or more growth factors. Growth factors generally are polypeptide hormones and include platelet-derived growth factor, epidermal

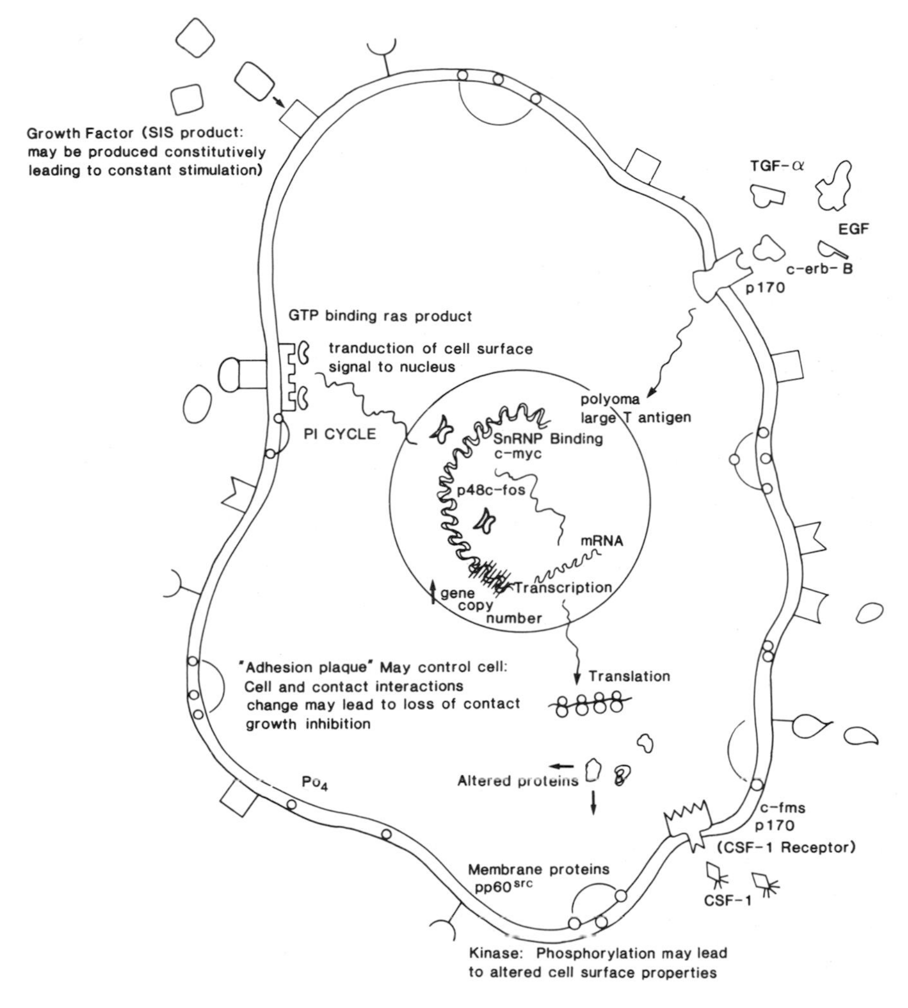

FIGURE 5.4. Proto-oncogene function in normal cells.

growth factor, insulin-like growth factors, T cell growth factor (interleukin 2), and colony-stimulating factor. Each of these hormones is known or suspected to have a corresponding cell surface receptor. Cells express one or more of these receptors depending on the cell type and stage of development. By mechanisms that remain to be elucidated, a growth factor's interaction with its receptor stimulates cell division in some cases and cell differentiation in others (for details see Chapter 8).

Nuclear proteins have numerous potential functions pertaining to development and differentiation. They could participate in DNA replication, or gene expression, or both. For example, they might act to open the DNA helix at particular promoter sites. Individual genes might encode specific nuclear binding proteins, enabling increased expression. Differentiated cells might possess sets of such proteins that in some manner determine which subset of the complete genome is expressed. These gene products could regulate transcription, processing of RNA, or gene structure (e.g., methylation of DNA). DNA binding proteins might enhance site-specific recombination; for example, generation of specific immunoglobulin genes from germ line precursors in B cells.

Cytoplasmic proto-oncogene proteins could potentially be involved with translational activities. Such proteins might act to stabilize mRNA or to increase its rate of degradation. They could act as protein initiation or elongation factors. They might subtly change ribosomes to favor one class of mRNAs over another. Other roles for cytoplasmic regulatory proteins include acting as second messenger intermediates between the cell surface and the nucleus.

Cellular membrane proteins include receptors for growth factors or differentiation signals. Membrane proteins could also act as channels or pores to permit or exclude entry of small molecules or ions, e.g., amino acids, glucose, or calcium. They might be involved with cellular "contact inhibition," the property of normal cells that prevents them overgrowing each other. They could be involved in maintenance of a proper cytoskeleton and hence cell shape and size.

Nuclear membrane proteins, especially those associated with small nuclear ribonuclear proteins, might be involved with RNA splicing or other steps in the processing of mRNA. They could permit or exclude "messengers" between cell surface and nucleus.

The first oncogene product to be characterized in detail was that encoded by the *src* gene of Rous sarcoma virus. Called pp60^{src}, it is a 60-kilodalton (kD) phosphoprotein with tyrosine kinase activity. This protein is associated with the cell plasma membrane. A number of independent oncogenes, among them *ros* and *abl* (see Table 7.1), also have been found to encode tyrosine kinase phosphoproteins. The phosphorylation of tyrosine by cellular kinases is relatively unusual; serine and threonine are the more frequent targets of cellular protein kinases. The tyrosine kinase activity of pp60^{src} increases dramatically in differentiated cells of various types, and it is hypothesized that this enzyme somehow regulates cellular differentiation.

The *ras*-like oncogenes, including Harvey-, Kirsten-, and N-*ras*, encode 21-kD proteins, which bind GTP and are localized to the inner plasma membrane. They may prove to be involved with "signal transduction" between environment (plasma membrane) and cell nucleus. The *ras* proteins share many characteristics with G proteins, proteins that link plasma membrane receptor molecules with various enzymes that generate cytoplasmic second messengers. (See discussion of transmembrane signaling, Chapter 8.)

The *myc* genes encode a nuclear protein that may be associated with the small nuclear ribonucleoproteins. The *fos* gene also encodes a nuclear protein, appearing within 1 hour of stimulation of quiescent cells by mitogens. These and other members of the nuclear proto-oncogene group probably play important roles in gene expression by regulating RNA processing, transcription, etc.

The product of the *sis* gene is a 28-kD analogue of platelet-derived growth factor. The *erb*-B product is a truncated version of the receptor for epidermal growth factor. The *fms* product is related to the receptor for colony-stimulating factor (CSF)-1. The *erb*-A product encodes a protein similar to the receptor for thyroid hormone.

Each of these proto-oncogene-coded functions could potentially play a role in normal cellular growth or differentiation as well as in neoplastic transformation. Here then is physical evidence in support of the oncogene hypothesis: Normal cells contain genes that, at least in their modified viral forms, are capable of inducing neoplastic transformation. These data are not the same as proving that the cellular proto-oncogenes themselves cause, or are involved in causing, cancer in intact organisms. However, it does provide a basis for further investigation into how such cellular oncogene-related sequences might be involved in neoplasia.

Activation of Cellular Proto-oncogenes and Human Cancer

The oncogene hypothesis proposes that normal genes involved in development or differentiation may be altered in such a way that their products transform the cell to neoplastic growth. Each of the normal mechanisms mentioned in the preceding section could conceivably be altered in one or more ways that would lead to malignant transformation. Potential mechanisms are numerous. Some of them are outlined here and evidence is provided to support a few. Details of individual oncogenes are presented in subsequent chapters.

Changes resulting in alteration of proto-oncogene (or protein) to transforming oncogene (or protein) could occur at the DNA, RNA, or protein level (Table 5.4).

DNA Changes

The gene itself might be changed by rearrangement, insertion, deletion, or point mutation. It could be amplified to an increased copy number. Examples include the following.

1. *int*: These sequences have been defined by their activation after nearby insertion of mouse mammary tumor viruses (see Chapter 11).
2. *myc*: Amplified *myc* genes have been noted in some tumor cell lines and in small cell carcinomas of the lung. The *myc* genes also are activated by insertion, e.g., induction of malignancy by avian leukosis virus (see Chapter 10).
3. *erb* B: The transforming *erb* B gene contains a deletion of 5′ and 3′ proto-*erb* B sequences. These sequences are thought to encode regulatory elements, the lack of which results in constitutive stimulation for growth and division (see Chapter 8).

TABLE 5.4. Potential mechanisms that change proto-oncogene or *onc*-protein to oncogenic gene or protein.

DNA changes
Rearrangement
Insertion/deletion
Point mutation(s)
Amplification → increased copy number
RNA changes
Strong promoter insertion → increased copy number
Enhancer → increased copy number
Processing mutation
Fusion message
Increased or decreased stability of mRNA
Increased or decreased ribosome affinity
Protein changes
Too much/too little
Altered function
Altered stability

4. *ras*: Proto-oncogene sequences can be isolated from normal cells, *ras*-containing virus-infected cells, and malignant cells or cell lines. These sequences have been cloned using bacterial vectors. It was observed that cloned viral *ras* genes and cloned malignant cell-derived *ras* genes are able to transform NIH 3T3 cells. Normal *ras* genes, however, are not transforming under the same conditions. The difference between transforming and non-transforming *ras* genes frequently involves single base pair changes at particular sites, as demonstrated by DNA sequence analysis (see Chapters 6 and 9).

RNA CHANGES

Increased levels of proto-oncogene mRNA might result from increased gene copy number, increased promotor usage, or insertion of a stronger promoter upstream from the gene coding regions (Fig. 5.5). Rearrangement of DNA sequences might also provide a strong promoter for proto-oncogene sequences. Displaced enhancer sequences could lead to increased mRNA expression. The mRNA itself might be processed differently; it may be part of a fusion message or be more or less stable than the unaltered mRNA. Examples include the following.

1. CML Ph^1 chromosome. Here translocation leads to an altered *abl* proto-oncogene mRNA (Fig. 5.6).
2. Burkitt's lymphoma cells. The *myc* gene is often translocated from chromosome 8 to one of the immunoglobulin gene loci: heavy chain (chromosome 14), κ light chain (chromosome 2), or λ light chain (chromosome 22), presumably putting *myc* in an actively transcribed region of DNA in B lymphocytes (see Fig. 6.4).

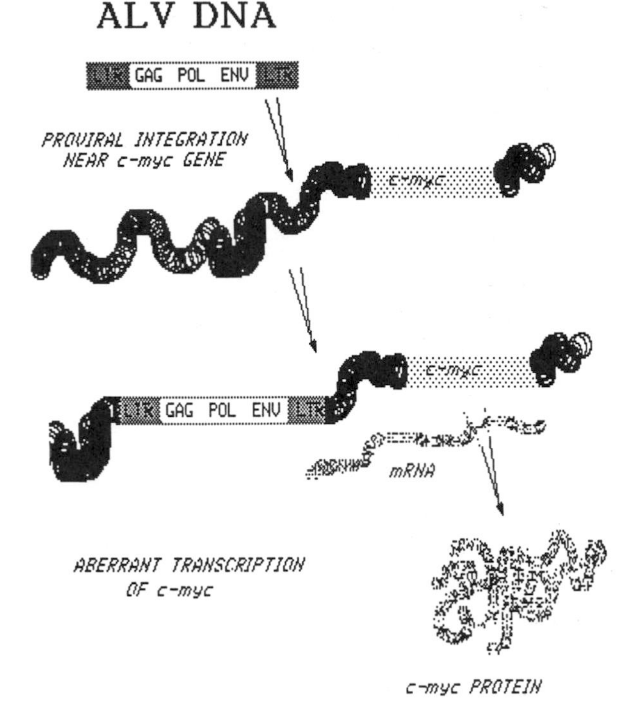

FIGURE 5.5. Insertion of a strong promoter upstream from a c-*onc* gene. Activation of several c-*onc* genes occurs by insertion of retroviral proviruses upstream from the gene. For example, c-*myc* is activated by insertion of avian leukosis virus in some avian hematopoietic malignancies.

Protein Changes

There could be too many or too few protein molecules as a result of altered mRNA or altered translation. The abnormal protein might be constitutively active; i.e., its rate of expression normal but its level unregulated. The protein product might be altered owing to mutation at the DNA level. Conceivable examples include the following.

1. Altered DNA binding leading to endless rounds of DNA replication and cell division
2. Altered growth factor (e.g., *sis* product), again leading to enhanced cell replication
3. Altered growth factor receptor (e.g., *erb* B product), rendering stimulation by exogenous growth factor unnecessary (again leading to increased cell division)
4. Altered membrane protein with loss of normal "contact inhibition," leading to disorganized growth in tissues
5. Altered cytoskeleton resulting in altered cell shape and perhaps cell–cell interaction parameters

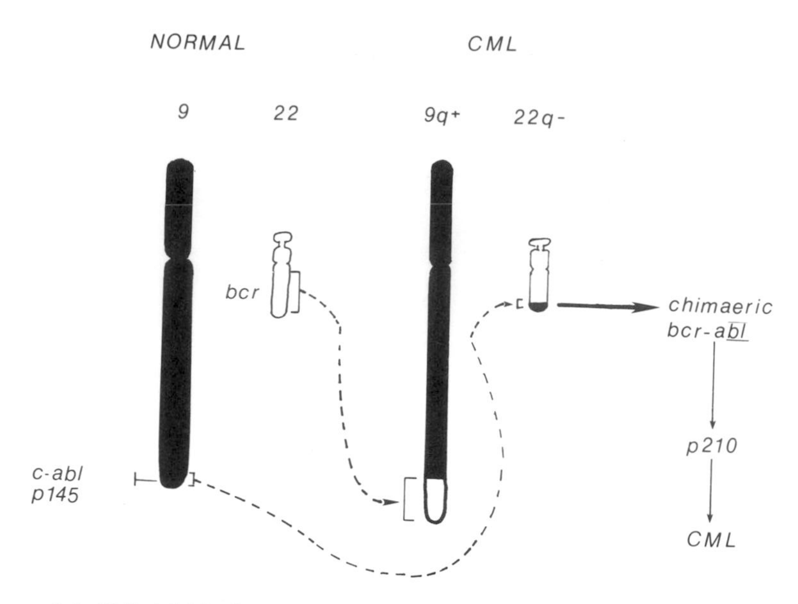

FIGURE 5.6. Philadelphia chromosome results from translocation of the *abl* gene on chromosome 9 to the *bcr* region of chromosome 22.

Functional Assays Define Oncogene Complementation Groups

Since the early days of tissue culture, scientists have known that cells isolated directly from organisms grow in vitro with only a limited life-span. These cells are called *primary cultures* (Chapter 2). However, during the growth of primary cell cultures an occasional subline may arise with an indefinite or "immortal" life-span; such immortal cells become *established cell lines*. It is assumed that established cell lines have acquired genetic changes that permit their indefinite growth in vitro, although exactly what the changes are have not been elucidated. Several of these cell lines do not form tumors in animals and thus are not "transformed," or "malignant." An example of such a cell line is NIH 3T3, which was derived from mouse fibroblasts.

Introduction of certain oncogenes, e.g., *ras*, into some cell lines such as NIH 3T3 immediately produces a fully transformed phenotype, i.e., a distinct cellular morphology, anchorage-independent growth, reduced requirement for serum, or tumor induction by cells injected into susceptible animals. However, transformation of primary rat embryo fibroblasts requires transfection by at least two different oncogenes. This difference between transformation of established cell lines versus transformation of primary cell cultures led to the concept of oncogene complementation groups (Table 5.5). One group of oncogenes is required for the "immortalization" or "establishment" of indefinite in vitro growth of primary cells, and a second group of oncogenes is required to produce the fully transformed phenotype encompassing the characteristics discussed above.

TABLE 5.5. Oncogene complementation groups.

"Immortalization"
myc Family, including v-*myc*, c-*myc*, N-*myc*, L-*myc*
Adenovirus EIA
Polyoma large T antigen
p53
v-*myb*
SV40 large T antigen
"Transformed phenotype"
ras Family, including H-*ras*, K-*ras*, N-*ras*
Adenovirus EIB (19-kD protein)
Polyoma middle T antigen
v-*src*
SV40 large T antigen

In Vitro Transformation Is a Multistep Process

Landmark experiments by Robert Weinberg's group at MIT showed that transfection of primary rat embryo fibroblasts by either the *myc* or *ras* gene alone was unable to transform the cells. However, when *myc* and *ras* genes were introduced together into the same cell, they "complemented" each other to produce the fully transformed phenotype. Expansion of this work revealed that DNA tumor virus transforming genes also were functional in these assays. Adenovirus E1A sequences and polyoma large T antigens provide *myc*-like immortalization functions, whereas adenovirus E1B genes and the polyoma middle T antigens were *ras*-like, providing the "transformed phenotype."

The implications of these observations are severalfold. First, cellular proto-oncogenes and independently evolved DNA tumor virus transforming genes appear to subserve similar cellular functions functions required for the normal growth and development of cells. Second, at least two and possibly more genes are required to establish the fully transformed phenotype. Although many "immortalization" genes such as *myc* have a nuclear location and several "transformed phenotype" genes such as *ras* are located elsewhere in the cell, the contribution of each *type* of gene product to the complete tumor cell phenotype is complex and not well understood.

In Vivo Tumorigenesis Is a Multistep Process

Cancer itself (in contrast to the transformation of cultured cells) is known to be a multistep process. Evidence supporting this view includes (1) the time interval between exposure to known carcinogens (e.g., chemicals, radiation) and the development of cancer; (2) the differential effects of chemicals "initiating" and "promoting" neoplasia; and (3) the "inheritance of cancer," which appears actually to be the inheritance of a *predisposition* to cancer; examples here include retinoblastoma and chromosome 13 deletions, Wilms' tumor and chromosome

11 alteration, and familial colon polyps where no known chromosomal aberration has yet been associated (see Chapter 6). In these cases cancer in the organism is frequent if not inevitable; however, every cell in the affected tissue does not *itself* undergo malignant transformation. This fact implies that a predisposition in each cell exists but an additional insult is required to transform any particular cell.

At this juncture, the oncogene hypothesis appears to explain neoplastic transformation better than any previous theory. Although some older theories included the concept that genetic alterations might underlie cancer, these theories were vague as to what specifically such alterations might be. The oncogene hypothesis posits that specific normal genes necessary for normal growth and differentiation become changed in particular definable ways (e.g., point mutation, insertion, amplification, deletion, rearrangement), leading to neoplastic transformation and hence tumorigenesis. Subsequent chapters outline each of the known oncogenes and their relations to human and animal malignancy. We first turn to a description of what is known of changes in human cellular proto-oncogenes that lead to activation of these genes in human cancers.

Bibliography

Alitalo K, Schwab M, Lin CC, et al. Homogeneously staining chromosomal regions contain amplified copies of an abundantly expressed cellular oncogene (c-myc) in malignant neuroendocrine cells from a human colon carcinoma. Proc Natl Acad Sci USA 80:1707, 1983.

Ar-Rushdi A, Nishikura K, Erickson J, et al. Differential expression of the translocated and untranslocated c-myc oncogene in Burkitt's lymphoma. Science 222:390, 1983.

Bargmann CI, Hung M-C, Weinberg RA. Multiple independent activations of the neu oncogene by a point mutation altering the transmembrane domain of p185. Cell 46:649, 1986.

Bishop JM. Cellular oncogenes and retroviruses. Annu Rev Biochem 52:301, 1983.

Bishop JM. Oncogenes and proto-oncogenes. Hosp Prac 52:301, 1983.

Blair DG, Oskarsson M, Wood TG, et al. Activation of the transforming potential of a normal cell sequence: a molecular model for oncogenes. Science 212:941, 1981.

Blick M, Westin E, Gutterman J, et al. Oncogene expression in human leukemia. Blood 64:1234, 1984.

Capon DJ, Seeburg PH, McGrath JP, et al. Activation of K-ras-2 gene in human colon and lung carcinomas by two different point mutations. Nature 304:507, 1983.

Cooper GM. Cellular transforming genes. Science 217:801, 1982.

Corcoran LM, Adams JM, Dunn AR, et al. Murine T lymphomas in which the cellular myc oncogene has been activated by retroviral insertion. Cell 37:113, 1984.

Dickson C, Smith R, Brookes S, et al. Tumorigenesis by mouse mammary tumor virus: proviral activation of a cellular gene in the common integration region int-2. Cell 37:529, 1984.

Eva A, Robbins KC, Andersen PR, et al. Cellular genes analogous to retroviral onc genes are transcribed in human tumor cells. Nature 295:116, 1982.

Huang C-C, Hay N, Bishop JM. The role of RNA molecules in transduction of the proto-oncogene c-fps. Cell 44:935, 1986.

Land H, Parada LF, Weinberg RA. Tumorigenic conversion of primary embryo fibroblasts requires at least two cooperating oncogenes. Nature 304:596–602, 1983.

Land H, Parada LF, Weinberg RA. Cellular oncogenes and multistep carcinogenesis. Science 222:771, 1983.

Marshall CJ, Hall A, Weiss RA. A transforming gene present in human sarcoma cell lines. Nature 311:671, 1982.

Müller R, Slamon DJ, Tremblay JM, et al. Differential expression of cellular oncogenes during pre- and postnatal development of the mouse. Nature 299:640, 1982.

Murphree AL, Benedict WF. Retinoblastoma: clues to human oncogenes. Science 223:1028, 1984.

Neel BG, Hayward WS, Robinson HL, et al. Avian leukosis virus-induced tumors have common proviral integration sites and synthesize discrete new RNAs: oncogenesis by promoter insertion. Cell 23:323, 1981.

Nusse R, Van Ooyen A, Cox D, et al. Mode of proviral activation of a putative mammary oncogene (int-1) on mouse chromosome 15. Nature 307:131, 1984.

Rowley J. Human oncogene locations and chromosome aberrations. Nature 301:290, 1983.

Ruley HE. Adenovirus early region 1A enables viral and cellular transforming gene to transform primary cells in culture. Nature 304:602–606, 1983.

Saito H, Hayday AC, Wiman K, et al. Activation of c-myc gene by translocation: a model for translocational control. Proc Natl Acad Sci USA 80:7476, 1983.

Slamon DJ, deKernion JB, Verma IM, et al. Expression of cellular oncogenes in human malignancies. Science 224:256, 1984.

Spector DH, Varmus HE, Bishop JM. Nucleotide sequences related to the transforming gene of avian sarcoma virus are present in DNA of uninfected vertebrates. Proc Natl Acad Sci USA 75:4102, 1978.

Turc-Carel C, Philip I, Berger M-P, et al. Chromosomal translocations in Ewing's sarcoma. N Engl J Med 309:497, 1983.

Westin EH, Wong-Staal F, Gelmann EP, et al. Expression of cellular homologues of retroviral onc genes in human hematopoietic cells. Proc Natl Acad Sci USA 79:2490, 1982.

Willecke K, Schafer R. Human oncogenes. Hum Genet 66:132, 1984.

Yunis JJ. The chromosomal basis of human neoplasia. Science 221:227, 1983.

6
Oncogenes and Human Cancers

Overview

Nearly all of the known retroviral oncogenes have proto-oncogene counterparts detectable in human DNA. Many of these proto-oncogenes, activated through a variety of mechanisms, have been associated with human cancers. Perturbations in the structure or expression of certain proto-oncogenes appear to be the general means by which this activation occurs.

Examples of structural changes are seen in single amino acid substitutions in the *ras* family of genes isolated from human leukemias, colon carcinomas, and bladder carcinomas. In chronic myelogenous leukemia, the translocation between chromosomes 9q and 22q results in the formation of a novel fusion gene termed *bcr-abl* whose gene product differs from that of the normal c-*abl* protein in size and in the ability to autophosphorylate.

Perturbations in the expression of a proto-oncogene can be due to the presence of abnormal regulatory elements as in the case of translocations involving c-*myc* seen in Burkitt's lymphomas, or due to amplification which augments expression by increasing the gene copy number. Amplifications of N-*myc* in neuroblastoma, and *erB-2/neu* in human breast cancer have been correlated with a more advanced stage and a poorer prognosis.

Recently, specific genetic elements which suppress tumorigenicity have been described. Several such elements have been localized to human chromosomes 11 and 13q leading to the conceptualization of cancer suppressor genes. In this paradigm, the loss of both alleles of such a suppressor gene, rather than the "activation" of a proto-oncogene, is responsible for tumorigenesis. In human retinoblastomas, the inactivation of a specific gene on chromosome 13q, called rb-1, is critical for neoplastic transformation.

In each case described, perturbations in any one gene may be involved in different phases (early or late) of oncogenesis depending on the disease context. Also, quite frequently, many genetic abnormalities are associated with a tumor that obscures the significance of any single genetic change. Furthermore, reliance on animal model systems to explain human malignancy must be taken with some caution since evidence exists implicating different oncogenes in the genesis of similar tumors depending on the species studied (man vs. mouse).

To this point, we have suggested some associations between oncogenes and the neoplastic state. What evidence do we have that endogenous transforming genes are involved in human cancer?

All of the retroviral oncogenes discovered thus far have cellular homologues (proto-oncogenes) that are thought to be involved in normal cellular function (see Chapter 5). If these normal genes are to cause cancer, their structure or their expression must be perturbed. Examples of such changes include the following:

1. Point mutations within the gene
2. Genetic rearrangements within the coding sequence of the gene
3. Genetic rearrangements outside the coding region
4. Amplification and/or overexpression of the gene

Each of these mechanisms results in the "activation" of one or another of the cellular proto-oncogenes that has been associated experimentally with human cancer. An additional mechanism for which there is preliminary experimental evidence is

5. Deletion of possible "anti-oncogenes."

Point Mutations: *ras* Gene and Human Neoplasia

The most notable class of oncogenes activated by point mutation is the *ras* family (Table 6.1). The first activated human oncogene was isolated from a human bladder carcinoma cell line, T24, by several researchers in 1982. Shih and Weinberg initially reported that a transforming principle present in the DNA of T24 cells could be transferred into NIH 3T3 cells by DNA transfection (Fig. 6.1). Specifically, genomic DNA isolated from the bladder carcinoma cell line would trans-

TABLE 6.1. Examples of *ras* mutations that transform in the 3T3 focus formation assay.

ras Allele	Source of allele	Codons 12	Codons 59	Codons 61	Focus formation
c-H-*ras*	Normal human	GGC Gly	GCC Ala	CAG Glu	No
c-H-*ras*	Bladder carcinoma lines	GTC Val	GCC Ala	CAG Gla	Yes
c-K-*ras*	Normal human	GGT Gly	GCA Ala	CAA Glu	No
c-K-*ras*	Lung carcinoma line	TGT Lys	GCA Ala	CAA Glu	Yes
N-*ras*	Normal human	GGT Gly	GCT Ala	CAA Glu	No
N-*ras*	Neuroblastoma line	GGT Gly	GCT Ala	AAA Lys	Yes

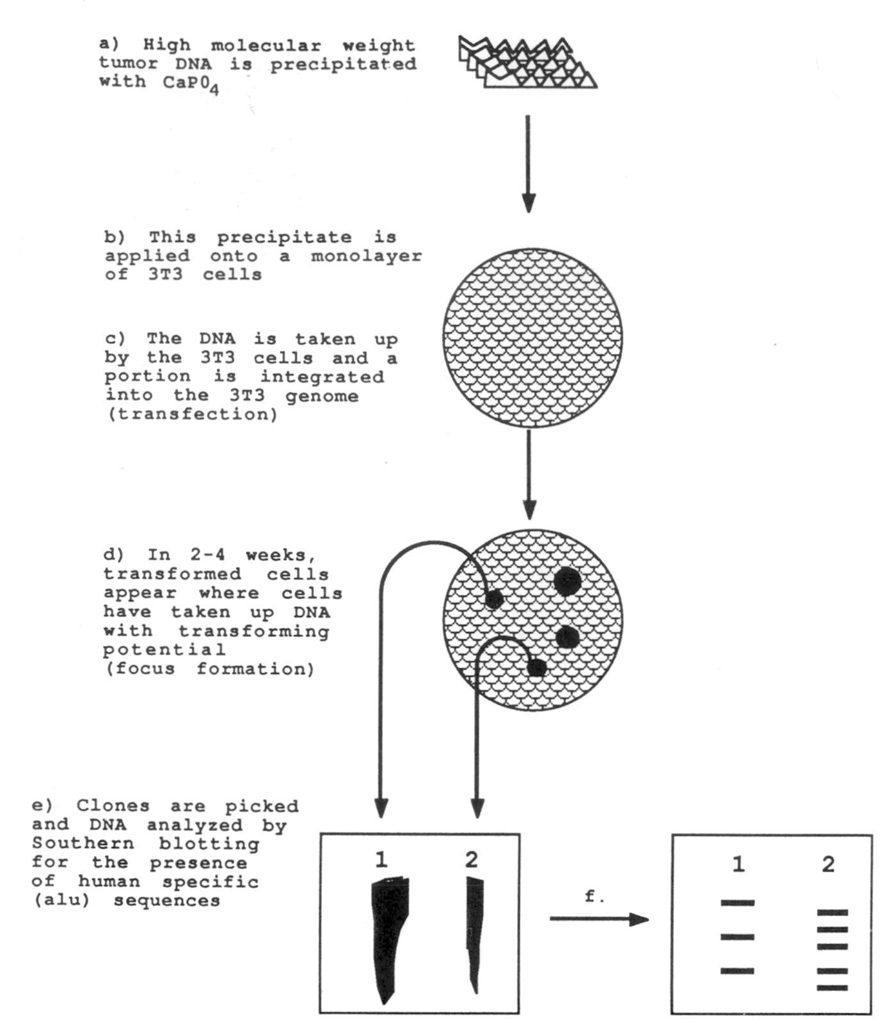

FIGURE 6.1. NIH 3T3 focus formation assay. Transfection is a method of introducing foreign DNA into cells in culture. Transfection of human tumor DNA and the isolation of resultant transformed NIH 3T3 cells was the means by which the first human transforming genes were isolated. After one round of transfection, transformed colonies have a significant amount of extraneous human DNA. A second cycle of transfection and focus selection eliminates this extraneous human DNA and enriches for transforming sequences.

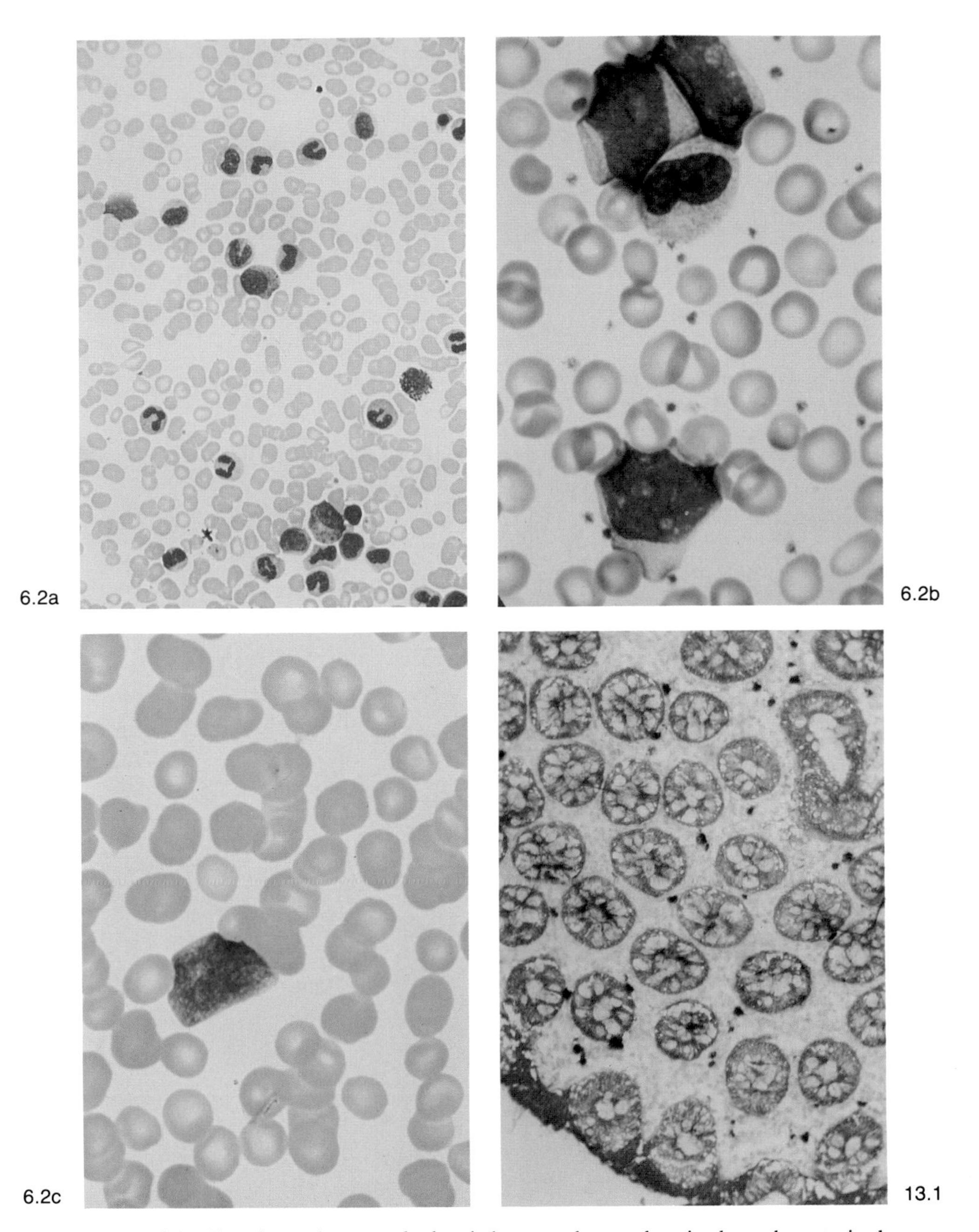

FIGURE 6.2. Chronic myelogenous leukemia has two phases: chronic phase characterized by high white blood cell counts of normal appearing granulocytic elements (a), and blast crisis, where primitive blast cells, either of myeloid (b) or lymphoid (c) lineage, predominate.

FIGURE 13.1. Immunoperoxidase staining of colon tissue by anti-CEA monoclonal antibodies. The dark staining areas identify the tumor cells. The levels of CEA released by these cells into serum is measured by sandwich ELISA and used to follow the clinical progression of gastrointestinal tumors. Photo generously provided by O. Gray.

Plate 2

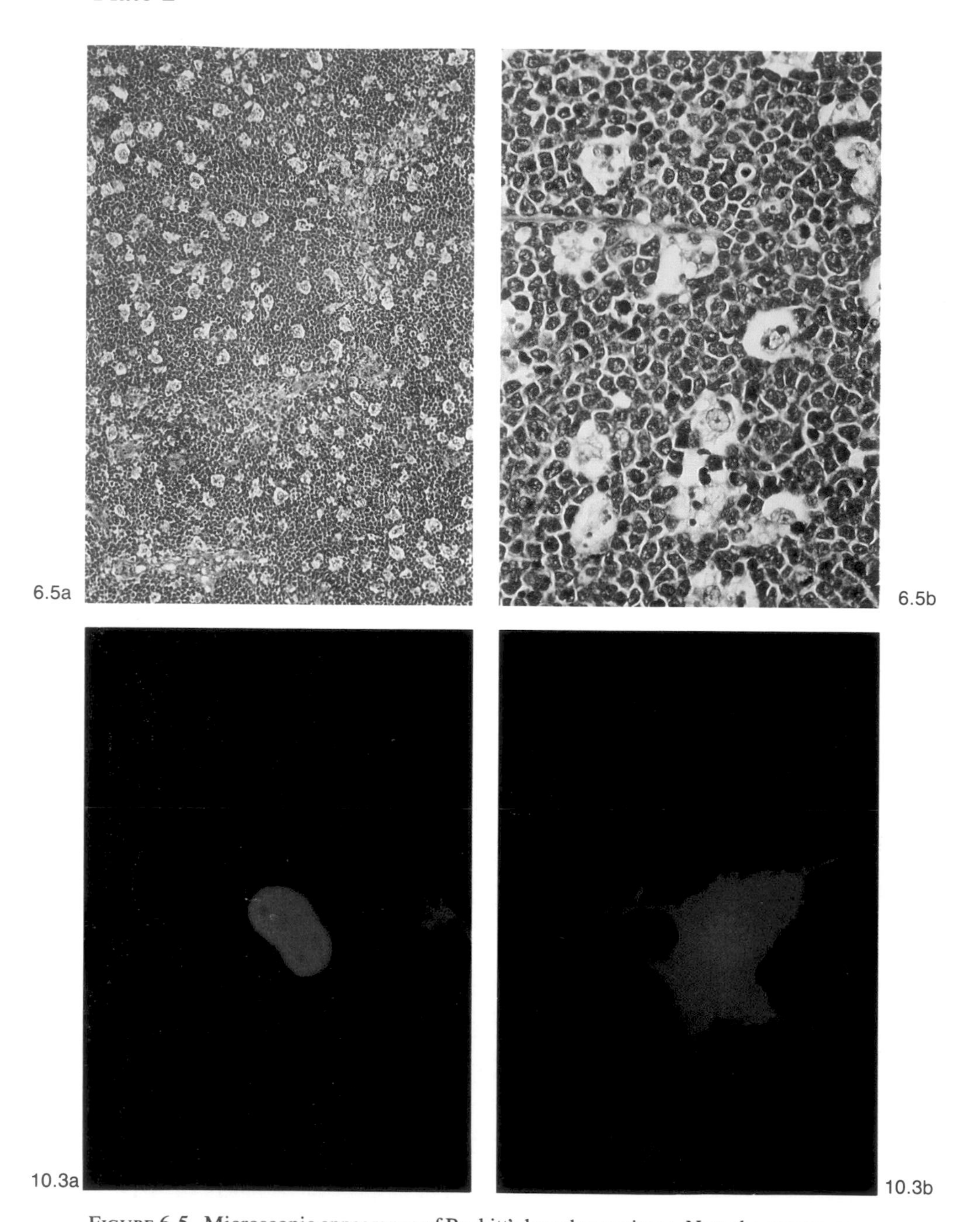

FIGURE 6.5. Microscopic appearance of Burkitt's lymphoma tissue. Note the monotonous population of malignant cells with large nuclei, and vacuoles in the cytoplasm.
FIGURE 10.3. Nuclear localization of c-*myc* is determined by specific amino acid sequences. Immunofluorescence of normal *myc* is localized to the nucleus (a), but a mutant *myc* with amino acid 262 to 372 deleted (within the third exon), now exhibits a diffuse cytoplasmic staining (b). This supports the notion that signals within the *myc* gene determine nuclear localization. (Photographs courtesy of Chi Dang and William Lee.)

form a small proportion of the 3T3 test cells. To prove that transformation was indeed due to the transfer of human DNA and not to a spontaneous event, Southern blots were performed on DNA from the transformed clones using an *alu* probe hybridizing only to human repetitive DNA sequences. This assay revealed abundant human-specific DNA in the transformed clones. The *alu* sequences served also as a convenient marker for cloning the transforming gene. (Details of the isolation of *ras* transforming genes are presented in Chapter 9.)

When sequenced, the human oncogene was found to be highly homologous to the Harvey *ras* viral oncogene that induces sarcomas and leukemias in susceptible murine strains. To elucidate the change responsible for conversion of this resident cellular gene into a transforming gene, the normal c-H-*ras* allele was examined. When both structural and functional analyses were completed, only a single nucleotide difference in the 12th codon of the *ras* gene was found; this difference changed amino acid 12 from glycine to valine (Table 6.1).

The implications of this finding are profound. Not only are there cellular homologues of viral oncogenes, but minor structural aberrations of these genes can lead to neoplasia.

The search then was on for more human cancer genes. Innumerable human tumor cell lines were screened in a similar fashion, and the findings were equally astounding. The cellular homologue of the viral Kirsten *ras* oncogene, c-K-*ras*, was activated at codon 12 in a lung and a colon carcinoma cell line. A similar mutation in codon 12 (glycine to arginine) was present in a primary lung tumor, but the mutation was not found in normal fibroblasts from the same patient. Similarly, an activated c-K-*ras* oncogene was found in one of five primary ovarian carcinomas by Feig and co-workers; this altered gene also was not found in the patient's normal fibroblasts.

The oncogene search led to discovery of previously unknown transforming genes. Brown and co-workers isolated a gene from a human fibrosarcoma cell line with a sequence closely related to the c-H- and c-K-*ras* genes. This gene was termed N-*ras* (no c-prefix was affixed because no viral homologue existed); the transforming allele was mutated at codon 61. Mutations in the N-*ras* gene, primarily in codons 12 and 13, often are found with acute myelogenous leukemia.

Although *ras* mutations are frequent, especially with myelogenous leukemias (approximately 40% of such tumors appear to contain them), they are by no means ubiquitous. Fujita et al. screened 23 primary bladder carcinomas and found transforming *ras* alleles in only two of them, both with mutations in codon 61. Feinberg et al. screened a variety of tumors for H-*ras* codon 12 lesions, and none of them appeared to have such a mutation. On the other hand it might be argued that the 3T3 transfection assay is an insensitive method to detect transforming genes because, at least in solid tumors, significant numbers of non-neoplastic stromal and inflammatory cells might dilute any positive signal from malignant cells. In addition, the K-*ras* gene is large, spanning over 45 kb so obtaining an intact gene for use in the transfection assay is technically difficult. Two groups have developed newer, more sensitive assays for detection of *ras* mutations. Bos et al. (1987; see also Rodenhuis et al.) used an oligomer hybridization assay able to detect codon 12 mutations specifically and found activated K-*ras* genes in 11 of 27 colorectal tumors and in 5 of 10 adenocarcinomas of the lung. Forrester et al. used an assay based on the ability of RNAse A to cleave single-base mismatches in RNA-RNA duplexes. Although their method could detect mutation in codons other than 12 (in contrast to the oligonucleotide hybridization technique) they succeeded in detecting activated K-*ras* codon 12 mutations in some 40% of the 66 colon tumors they examined. Still, no human tumor has been associated exclusively with mutations of one of the *ras* alleles, although relatively more epithelial tumors carry H- and K-*ras* lesions and N-*ras* lesions predominate in hematopoietic neoplasms. Activation of each of the three *ras* proto-oncogenes by point mutation has been detected in acute leukemias; however, a unique clinical presentation has not been associated with activation of a particular *ras* gene. Other than minor structural differences, the H-, K-, and N-*ras* genes are remarkably similar (see Chapter 9). Therefore it may be argued that regardless of which *ras* oncogene(s) is activated, the result is the same.

It is conceivable that *ras* gene activation is the critical event necessary for the development of a full-blown malignant picture; however, experimental evidence suggests that this supposition is probably not true. Myelodysplasia, or preleukemia, is characterized by abnormal appearing white blood cells and reduced numbers of each blood cell type (cytopenias). This disorder can persist for months to years; acute leukemia subsequently develops in 20 to 40%. Two groups have shown that mutations in the *ras* proto-oncogenes can be found in the preleukemic phase of myelodysplasia (in total, 5 of 12 patients). In one patient, the mutation was detected in the blood cells 1.5 years prior to leukemic conversion. These data suggest that mutations in the *ras* genes may be involved in initiation of human leukemia. Because of the considerable delay between the induction of the genetic lesion and the onset of leukemic transformation, it is hypothesized that other genetic factors must work in conjunction with *ras* to induce acute leukemia. Further evidence that mutant *ras* alleles are involved in the earliest phases of neoplastic transformation comes from work examining premalignant lesions of the colon: mutations of the K-*ras* gene were detected in five of six human colonic adenomas that were found adjacent to carcinomas.

Activated *ras* genes appear not to be necessary for maintenance of the transformed phenotype. Using a focus-forming assay in 3T3 cells. Albino et al. analyzed the transforming potential of a variety of melanoma cell lines derived from primary and metastatic tumor sites in the same patient. An activated H-*ras* oncogene was found in only one of five cell lines derived from this patient. Thus *ras* activation has an important role in tumorigenesis, but in this case, maintenance of malignancy did not require *ras* activation.

Table 6.2 lists tumor types from which activated *ras* alleles have been isolated. Several general statements may be made: (1) Only specific *ras* mutations are active in the 3T3 transfection assay; such lesions affect codons 12, 13, 59, 61, and 63 as determined by in vitro mutagenesis of the normal H-*ras* allele. (2) Activated *ras* oncogenes can be isolated from virtually any cell type and even in

TABLE 6.2. Human tumors exhibiting transforming *ras* genes.

Tumor type	Origin of cells
c-H-*ras*-1	
Bladder carcinoma	Cell line
Bladder carcinoma	Primary tissue
Lung carcinoma	Cell line
Melanoma	Cell line
Mammary carcinosarcoma	Cell line
Acute myelogenous leukemia	Primary tissue
c-K-*ras*-2	
Lung carcinoma	Cell line
Lung carcinoma	Primary tissue
Colon carcinoma	Cell line
Colon carcinoma	Primary tissue
Pancreatic carcinoma	Cell line
Gallbladder carcinoma	Cell line
Rhabdomyosarcoma	Cell line
Ovarian carcinoma	Cell line
Ovarian carcinoma	Primary tissue
Gastric carcinoma	Primary tissue
Acute lymphocytic leukemia	Cell line
Acute myelogenous leukemia	Primary tissue
Myelodysplasia	Primary tissue
Renal cell carcinoma	Primary tissue
Bladder carcinoma	Cell line
N-*ras*	
Neuroblastoma	Cell line
Burkitt's lymphoma	Cell line
Fibrosarcoma	Cell line
Rhabdomyosarcoma	Cell line
Promyelocytic leukemia	Cell line
Acute myelogenous leukemia	Primary tissue
Melanoma	Cell line
T cell leukemia	Cell line
Chronic myelogenous leukemia	Primary tissue
Myelodysplasia	Primary tissue

premalignant stages. (3) No normal cells harbor the transforming genetic lesions. (4) Though *ras* activation may play an important role in neoplastic transformation, its continued presence appears not to be necessary for maintenance of the transformed state. (5) The 3T3 transfection assay biases for discovery of *ras* family oncogenes.

DOES *ras* ACTIVATION CAUSE CANCER?

The presence of activated *ras* genes only in tumor DNA and not in normal DNA from the same individual suggests that *ras* has a role in tumorigenesis. Nevertheless, the possibility remains that such *ras* mutations are epiphenomena. It is well known that transformed cells have higher mutation rates than do most normal cells. Thus it is possible that *ras* lesions are "bystander aberrations" that have no real role in tumor progression. Both circumstantial and direct evidence, however, argue against this possibility. Even though incidental mutation might explain activated *ras* alleles in tumor cell lines, it is difficult to reconcile such an explanation with evidence that a high proportion of primary tumor cells (from freshly biopsied specimens) also show activated *ras* alleles. Random genetic lesions that impart no selective advantage should be acquired and lost at similar frequencies; thus a population of cells would show no net gain in the frequency of such a lesion. The frequency with which transforming *ras* alleles are seen in human tumors (between 15 and 40% depending on the assay used and the disease studied) and the clonality of the individual lesions, however, argue strongly that *ras* mutations impart a distinct selective advantage to tumor cells.

Direct evidence suggesting that an activated *ras* gene is important in human tumorigenesis comes from work by Harris et al. A transforming v-H-*ras* gene was introduced into primary normal human bronchial epithelial cells by protoplast fusion. When assayed for transformation by plating in soft agar, a small number of colonies grew. The soft agar colonies appeared phenotypically transformed and gave rise to tumors when injected into nude mice; furthermore, the resultant cell lines produced the v-H-*ras* protein. It is of interest that passage of these cells in nude mice for even a short period of time resulted in progressive and massive chromosomal rearrangements manifested by hyperploidy and marker chromosomes. These data suggest that v-H-*ras* might contribute to tumor progression by augmenting genetic instability, thus encouraging further critical mutations.

One mechanism of achieving this "genetic instability" is by increasing the probability of cell fusion events. Wong and co-workers observed that *ras*-transformed cells had up to 100-fold greater rates of cell fusion than their nontransformed counterparts. The spontaneously fused cells were uniformly hyperdiploid, containing an increased number of chromosomes. Similar spontaneous fusion events have been reported to increase the virulence of experimental murine tumors.

Aberrant *ras* genes are associated with a number of human cancers; however, their role in oncogenesis is still unclear. Current evidence suggests that *ras* genes contribute both to the initiation of malignant transformation and to the progression of the malignant disease (see Chapter 9).

Genetic Rearrangements

Cytogeneticists were the first to note consistent chromosomal translocations associated with a variety of tumors. It was supposed that genetic rearrangements resulted from these translocations and that the rearrangements were related to the pathogenesis of the tumor. More recently, involvement of known protooncogenes in chromosome rearrangements has been inferred by the proximity of chromosomal breakpoints to the known loci of particular oncogenes (e.g., *myc*, *abl*, *ets*). In addition, new oncogenes have been revealed by cloning and mapping of tumor-associated chromosome breakpoints (e.g., *bcl*-1 and *bcl*-2). In some cases translocations occur outside the protein coding domain of the oncogene (*myc*), whereas in others the genetic rearrangement leads to formation of hybrid genes (*bcr/abl*). Thus the normal structure of a resident proto-oncogene may be perturbed by chromosomal rearrangements, and this perturbation may result in neoplastic transformation.

Generation of Fusion *onc*-Proteins: Philadelphia Chromosome and CML

An example that well illustrates how formation of a hybrid oncogene may play a critical role in tumorigenesis is seen in chronic myelogenous leukemia (CML). This disorder is characterized by a chronic phase during which the patient's mature blood granulocytes are greatly increased (Fig. 6.2; see Plate 1). The spleen and liver may be enlarged owing to infiltration by malignant cells, and normal bone marrow elements are often replaced, leading to anemia and thrombocytopenia. The chronic phase is easily controlled using simple and relatively benign chemotherapeutic measures. However, at a median time of 3 years after diagnosis, all patients invariably enter a blast crisis where immature malignant cells (blasts) emerge, and the clinical picture resembles that of acute leukemia (Fig. 6.2; see Plate 1). Although CML presents initially as a malignant proliferation of myeloid and granulocytic cells, one-third of blast crisis leukemias are lymphoid, rather than myeloid. This finding suggests that CML results from transformation of a very primitive hematopoietic stem cell.

In 1960 Nowell discovered an abnormal chromosome in the white blood cells of patients with CML, now called the Philadelphia chromosome (Ph^1) after the city in which it was discovered (Fig. 6.3). This chromosome originally was thought to be a deletion (or loss) of a portion of the long arm of chromosome 22, but when better chromosomal banding techniques were developed the abnormality was found to be a reciprocal exchange or translocation of genetic material between chromosomes 9 and 22. The 9:22 translocation, designated t(9:22)[1], is present in 90 to 95% of patients with CML and may be causally related to the disease.

The first clue that oncogenes might be involved in the Ph^1 translocation was the discovery that the translocated c-*abl* proto-oncogene, normally resident on chromosome 9, mapped to an area close to the usual breakpoint in chromosome 22.

[1]Cytogenetic nomenclature: There are 22 human chromosomes designated 1 to 22 and sex chromosomes designated X and Y. The long arms are designated q and the short arms p. Each chromosome has multiple bands as seen by various staining procedures. These bands are referred to by numbers after the arm designation, e.g., 8q24. Translocations are referred to by t with a colon between the translocated portions of the chromosomes, e.g., t(9:22).

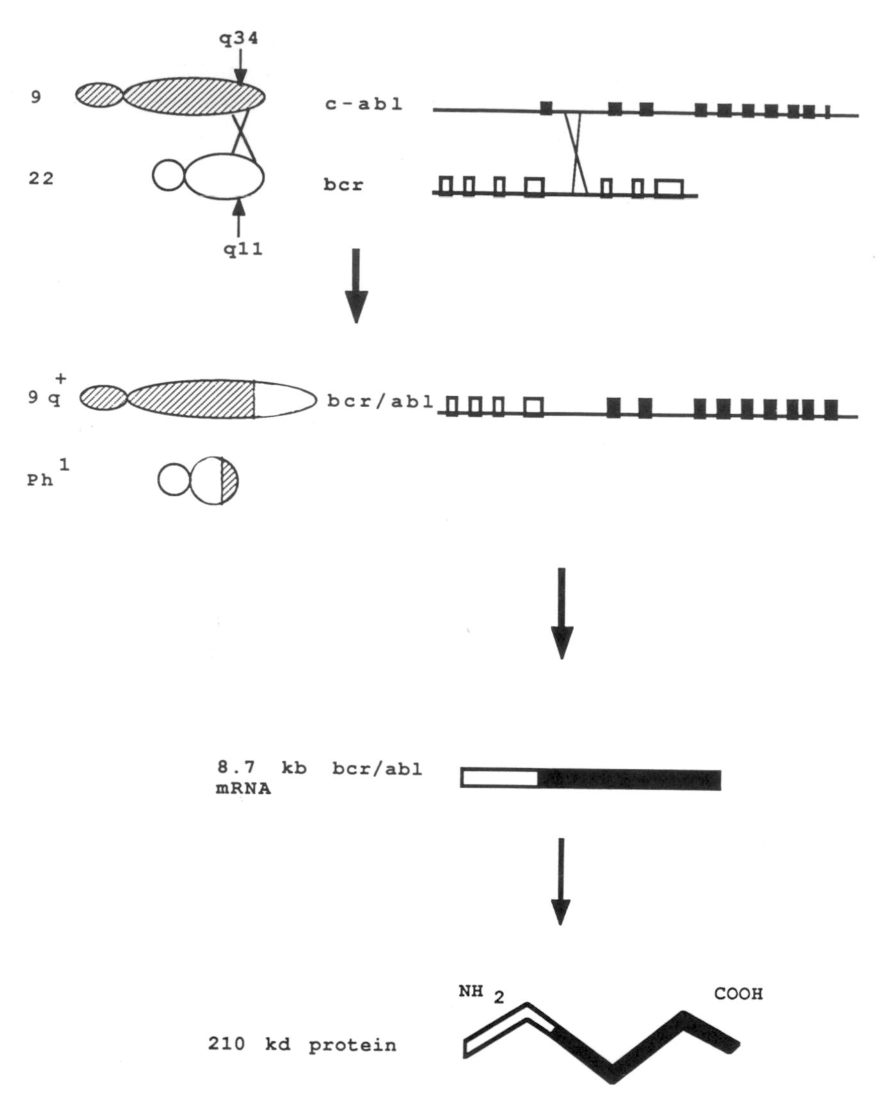

FIGURE 6.3. The *bcr/abl* fusion mRNA is generated by translocating the c-*abl* gene from its location in chromosome 9 to a new location in chromosome 22. The translocation puts c-*abl*, which has been truncated at its 5′ exons, downstream from the 5′ exons of a truncated *bcr* gene. The *bcr/abl* fusion mRNA is produced after splicing of exons over large distances (>100 kb) and results in a hybrid protein.

Heisterkamp and co-workers were the first to clone the breakpoint region; their findings quickly became among the most interesting in molecular genetics. Knowing that the c-*abl* locus is transferred during the 9:22 translocation, these scientists used a v-*abl* probe to isolate clones from the 5′ end of the gene. The 5′ clones subsequently were employed as probes to see if rearrangements were present in DNA isolated from white blood cells of CML patients. In one patient the 5′ *abl* region did appear to be rearranged by restriction enzyme mapping. A fragment representative of this region then was cloned for further analysis. Did this novel clone contain the exact chromosomal breakpoint involved in CML? To answer this question, fragments of the isolated clone were used to probe DNA from mouse/human somatic cell hybrids.

Somatic cell hybrids are mouse cells fused with human cells and contain both mouse and human chromosomes. During passage of the hybrid cells, human chromosomes are preferentially and randomly lost. It thus is possible to isolate hybrid clones containing only one or a few human chromosomes. A set of these clones can be isolated representing each human chromosome uniquely. It then is possible to determine from which human chromosome(s) a cloned DNA originates by using the clone as a probe of Southern blots containing DNA from the set of mouse/human hybrids. Using such a technique, Groffen and co-workers proved they had isolated the exact translocation breakpoint of chromosomes 9 and 22 in CML. Comparing the number of such breakpoint clones, they found that the translocation site on chromosome 9 was variable. In fact, only one patient had a chromosome 9 breakpoint detectably close to the c-*abl* gene. However, 17 of 19 CML patients had their chromosome 22 breakpoints localized to a 5.8-kilobase (kb) region. If one considers that the human genome is 3×10^6 kb in length, 5.8 kb represents 0.0002% of the human genome. The localization of the chromosome 22 translocation breakpoints within such a small area argues that this region plays a critical role in CML.

The CML translocation breakpoint so precisely inserted into chromosome 22 suggested that the expression of c-*abl* (or a closely linked gene) might be altered in some way in CML. Simultaneously, several scientists reported the presence of an abnormal c-*abl* mRNA in CML cells. Normal c-*abl* transcripts comprise two bands of 6 and 7 kb, on Northern blots. CML cells, on the other hand, express an 8.7-kb *abl* transcript. Somatic cell hybrids containing the Philadelphia chromosome also express the abnormal 8.7-kb transcript, whereas hybrids bearing a normal chromosomal 22 do not. Cloning the cDNA corresponding to the abnormal *abl* transcript revealed that the 5′ sequences of the mRNA originated from chromosome 22, whereas the 3′ mRNA sequences consisted of *abl*, truncated at its 5′ end. The final gene product is therefore a fusion mRNA derived from a chromosome 22 gene, termed *bcr* (for breakpoint cluster region), and a foreshortened c-*abl* gene translocated to chromosome 22.

The CML cells synthesize a fusion *abl* protein of 210 kD in addition to or instead of the normal c-*abl* protein of 145 kD. Furthermore, the fusion protein shows increased autophosphorylation compared to the normal *abl* protein.

The finding of a fusion *abl* protein in CML cells recalls the original isolation and characterization of retroviral *abl*. The Abelson murine leukemia virus is replication-defective and carries the v-*abl* oncogene instead of some viral structural gene sequences. It efficiently induces T cell lymphoma/leukemia in susceptible mice. The transforming v-*abl* product is translated from a fusion mRNA between viral *gag* sequences at the 5′ end and truncated *abl* sequences at the 3′ end. The v-*abl* protein also has increased autophosphorylation compared with

TABLE 6.3. Comparison of the properties of c-*abl*, virally induced v-*abl*, and *bcr*/*abl*

Property	c-*abl*	v-*abl*	*bcr*/*abl*
mRNA	Normal	*gag*/*abl* fusion	*bcr*/*abl* fusion
abl	Normal	5′ and 3′ truncation	5′ truncation
Autophosphorylation	No	Yes	Yes
Transforming activity in 3T3 assay	No	Yes	No

the normal c-*abl* protein. Furthermore, the ability to autophosphorylate correlates with the ability to transform. Thus activation of murine c-*abl* by viral transduction is essentially the same as activation of human c-*abl* by chromosomal translocation in CML (Table 6.3).

Despite the similarity of the p210 *bcr-abl* protein to the v-*abl* protein, *bcr-abl* is unable to transform NIH/3T3 fibroblasts whereas v-*abl* does so easily. When viral gag sequences are placed in the 5′ region of the *bcr-abl* fusion gene, the resultant hybrid gene can now transform fibroblasts. This suggests that the N-terminus of the fusion protein involving c-*abl* is important for determining the range of cells this gene can transform. This point is underscored by the discovery of an aberrant *bcr-abl* fusion gene in patients with Philadelphia chromosome positive acute lymphocytic leukemia. These patients differ from CML patients in that they present as de novo acute lymphocytic leukemia (i.e., without a chronic phase) though they carry the t(9q;22q). Subsequent work determined that in this disease, a smaller (190 kD) fusion protein is consistently found which is the result of a shorter *bcr* contribution to the 5′ portion of the fusion protein.

CML and Other Oncogenes

In addition to the *abl* gene, evidence suggests that other transforming genes may be involved in the pathogenesis of both murine hematological malignancies (Abelson MuLV) and CML. If DNA from T cell lymphomas induced by Abelson-MuLV is transfected into NIH 3T3 cells, foci of transformed cells appear. However, no v-*abl* sequences are detectable using DNA hybridization analysis. Comparing CML DNA from cells at the two stages of the disease, it was found that activated *ras* genes were present in three of 6 cases of CML in blast crisis, whereas in chronic CML only two of 27 cases showed *ras* gene activation (based on research by three independent groups: 0/13, 1/6, 1/8 cases). These results imply that in addition to c-*abl*, other activated cellular oncogenes occur in the transformation pathway(s) leading to CML.

ALTERED REGULATION FOLLOWING REARRANGEMENT: C-*myc* GENE

Involvement of the c-*myc* proto-oncogene in human cancers is an example of gene activation by translocation without perturbation within the coding sequences. In this situation it is probable that control regions for *myc* expression have been altered and appropriate expression abrogated.

With mammalian neoplasia, particularly human and murine B cell lymphomas, *myc* has been localized to chromosome areas consistently involved in translocations (Fig. 6.4). In many animals there is firm association of *myc* with neoplasia.

The most thoroughly studied human disease involving *myc* is Burkitt's lymphoma. Its animal counterpart is murine plasmacytoma. Burkitt's lymphoma is a tumor of lymph nodes involving B lymphocytes, the immunoglobulin-producing cells (Fig. 6.5; see Plate 2). It is characterized by massive enlargement of lymph nodes due to infiltration with malignant B cells. The tumor frequently spreads to bone marrow and the central nervous system. The *growth fraction*, or percentage of dividing cells, is high, and treatment of the disease with aggressive chemotherapy can be curative. Burkitt's lymphoma was described first in African patients. Differences between African and non-African Burkitt's lymphoma, e.g., site of disease at presentation and other discriminating factors, led to two subcategories: the African and non-African varieties. For example, African children frequently present with jaw masses (Fig. 6.6), whereas non-African (American) children almost always have their primary disease in the abdomen. Burkitt's lymphoma is now associated with the acquired immune deficiency syndrome (AIDS) as well.

Study of Burkitt's lymphoma has been facilitated by in vitro studies of Burkitt cells. Consistent cytogenetic abnormalities are found. Most Burkitt's lymphoma cells carry a reciprocal translocation between chromosome 8q24 and chromosomes 2, 14, or 22. Immunoglobulin light chain loci are on chromosomes 2p11(κ) and 22q11(λ), whereas immunoglobulin heavy chains localize to chromosome 14q32. The human c-*myc* proto-oncogene is on chromosome 8q24. These areas correspond precisely to the translocations involved in Burkitt's lymphomas. This fact prompted investigators to determine if in fact c-*myc* was juxtaposed to an immunoglobulin locus in these translocations. Analysis by Southern blotting and by cloning and sequencing of the pertinent breakpoints showed that the human c-*myc* locus and the immunoglobulin heavy chain genes often are translocated such that they lie "head to head," i.e., the 5′ portion of the *myc* genes abuts the 5′ portion of the heavy chain "switch" region (Fig. 6.7). More recent evidence implicates the J_H (heavy chain "joining") region on chromosome 14 in the t(8;14) translocation. Many of these translocations place the rearranged *myc* allele adjacent to immunoglobulin enhancer elements, thus interrupting the normal transcriptional control mechanisms of the *myc* gene.

"Dysregulation" of the Translocated c-*myc* Gene

With Burkitt's lymphomas and murine plasmacytomas, there often are deletions or point mutations involving the first (noncoding) exon and surrounding noncoding regions of the c-*myc* gene (see Chapter 10). With murine plasmacytomas such abnormalities were seen in 24 of 25 tumors where the gene was mapped. In humans, all of seven primary American Burkitt's lymphomas harbored gross rearrangements of c-*myc* involving the first exon, first intron, and 5′ flanking region; in all of eight fresh African Burkitt's lymphomas, there was evidence for

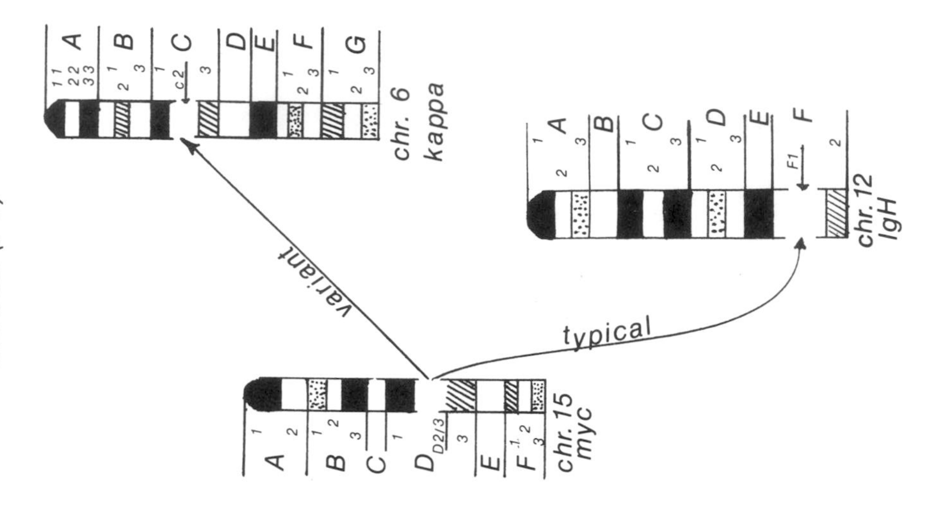

MOUSE (PC)
chr. 6
kappa
variant
chr. 15
myc
typical
chr. 12
IgH

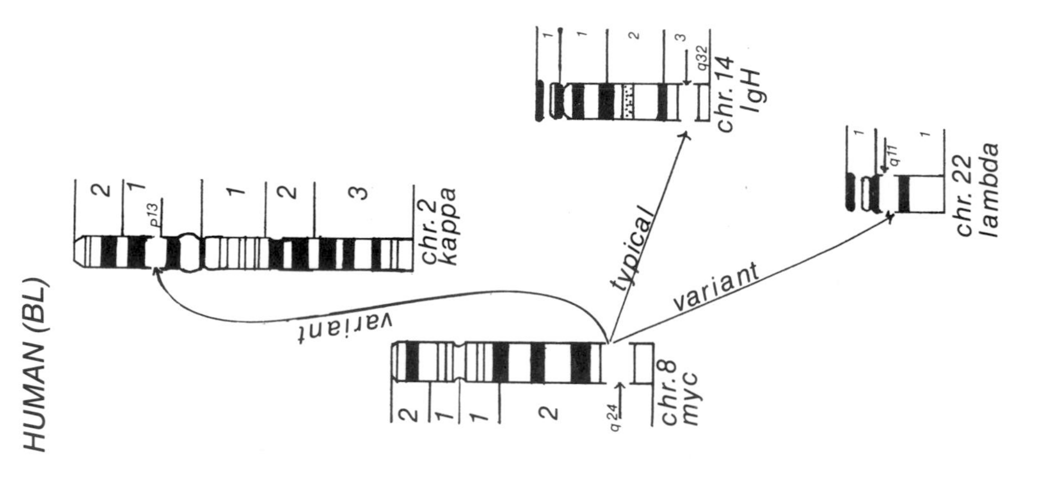

HUMAN (BL)
chr. 2
kappa
p13
variant
typical
chr. 8
myc
q24
chr. 14
IgH
q32
variant
chr. 22
lambda
q11

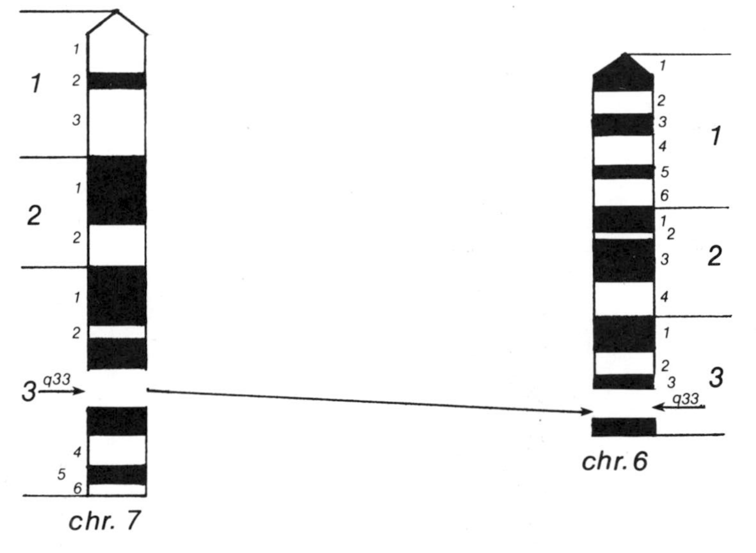

FIGURE 6.4. These c-*myc* translocations are found in human Burkitt's lymphoma, mouse plasmacytomas, and rat plasmacytomas.

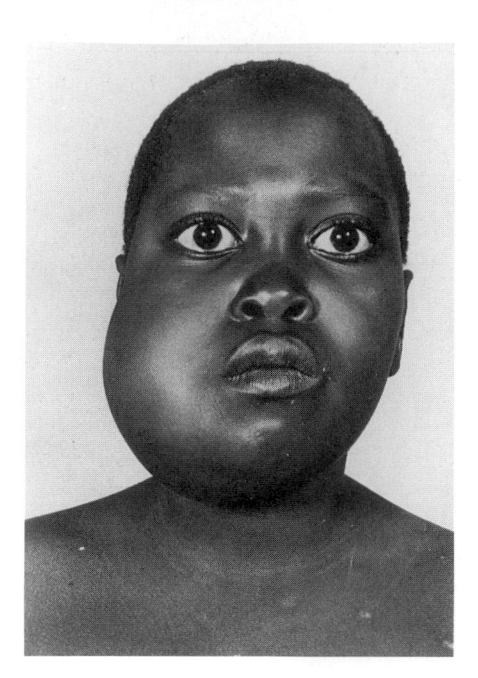

FIGURE 6.6. This child has the common presenting finding for African Burkitt's lymphoma, a jaw mass.

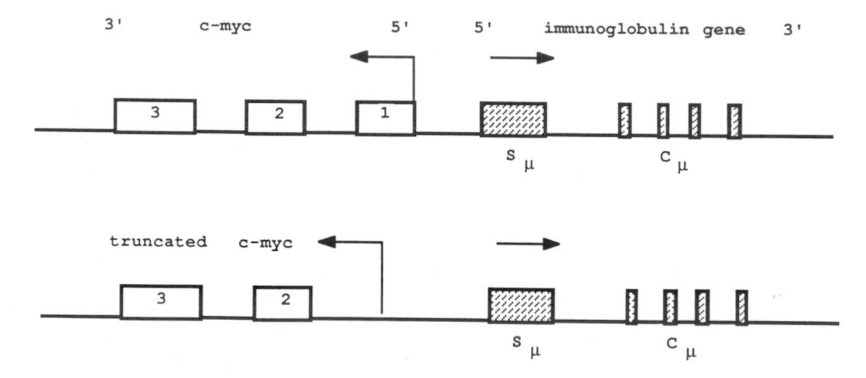

FIGURE 6.7. Configuration of the c-*myc* gene involved in the t(8:14) translocation. The arrows show the transcriptional orientation of the genes.

mutations within the same region. The immediate question is: Does translocation of the c-*myc* gene and the associated changes in the 5′ portion of the gene stimulate its transcription?

Analysis of RNA shows increased expression of *myc* in tumor cells compared with normal lymphocytes, but expression is not uniformly higher than the increase seen in Epstein-Barr virus immortalized (but nontumorigenic) B lymphocytes that do not carry the *myc* translocation. Thus elevated transcription of *myc*, though important, cannot completely explain B cell transformation. It has been postulated that translocation might bring cryptic enhancers into close proximity with the *myc* promoter, allowing increased rates of transcription. However, there are several examples where the distance between the immunoglobulin genes and the translocated c-*myc* locus is so great that enhancer effects are less likely. In addition, protein analysis shows no difference between translocated and natural *myc* alleles, a finding corroborated by DNA sequence data. Thus there are no constituent structural mutations within the coding region that activate c-*myc*.

There is growing evidence that normal regulation of c-*myc* expression is disturbed by the translocation. For example, in cells with the 8:14 translocation, there remains an untranslocated *myc* allele on the uninvolved chromosome 8. However, in most cell lines tested, only the translocated *myc* gene is expressed, the normal allele being silent. This result suggests that the translocated *myc* allele may be constitutively transcribed, resulting in suppression of the normal allele.

There also is evidence that normal translation mechanisms are disturbed upon translocation. The normal c-*myc* gene has three exons, of which the second and third encode the protein (Fig. 6.8). The full-length (three exon) *myc* mRNA has sequences in the first exon extending into the second exon that are complementary to each other. These regions could form hairpin loops by hydrogen bonding. Some such hypothetical loops are "strong" enough to prevent read through by ribosomes; this effect would inhibit translation of the mRNA into protein. When efficiency of translation is compared in vitro, the three-exon *myc* mRNA is ten times *less* efficient than the two-exon mRNA with deleted first exon. There also is growing evidence for sequences within and around the first exon that act as repressors of transcription and as sites for premature termination of the messenger RNA (i.e., incomplete and foreshortened mRNAs are generated). Thus elements within the first *myc* exon and surrounding regions potentially may act to (1) decrease the rate of transcription, (2) stall or stop transcription once it has begun, and (3) stall or stop translation of mature mRNA. It therefore is conceivable that translocation of *myc*, commonly accompanied by deletions and point mutations in and around the first exon, induces dysregulated *myc* expression, and thus may have an important role in tumor induction.

Association does not, of course, mean causation. To prove that deregulation of *myc* indeed induces cancer would require the controlled introduction of a *myc* mutation into an experimental animal. Such an experiment has in fact been performed by inserting a number of *myc* gene constructions into the germ line DNA of mice. In an elegant study, Adams et al. created transgenic mice (see Chapter 12) carrying the mouse c-*myc* gene associated with different transcriptional

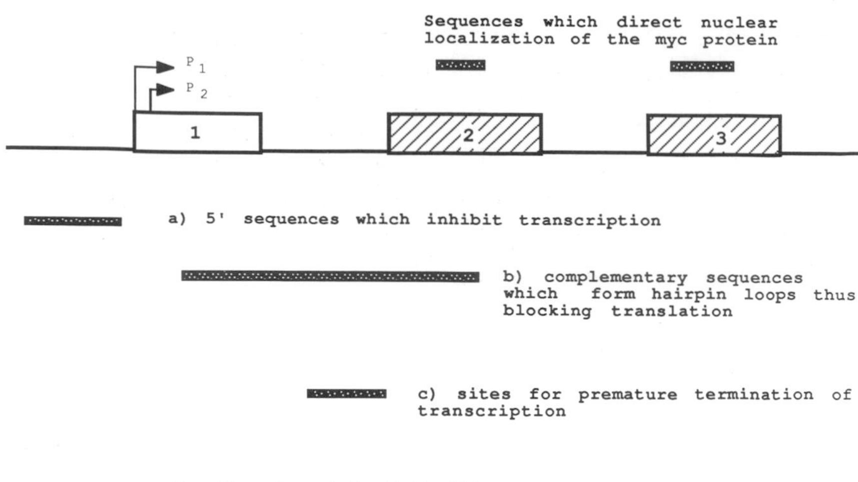

FIGURE 6.8. Elements of the c-*myc* gene that might control its expression. P_1 and P_2 denote the two promoters used by the gene. The hatched boxes represent the coding exons (2 and 3), and the open box represents the noncoding first exon. The possibilities for control of c-*myc* expression include: (a) blocking transcriptional initiation; (b) blocking translation of mature mRNA by hairpin loop formation; (c) sites for premature termination of transcription (attenuation); and (d) altered protein half-life.

control regions including an immunoglobulin enhancer region, a mouse retroviral long terminal repeat (LTR) enhancer, the SV40 enhancer, the metallothioneine control region, and the normal *myc* upstream sequences. These scientists found that 13 of 15 transgenic mice bearing an immunoglobulin-enhanced c-*myc* gene developed aggressive B cell lymphomas, whereas none of five mice carrying the *myc* gene with normal upstream sequences developed tumors. In addition, 3 of 21 mice bearing the SV40/*myc* construction developed other tumors: a lymphosarcoma, a renal cell carcinoma, and a fibrosarcoma.

The fact that different enhancer-driven *myc* constructions induce different types of cancer suggests that deregulated expression of *myc* within a given tissue type is important for tumor formation. Furthermore, each B cell tumor is of clonal origin, i.e., derived from an individual progenitor cell expressing only the introduced *myc* gene. The latter finding is of fundamental importance. If the deregulated *myc* gene provides the sole transforming event, one would expect the resultant tumors to be multiclonal because the introduced *myc* gene is present in every B cell of the transgenic animal. Instead, all lymphomas studied gave unique immunoglobulin gene rearrangement patterns, pointing to a monoclonal origin for each tumor. This result implies that *myc* deregulation is necessary but in and of itself not sufficient to induce lymphomas. A second genetic event must occur for tumors to form, a finding consistent with previous in vitro work showing that at least two oncogenes must be present for transformation of primary cells in culture. These data support the hypothesis that cooperation between activated oncogenes is important for tumorigenesis (Chapter 5).

The c-*myc* gene is not the only gene perturbed by translocation in human neoplasia. The *bcl*-1 and *bcl*-2 genes were discovered by translocation in non-Burkitt lymphomas (see Chapter 11). In addition, a specific translocation t(11;22)(q23;q11-12) is the hallmark of Ewing's sarcoma and neuroepitheloma. Systematic study of translocations in human neoplasia may well reveal altered regulation of other proto-oncogenes as a result of chromosome rearrangement.

Gene Amplification and Overexpression

Either an increased transcription rate or an increased amount of the DNA template can result in gene overexpression. An increase in gene copy number, termed *gene amplification*, appears to be a common mechanism of augmenting gene expression in tumor cells. Gene amplification occurs in two forms defined by chromosome analysis: double minutes (DMs) or homogeneous staining regions (HSRs) (Fig. 6.9). DMs appear as small extrachromosomal particles that have no centromere and therefore distribute randomly at mitosis. HSRs, on the other hand, appear as monotonous regions in existing chromosomes, often elongating the arms on which they reside. That these structures contain amplified DNA sequences first was established during studies of cancer cells and drug resistance; the potential role of gene amplification in tumorigenesis was demonstrated later. One well-studied example shows that amplification of the N-*myc* gene appears to play an important role in human neuroblastoma (Table 6.4).

Neuroblastoma is the most common extracranial solid tumor of childhood, arising from neuroectoderm and consisting of anaplastic sympathetic ganglion cells. The most common clinical presentation is an abdominal mass in an infant due to a tumor growing from the adrenal medulla. The disease is staged by size and extent of tumor burden from I (the least) to IV (the most). Stage III denotes a large abdominal mass crossing the midline of the body, and stage IV implies metastatic disease, most frequently to bone marrow, skin, bone, and liver. Stage IVS is a subclassification indicating a small primary lesion but with metastatic disease to organs other than bone. The curious propensity for stage IVS disease (as well as for some stage I and II disease) to regress spontaneously justifies its separation into a separate category. Stage I and II disease can be treated surgically with a reasonably good prognosis. However, stages III and IV have a poor prognosis, as these tumors are resistant to most forms of chemotherapy and radiation therapy.

During screening of cell lines for overexpression of oncogenes, some neuroblastoma lines were found to express high levels of a *myc*-like gene not previously described. This new gene, called N-*myc*, was localized to chromosome 2p23-24. On examination of tumor genomes, multiple copies of this gene denoting gene amplification frequently were observed. N-*myc* was cloned, and sequenced revealing homology with c-*myc*. The two genes have a similar

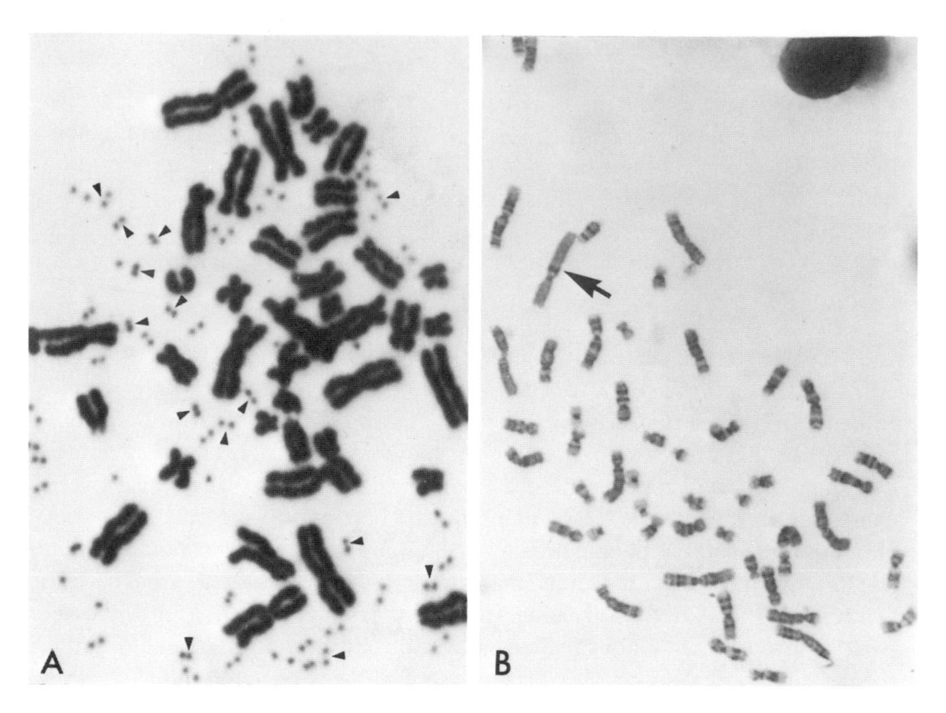

FIGURE 6.9. Double minute (DM) and homogeneously staining region (HSR) chromosomes in human colon carcinoma cells. (A) Metaphase spread from human colon carcinoma cell line (COLO 320 DM) showing numerous DM chromosomes. Some of the DMs are indicated by small arrowheads. (B) Metaphase spread from a human carcinoma cell line showing a chromosome with HSR (arrow). (Kindly provided by Dr. C.C. Lin)

three-exon structure and appear to have similar functions judged by the ability of either when activated by a strong promoter to complement an activated *ras* gene in the primary rat embryo fibroblast transformation assay (see Chapter 10).

Amplification of the N-*myc* gene was observed in 22 of 27 neuroblastoma cell lines. This frequency of N-*myc* amplification prompted scientists to screen primary neuroblastoma tissue. They found a strong correlation between N-*myc* amplification and the stage of the disease. For example, none of 8 stage I patients and 2 of 16 stage II patients had N-*myc* amplification, whereas 13 of 20 stage III tumors and 19 of 40 stage IV tumors showed N-*myc* amplification. In a more recent study, N-*myc* amplification was shown to be as important a determinant of patient survival as the stage of the disease. When stage II patients were segregated into those with and without N-*myc* amplification, all those with amplification died within 2 years, whereas 80% of those without amplification survived.

N-*myc* amplification occurs predominantly in neuroblastomas; occasional retinoblastomas and a few small cell lung carcinomas also contain amplified N-*myc* (Table 6.4). Thus N-*myc* amplification can be used to distinguish one

TABLE 6.4. Amplification of proto-oncogenes in human tumor cells.

Proto-oncogene	Tumor
c-*abl*	Chronic myelogenous leukemia (K562)
c-*erbB*	Epidermoid carcinoma (A431) Squamous carcinoma Glioblastoma
c-*erbB-2*	Adenocarcinoma of the salivary gland Gastric carcinoma Mammary carcinoma
c-*ets*-1	Acute myelomonocytic leukemia
c-*myb*	Adenocarcinoma of colon Acute myelogenous leukemia
c-*myc*	Promyelocytic leukemia (H160) Colon APUDoma (COLO 320) Small cell carcinoma of the lung Carcinoma of the breast (SKBr-3) Carcinoma of the breast Gastric adenocarcinoma
L-*myc*	Small cell carcinoma of the lung
N-*myc*	Neuroblastoma Small cell carcinoma of the lung Retinoblastoma
K-*ras*	Carcinoma of the lung Gastric carcinoma
N-*ras*	Mammary carcinoma (MCF-7)

tumor type from another. Neuroblastoma is often difficult to distinguish histologically from lymphoma and Ewing's sarcoma. Amplification of N-*myc* in small round tumor cells in appropriate clinical situations suggests a diagnosis of neuroblastoma. The diagnostic and prognostic value of N-*myc* gene copy number may make this determination an important clinical laboratory test for patients suspected of having neuroblastoma.

Amplification of c-*myc* is seen in a broader spectrum of human tumors than is N-*myc* amplification (Table 6.4). However, no tumor type or cell line shows as close a correlation between gene copy number and stage of disease as does N-*myc* and neuroblastoma. Occasional cell lines derived from human small cell carcinoma of the lung (SCCL) exhibit c-*myc* (and N-*myc*) amplification. Usually such amplification is correlated with a specific subtype, the large cell variant, that is more aggressive and less responsive to conventional chemotherapy and irradiation. Study of primary SCCL tissue, on the other hand, does not support a correlation of *myc* amplification with tumor stage. Wong and co-workers examined *myc* amplification in primary and metastatic tissue from SCCL. They found three or more gene copies in only 5 of 45 patients. It was of interest that metastatic SCCL tissue showed the same amplification as did its primary tumor.

Amplification or rearrangement of c-*myc* was described in 32% of primary breast carcinomas at presentation but appeared to be correlated only with the patient's age and not with tumor characteristics such as tumor grade or the presence of estrogen receptors. Overexpression of c-*myc* in human melanoma cell lines was inversely correlated with the expression of MHC class I antigens and with more virulent behavior (see Chapter 10).

Amplification of genes other than those of the *myc* family also has been associated with tumor aggressiveness. Using an antibody to the *ras* product p21, Schlom found a correlation between the histologic grade of pathological specimens of prostatic tumor tissue and the level of *ras* gene expression (see Viola et al.). Histologic grade in prostatic carcinoma is one of the best indicators of tumor aggressiveness and ultimately of patient survival. Amplification of the *neu* (*erb* B-2 = HER-2) gene had been noted in human breast cancer cell lines. Subsequently Slamon and co-workers examined primary breast cancer tissue and found *neu* gene amplification in 30% of tumors. This amplification was significantly correlated with both overall survival and time to relapse in breast cancer patients. Among the parameters classically regarded to be of prognostic significance in breast cancer—tumor size, patient age, presence or absence of estrogen and progesterone receptors, and metastasis to regional lymph nodes—amplification of *neu* was best correlated with the number of positive lymph nodes. Using multivariate analysis, these researchers demonstrated that *neu* gene amplification was an independent and powerful prognostic indicator of disease state in breast cancer. Amplification of *neu* also has been noted in a human salivary gland adenocarcinoma. Examples of other oncogene amplifications and their associated neoplastic states are presented in Table 6.4.

Despite the association of oncogene amplification, with stage and prognosis of some tumors, there are no data to support a causal role for oncogene amplification as an initiator of neoplasia. In fact, it is more likely that amplification plays a role in cancer progression rather than initiation as gene amplification has not been substantiated in normal cells.

Deletion of Suppressor Genes

Oncogenes are considered to act in a dominant manner if expression of one transforming allele contributes to the neoplastic phenotype. The genes we have so far discussed are of this type. There is experimental evidence for another class of genes whose expression inhibits transformation. *Loss* of such genes then may lead to neoplastic conversion.

The existence of cancer "suppressor genes" was suggested first by somatic cell hybridization experiments. When normal cells are fused with malignant cells, the resultant hybrids appear benign, suggesting that malignancy is actually a recessive trait. When certain chromosomes are lost in such hybrids, the ability to induce tumors is restored. In such experiments, the presence of human chromosomes 11 appears to suppress the transformed phenotype. The existence of genes capable of suppressing tumorigenicity has been suggested by other lines of experimentation. When v-H-*ras* is introduced into hamster embryo fibroblasts, malignant cell lines rarely form. However, emergence of rare tumorigenic lines always is associated with loss of a hamster chromosome 15. Similarly, when transforming EJ-*ras* genes are transfected into normal human diploid fibroblasts, no transformation is seen.

Retinoblastoma

Retinoblastoma, a relatively common malignant tumor of the eye in children (seen in approximately one per 20,000 live births) is one human malignancy where loss of a suppressor gene may be a critical step in tumorigenesis. There is considerable interest in the disease because of its hereditary nature. Neoplastic cells in the retinoblastoma arise from the fetal retinal layer. Treatment is generally by enucleation (removal of the affected eye); however, some patients have received radiotherapy, an efficacious treatment for small tumors.

Hereditary and sporadic cases of retinoblastoma occur. In the hereditary form (approximately 40% of cases), peripheral blood lymphocytes carry a deletion at chromosome 13q14, and individuals pass this deletion to their progeny. Retinoblastoma classically is transmitted as an autosomal dominant trait, but on the molecular level the disease appears to result from loss of a normal "suppressor" allele. In the sporadic form, no clear germ line mutation is seen, but the tumor cells frequently exhibit abnormalities in chromosome 13.

The natural histories of hereditary and sporadic retinoblastoma are different. The hereditary disease presents earlier in life, it commonly has bilateral ocular involvement, and patients have a higher risk for developing secondary tumors such as osteosarcomas. The sporadic form presents later, it is localized to one eye, and patients have no greater risk for secondary tumors compared to the general population. The clinical data are explained best by a "two-hit" hypothesis of transformation proposed by Knudson. This hypothesis states that sequential mutation of at least two independent genes is necessary for transformation of normal retinoblasts to their malignant counterparts. In the hereditary form, the first "hit" is inherited as the chromosome 13 deletion, and the second occurs randomly in susceptible retinal cells. Because cells in both retinas already carry the first mutation, the risk of tumor formation is equal and high for both eyes. In the sporadic form, two chance mutations must occur within the same retinoblast. Because either mutation alone is unlikely, there is an extremely low probability for both events to occur simultaneously in the same cell. Current evidence indicates that the two "hits" appear to be successive functional losses of both alleles of a retinoblastoma-associated gene, termed rb-1, located on chromosome 13q14 (Fig. 6.10).

A karyotypic assignment for a chromosomal deletion represents only a gross approximation of its molecular location. Isolated deletions encompassing many

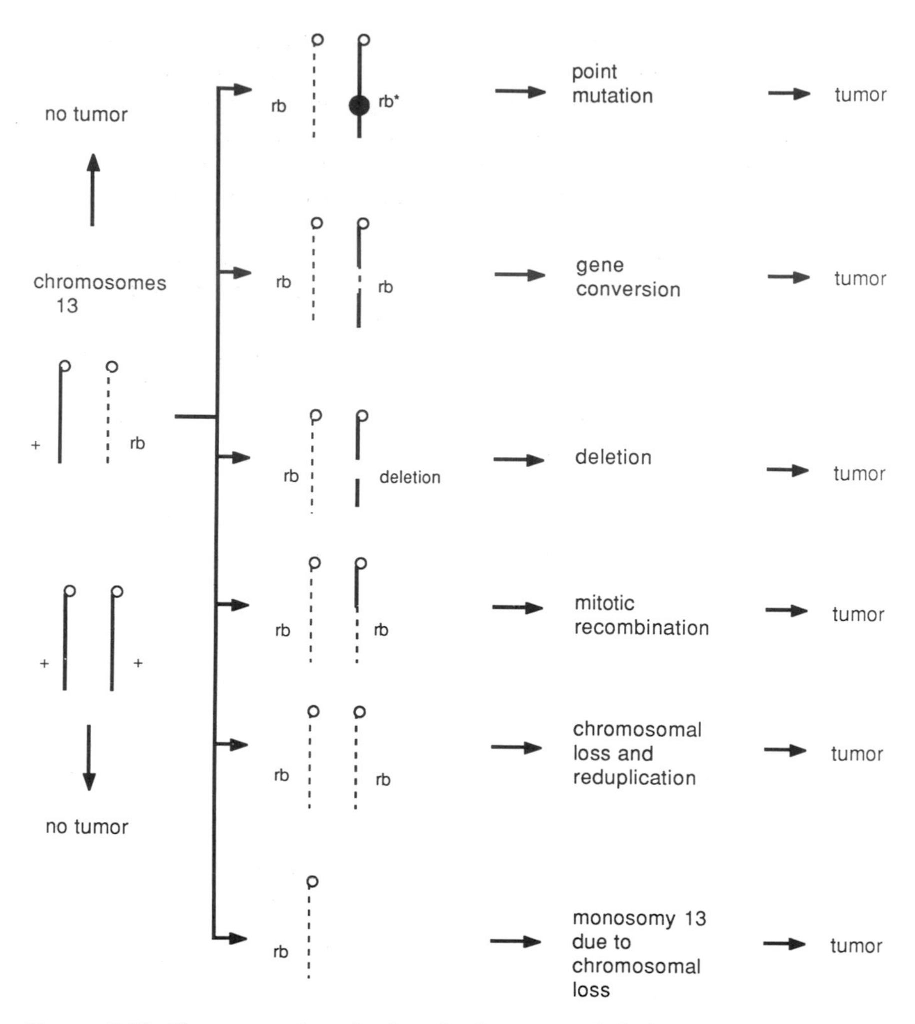

FIGURE 6.10. Chromosomal mechanisms leading to neoplasia in retinoblastoma. Solid lines represent chromosomes carrying the normal (wild-type) rb-1 allele, designated "+." Interrupted lines denote chromosomes harboring the abnormal rb-1 allele. Any situation where only the abnormal alleles remain would lead to tumor formation.

kilobases of DNA can occur without any noticeable change in karyotype. Thus in many patients with retinoblastoma, no gross abnormalities are seen in chromosome 13 despite biochemical evidence for deletions in one or both of the chromosomes. Biochemical evidence takes advantage of the close physical linkage between the retinoblastoma locus and a gene encoding the enzyme, esterase D.

In most cases a deletion at chromosome 13q14 is accompanied by a reduction in esterase D activity. One patient with sporadic retinoblastoma had a grossly normal karyotype but exhibited only one-half the esterase D activity compared to similar cells from her parents. The tumor cells containing one copy of chromosome 13 had no detectable esterase D activity. This result suggests that the loss of both rb-1 alleles, linked to esterase D, is important in the development of retinoblastoma.

Isolation of DNA probes specific for certain regions of chromosome 13 has aided elucidation of the mechanism of rb-1 gene loss. Cavenee and colleagues used chromosome 13 restriction fragment length polymorphism (RFLP) analysis to distinguish between individual chromosomes 13. The principle of RFLP analysis is discussed in Chapter 13. Using this technique, these scientists found that retinoblastomas frequently showed loss of both normal chromosomes 13 with reduplication of the abnormal chromosome containing the rb-1 defect. Reduplication of the abnormal chromosome 13 appeared to be a frequent mechanism to reveal the recessive retinoblastoma-associated gene. Mitotic recombination generating two copies of the abnormal rb-1 allele with concomitant loss of the normal allele is another mechanism leading to the same result.

Subsequently a gene mapping to 13q14 has been cloned and analyzed and found to correspond to the postulated rb-1 locus. A 1.5 kb DNA probe from a chromosome 13 λ library (see Chapter 2) detected a deleted segment in 3 of 37 retinoblastomas. Chromosome "walking" techniques allowed isolation of some 30 kb of surrounding DNA. One small fragment from this area, used as a probe, detected a conserved sequence in murine and human DNA. This probe also detected a 4.7 kb transcript from normal retinal cells and cells from other tumor types but not from retinoblastoma cells containing the 13q14 deletion. Further analysis revealed three types of deviant restriction fragments in DNA obtained from various retinoblastomas: (1) absent fragment corresponding to the 4.7 kb transcript, (2) underrepresented fragment, (3) deleted or rearranged fragment. The earlier work was confirmed by another group who found that 16 of 40 retinoblastomas contained structural changes—internal deletions corresponding to truncated transcripts in the rb-1 gene. The remaining tumors had absent transcripts or abnormally expressed transcripts verifying that the rb-1 locus is that associated with retinoblastoma. Interestingly, similar aberrations in the rb-1 gene are seen in osteosarcomas, another cancer that retinoblastoma patients are prone to develop.

Knowledge of the retinoblastoma-associated locus has already allowed prenatal diagnosis by RFLP analysis for risk of retinoblastoma (see Chapter 13).

Curiously, homozygous gene deletion is associated relatively commonly with oncogenesis in Drosophila. For example, deletion or insertional inactivation of a gene called "lethal (2) giant larvae" (l(2)gl) results in malignant neuroblastomas and tumors of imaginal discs, similar to human retinoblastoma. It might be postulated that l(2)gl in Drosophila and rb-1 in humans control normal cell proliferation and perhaps post-mitotic differentiation of the visual apparatus.

Wilms' Tumor

The clinical and molecular aspects of Wilms' tumor (nephroblastoma), another childhood neoplasm, are remarkably similar to those of retinoblastoma. Wilms' tumor is an embryonal neoplasm presenting as a kidney mass in young children. There appear to be both hereditary and sporadic forms of the disease. The sporadic tumors usually are unilateral; the hereditary tumors present at an earlier age and are bilateral and multifocal. In addition, hereditary Wilms' tumors often are associated with aniridia (lack of irises) and mental retardation, a constellation of abnormalities called the aniridia–Wilms' tumor syndrome. Affected individuals harbor a germ line deletion on the short arm of chromosome 11, band p13, which is seen also in Wilms' tumor cells themselves. The Beckman–Weidemann syndrome is a related disorder where patients develop Wilms' tumors, hepatoblastomas, rhabdomyosarcomas, and adrenal carcinomas in association with hemihypertrophy. These individuals also carry constitutive abnormalities on chromosome 11, region 11p13-11p15. RFLP analysis of tumor cells characteristically shows loss of segments on chromosome 11p. Thus with Wilms' tumor the loss of an important normal gene on chromosome 11p appears to be associated with, and may be implicated in, the pathogenesis of the disease.

Several additional studies have associated deletions of particular chromosome segments and human neoplasia. Fortuitous detection of a chromosome 5 deletion in a patient with familial adenomatous polyposis (FAP), a disease that predisposes to adenocarcinoma of the colon led to mapping of an FAP associated gene to chromosome 5. Analysis of a family with FAP showed deletion of a 4.4 kb fragment localizing to 5q21-q22 in every affected individual. Extension of this work to sporadic colon cancers showed at least 20% of these tumors lost a 5q allele present in matched normal tissue. This data points to a gene whose homozygous loss may predispose to colon carcinoma.

Nine patients with small cell lung cancer were shown to have lost alleles at 3p21. Loss of 3p alleles also occurred in 11 of 11 evaluable patients with renal cell carcinoma. Chromosome 3p has been associated with translocations in renal cell carcinoma [t(3;11) and t(3;8)] and with deletions and translocations in melanoma.

Tables 13.1 and 13.2 list a number of other tumors in which chromosomal deletions have been identified. How important the loss of this genetic information is in the pathogenesis of these neoplasms remains to be elucidated.

A Study in Molecular Oncogenesis: Breast Cancer

The consistency of the genetic abnormalities in hematological malignancies such as CML and Burkitt's lymphoma are dramatic. However, most human cancers are of epithelial origin, i.e., breast, colon, and most lung cancers. With these tumors, no single genetic perturbation emerges as the critical transforming event. An analysis of studies on mammary carcinogenesis is instructional for

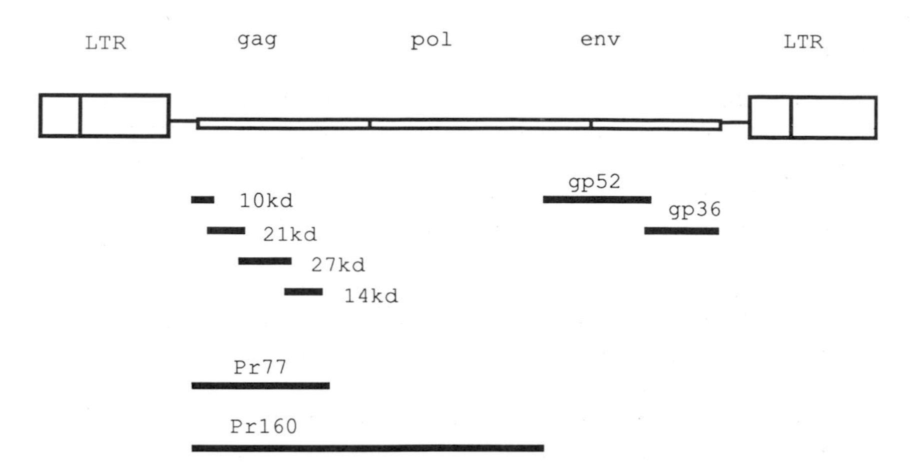

FIGURE 6.11. Proviral structure of the mouse mammary tumor virus.

it highlights two concepts that may be important in interpreting the pertinent scientific literature. First, a variety of genetic insults may produce the same neoplastic phenotype, and second, caution should be exercised when extrapolating data from mouse models to a human disease.

Models of murine mammary tumors, for example, suggest that a number of oncogenes could be involved in the transformation process. Some mouse mammary cancer is virally induced. Murine mammary tumor viruses (MMTVs) are type B retroviruses. A prototypic MMTV genome structure is depicted in Figure 6.11. As for other retroviruses, the proviral LTRs carry promoter, enhancer, and polyadenylation functions. In many mouse strains infection with the virus occurs via milk and is passed from mother to offspring nongenetically. In a few strains an MMTV provirus is integrated within the genome and transmitted genetically. In either case, tumors occur after a long latency period and arise from a single infected cell.

The MMTV carries no known transforming gene. In a susceptible mouse strain called C3H, 18 of 26 tested tumors appeared to have MMTV DNA integrated at a common site within the host genome. Cloning of this common area revealed a new gene, designated *int*-1, that was transcriptionally activated by the close proximity of an MMTV LTR enhancer (see Chapter 11). The *int*-1 gene is expressed only in murine mammary tumors induced by MMTVs, not in normal mammary glands. This finding suggests that *int*-1 activation by MMTV integration must play a role in murine mammary tumorigenesis.

Chemical carcinogens also have revealed endogenous genes important for mammary gland transformation. *N*-Nitroso-*N'*-methylurea (NMU) methylates the 7-nitrogen and 6-oxygen positions of deoxyguanosine causing guanine-to-adenine conversions. A single injection of NMU in 50-day-old rats induces mammary carcinomas at high frequency. These tumors also are hormone-dependent in that castrated female rats rarely develop them. Zarbl and colleagues investigated which oncogenes could be activated in these tumors using an NIH 3T3 transfection/focus formation assay. Transformation of 3T3 cells occurred in 83% of the tumors examined; in each case the activated oncogene was c-H-*ras* altered by a G-to-A change in codon 12. Normal mammary epithelial cell DNA was unable to induce transformation of 3T3 cells. This finding suggests that genes of the *ras* family play a role in murine mammary neoplasia.

The introduction of oncogenes into germ line tissue of mice is a powerful tool to probe oncogene function in the intact animal. As mentioned above, *myc* genes under transcriptional control of an MMTV promoter and enhancer have been constructed. The MMTV enhancer unit is hormonally sensitive: In the presence of steroid hormones, transcription is increased. An introduced *myc* gene con-

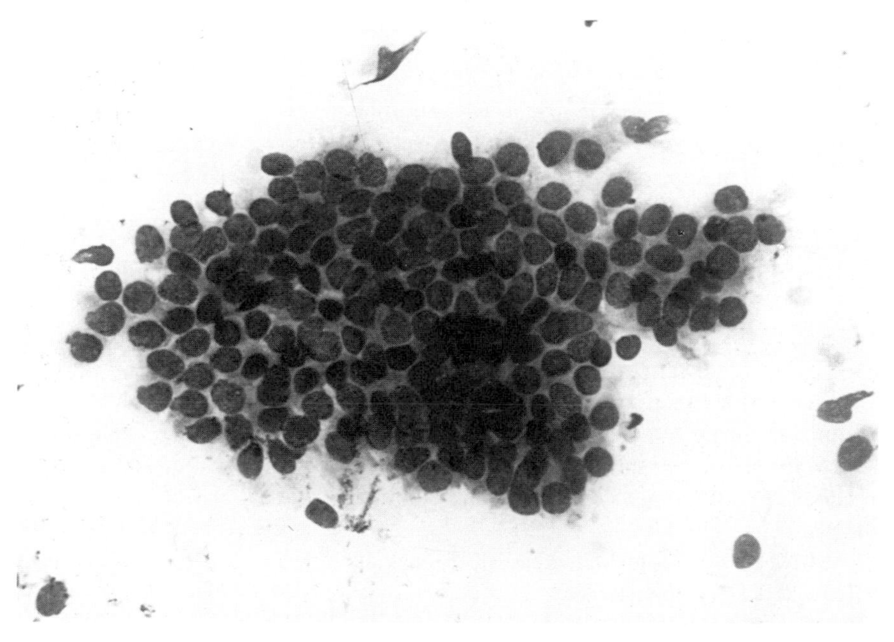

FIGURE 6.12. Pathology of human breast cancer. (A) Fine needle aspiration (FNA) sample from benign human breast tissue. The epithelial cells are small, uniform, and arranged in a monolayered sheet. (×100) (B) FNA sample from human breast cancer. The epithelial cells are large compared to the benign breast cells in A. There is variation in size and shape. The arrangement within cell clusters is irregular. (×100) (C) Histologic section from benign human breast tissue. The epithelium is arranged as well defined units surrounded by fibrous stroma. Numerous small ducts are present (arrowheads). One duct is leading into numerous lobules (arrows). (×20) (D) Histologic section from human breast cancer. Numerous nests of cancer cells (arrowheads) are infiltrating in a random fashion throughout the pale-staining stroma. One larger nest (arrow) shows tumor cells at the edges and the necrotic debris in the center. (×20) (A–D kindly provided by Dr. Britt-Marie Ljung)

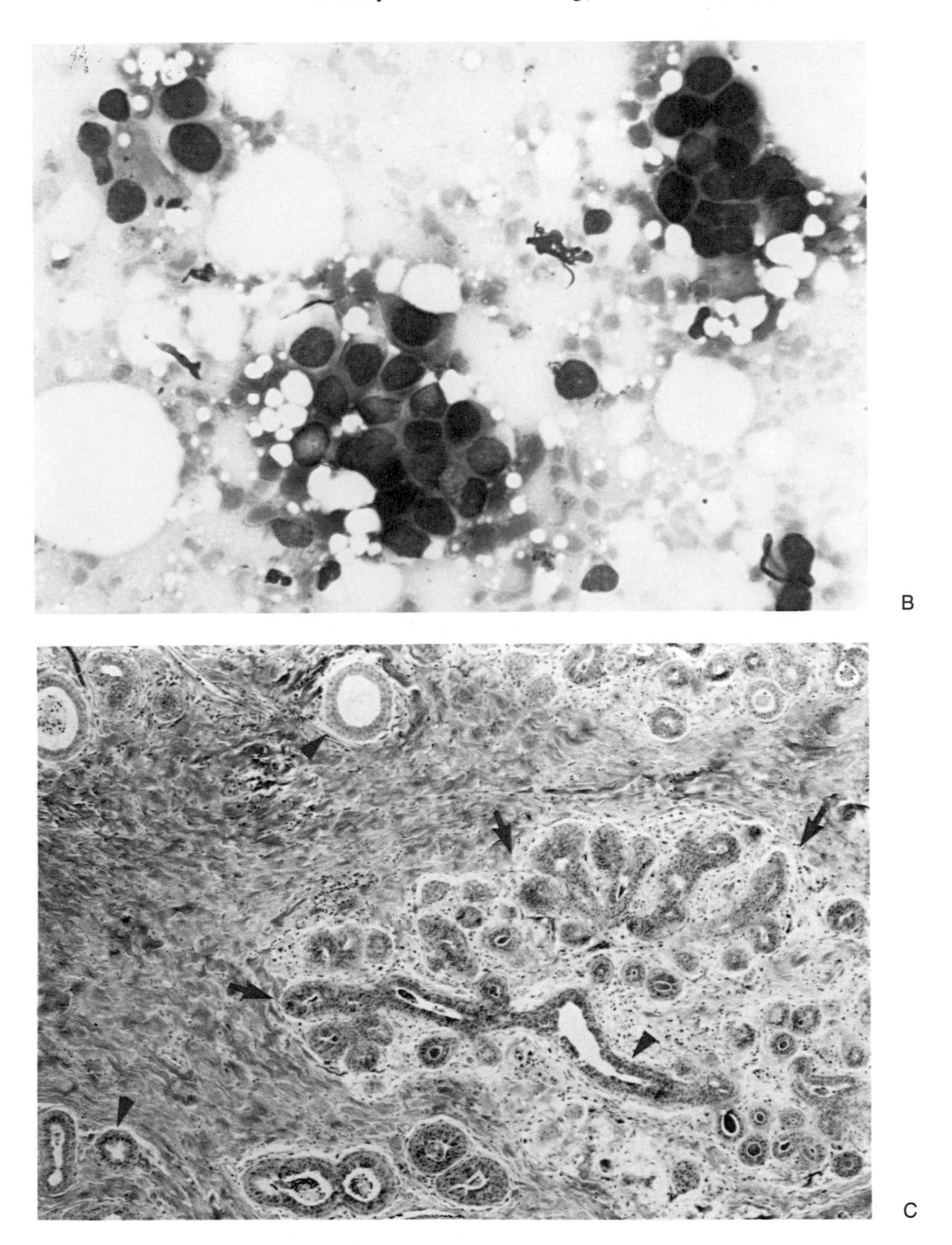

FIGURE 6.12. *Continued.*

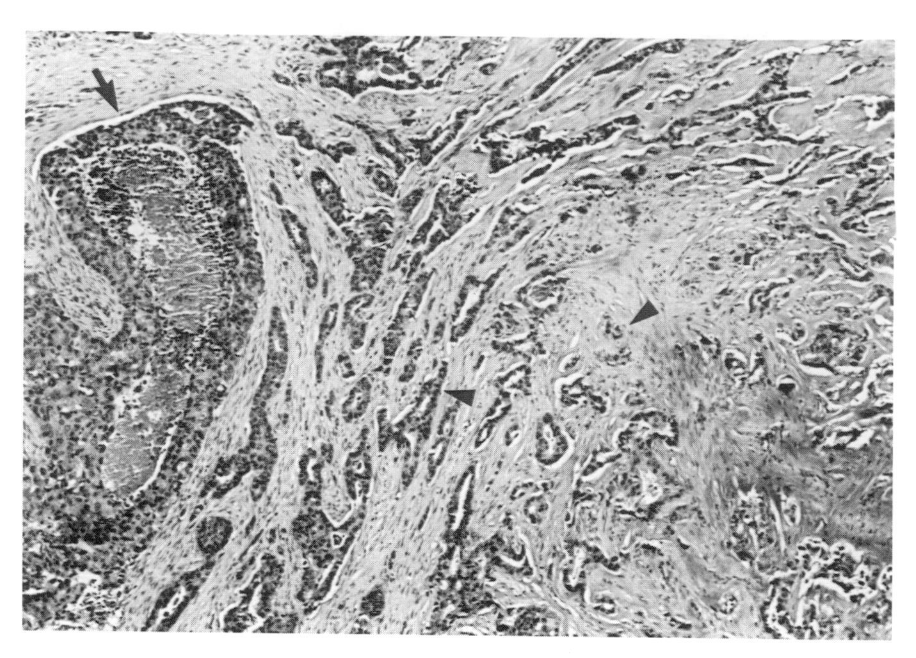

FIGURE 6.12. *Continued.*

trolled by a MMTV LTR would be predicted to be influenced by steroid hormones. Pregnancy exposes female animals to increased steroid hormone levels. It was found that MMTV LTR-*myc* transgenic mice develop mammary carcinomas at high frequency after their second or third pregnancy, and all of the tumors expressed the introduced *myc* gene. The long latency period prior to mammary cancer development implies that additional genetic events are required, presumably in cooperation with *myc* expression, for these tumors to form.

Animal models have implicated *myc*, *ras*, and *int*-1 in the genesis of mammary carcinomas. The role of these oncogenes in initiating human breast cancers is less certain (Fig. 6.12); however, perturbations of *erbB*-2/*neu*, and the loss of a putative suppressor gene on chromosome 11p in this disease are emerging as important associations.

Escot and co-workers studied the c-*myc* locus in 121 human primary breast carcinomas. They found that 32% of these tumors had a 2- to 15-fold amplification of c-*myc*. However, perturbations in *myc* did not impact on the prognosis of these patients, nor did it correlate with unfavorable clinical parameters. Since gene amplification has not been observed in non transformed tissue, it is highly unlikely, that *myc* amplification was the initiating event. Neither amplification nor mutations of the *ras* genes have been detected in primary breast cancer cells, and *int*-1 is not expressed in human breast cancers. Thus the proto-oncogenes implicated in the genesis of murine mammary carcinomas appear not to be important in initiating human breast cancer.

Two genetic aberrations, however, are emerging as significant lesions in the human disease: *erbB-2/neu* amplification, and loss of heterozygosity at the c-H-*ras* locus. Slamon and colleagues observed that the *erbB-2/neu* locus is amplified in 30% of primary breast cancers and that amplification is associated with a worse prognosis. Theillet and co-workers found a loss of a H-*ras* allele in breast tumor tissue in 14 of 51 patients associated with high grade lesions lacking hormone receptors. In these examples, the deletion of a H-*ras* gene itself is probably not the important event. It is hypothesized that the loss of a putative tumor suppressor gene positioned close to the H-*ras* locus on chromosome 11p is the critical genetic lesion. Thus, in human breast cancer, no single genetic abnormality predominates which may reflect the complex etiologic factors involved in the human disease.

Bibliography

Point Mutations

Albino AP, Lestrange R, Oliff A, et al. Transforming ras genes from human melanoma: a manifestation of tumor heterogeneity? Nature 308:69, 1984.

Balmain A, Ramsden M, Bowden G, et al. Activation of the mouse cellular Harvey ras gene in chemically induced benign skin papillomas. Nature 307:658, 1984.

Bargmann C, Hung ML, et al. Multiple independent activations of the neu oncogene by a point mutation altering the transmembrane domain of p185. Cell 45:649, 1986.

Bos JL, Toksoz D, Marshall CJ, et al. Amino acid substitutions at codon 13 of the N-ras oncogene in human acute myeloid leukemia. Nature 315:726, 1985.

Bos J, Fearon E, Hamilton S, et al. Prevalence of ras gene mutations in human colorectal cancers. Nature 327:293, 1987.

Brown R, Marshall CJ, Pennie SG, et al. Mechanism of activation of an N-ras gene in the human fibrosarcoma cell line HT 1080. EMBO J 3:1321, 1984.

Fasano O, Aldrich T, Tamanoi F, et al. Analysis of the transforming potential of the human H-ras gene by random mutagenesis. Proc Natl Acad Sci USA 81:4008, 1984.

Feig LA, Bast RC, Knapp RC, et al. Somatic activation of ras gene in a human ovarian carcinoma. Science 223:698, 1984.

Feinberg AP, Vogelstein B, Droller MJ, et al. Mutations affecting the 12th amino acid of the c-H-ras oncogene product occurs infrequently in human cancer. Science 230:1175, 1983.

Forrester K, Almoguera C, Han K, et al. Detection of high incidence of K-ras oncogenes during human colon tumorigenesis. Nature 327:248, 1987.

Fujita F, Yoshida O, Yuasa Y, et al. H-ras oncogenes are activated by somatic alterations in human urinary tract tumors. Nature 309:464, 1984.

Hirai H, Kobayashi Y, Mano H, et al. A point mutation at codon 13 of the N-ras oncogene in myelodysplastic syndrome. Nature 327:430, 1987.

Harris CC, Lechner JF, Yoakum GH. In vitro studies of human lung carcinogenesis. Carcinog Compr Surv 9:257, 1985.

Liu E, Hjelle B, Morgan R, Hecht F, Bishop JM. Mutations of the Kirsten-ras proto-oncogene in human preleukemia. Nature 330:186, 1987.

Rodenhuis S, van de Wetering M, Mooi W, et al. Mutational activation of the K-ras oncogene: a possible pathogenetic factor in adenocarcinoma of the lung. New Engl J Med 317:929, 1987.

Santos E, Martin-Zanca D, Reddy EP, et al. Malignant activation of a k-ras oncogene in lung carcinoma but not in normal tissue of the same patient. Science 223:661, 1984.

Shih C, Padhy LC, Murray MJ, et al. Transforming genes of carcinomas and neuroblastomas introduced into mouse fibroblasts. Nature 290:261, 1981.

Taparowski EH, Suard Y, Fasano O, et al. Activation of T24 bladder carcinoma transforming gene is linked to a single amino acid change. Nature 300:762, 1982.

Wong D, Liu E, Cadman E. Spontaneous transfer of drug resistance genes in cultured mammalian cells. Clin Res 33:460A, 1985.

Genetic Rearrangements Within the Body of the Gene: CML

Bolin RW, Robinson WA, Sutherland J, et al. Busulfan vs. hydroxyurea in the long-term therapy of CML. Cancer 50:1683, 1982.

Collins S, Kubowski I, Miyoshi I, et al. Altered transcription of c-abl oncogene in K 563 and other chronic myelogenous leukemia cells. Science 225:72, 1984.

Daley G, McLaughlin J, Witte O. The CML-specific p210 bcr/abl protein, unlike v-abl, does not transform NIH/3T3 fibroblasts. Science 237:532, 1987.

De Klein A, Guerts van Kessel A, Grosveld G, et al. A cellular oncogene is translocated to the Philadelphia chromosome in chronic myelogenous leukemia. Nature 300:765, 1982.

Groffen J, Stephenson JR, Heisterkamp N, et al. Philadelphia chromosomal breakpoints are clustered within a limited region bcr, on chromosome 22. Cell 36:93, 1984.

Heisterkamp N, Stephenson JR, Groffen J, et al. Localization of the c-abl oncogene adjacent to a translocation breakpoint in chronic myelogenous leukemia. Nature 306:239, 1983.

Hermans A, Heisterkamp N, von Lindern M, et al. Unique fusion of bcr and c-abl genes in Philadelphia chromosome positive acute lymphoblastic leukemia. Cell 51:33, 1987.

Hirai H, Tanaka S, Azuma M, et al. Transforming genes in human leukemia cells. Blood 66:1371, 1985.

Konopka JB, Watanbe SM, Witte ON. An alteration of the human c-abl protein in K 562 leukemia cells unmasks associated tyrosine kinase activity. Cell 37:1035, 1984.

Liu E, Hjelle B, Bishop JM. Transforming genes in chronic myelogenous leukemia. Proc Natl Acad Sci USA 85:1952, 1988.

Pryines R, Foulkes JG, Rosenberg N, et al. Sequences of the A-MuLV protein needed for fibroblast and lymphoid cell transformation. Cell 34:569, 1983.

Shtivelman E, Lifshitz B, Gali RP, et al. Fused transcript of abl and bcr genes in chronic myelogenous leukemia. Nature 315:550, 1985.

Genetic Rearrangement Outside the Coding Region of the Proto-oncogene: *c-myc*

Alitalo K, Bishop JM, Smith DH, et al. Nucleotide sequence of the v-myc oncogene of avian retrovirus MC29. Proc Natl Acad Sci USA 80:100, 1983.

Alitalo K, Saksela K, Winquist R, et al. Amplification and aberrant expression of cellular oncogenes in human colon cancer cells. In: Genes and Cancer. New York: Alan R. Liss, 1984, pp. 383–397.

Bentley DL, Groudine M. A block to elongation is largely responsible for decreased transcription of c-myc in differentiated HL60 cells. Nature 321:702, 1986.

Chung JH, Reed R, Swan E, et al. *cis*Acting regulatory sequences of human c-myc. Presented at the Second Annual Meeting on Oncogenes, Hood College, Frederick, MD.

Croce C. Role of chromosome translocations in human neoplasia. Cell 49:155, 1987.

Dalla-Favera R, Bregni M, Erickson J, et al. Human c-myc onc gene is located on the region of chromosome 8 that is translocated in Burkitt lymphoma cells. Proc Natl Acad Sci USA 79:7824, 1982.

Dalla-Favera R, Martinotti S, Gallo RC, et al. Translocation and rearrangement of the c-myc oncogene locus in human undifferentiated B-cell lymphocytes. Science 219:963, 1983.

Dani Ch, Blanchard JM, Piechaczyk M, et al. Extreme instability of myc mRNA in normal and transformed human cells. Proc Natl Acad Sci USA 81:7046, 1984.

Erikson J, ar-Rushdi A, Drwinga HL, et al. Transcriptional activation of the translocated c-myc oncogene in Burkitt lymphoma. Proc Natl Acad Sci USA 80:820, 1983.

Hann SR, Abrams HD, Rohrschneider LR, et al. Proteins encoded by v-myc and c-myc oncogenes: identification and localization in acute leukemia virus transformants and bursal lymphoma cell lines. Cell 34:789, 1983.

Hay N, Bishop JM, Levens D. Regulatory elements that modulate the expression of human c-myc. Genes and Development 1:659, 1987.

Leder P, Battey J, Lenoir G. Translocation among antibody genes in human cancer. Science 222:765, 1983.

Lee WMF, Schwab M, Westaway D, et al. Augmented expression of normal c-myc is sufficient for code transformation of rat embryo cells with a mutant ras gene. Mol Cell Biol 12:3345, 1985.

Lombardi L, Newcomb E, Dalla-Favera R. Pathogenesis of Burkitt lymphoma: expression of an activated c-myc oncogene causes the tumorigenic conversion of EBV-infected human B lymphoblasts. Cell 49:161, 1987.

Mellon P, Pawson A, Bister G, et al. Specific RNA sequences and gene products of MC29 avian acute leukemia virus. Proc Natl Acad Sci USA 75:5874, 1978.

Mougneau E, Lemieux L, Rassoulzadegan M, et al. Biological activities of v-myc and rearranged c-myc oncogenes in rat fibroblast cells in culture. Proc Natl Acad Sci USA 81:5758, 1984.

Saito H, Hayday AC, Wiman K, et al. Activation of the c-myc gene by translocation: a model for translational control. Proc Natl Acad Sci USA 80:7476, 1983.

Taub R, Kirsch I, Morton C, et al. Translocation of the c-myc gene into the immunoglobulin heavy chain locus in human Burkitt's lymphoma and murine plasmacytoma cells. Proc Natl Acad Sci USA 79:7837, 1982.

Watt R, Stanton LW, Marcu KB, et al. Nucleotide sequence of cloned cDNA of human c-myc oncogene. Nature 303:725, 1983.

Gene Amplification and Overexpression

General

Schmike RT. Gene amplification in cultured mammalian cells. Cell 37:705, 1984.

Schmike RT. Summary. In: Schmike RT (ed), Gene Amplification. Cold Spring Harbor, NY: Cold Spring Harbor Laboratory, 1982, p. 317.

Stark GR, Wall GM. Gene amplification. Annu Rev Biochem 53:447, 1984.

c-myc

Collins S, Grondine M. Amplification of endogenous myc-related DNA sequences in a human myeloid leukemia cell line. Nature 298:679, 1982.

Little CD, Nau MM, Carney DN, et al. Amplification and expression of the c-myc oncogene in human lung cancer cell lines. Nataure 306:194, 1983.

Sikora K, Evans GI, Stewart J, et al. Detection of the c-myc oncogene product in testicular cancer. Br J Cancer 52:171, 1985.

Watson JV, Stewart J, Evan GI, et al. The clinical significance of flow cytometric c-myc oncoprotein quantitation in testicular cancer. Br J Cancer 53:331, 1986.

Wong AJ, Ruppert JM, Eggleston J, et al. Gene amplification of c-myc and N-myc in small cell carcinoma of the lung. Science 233:461, 1986.

N-myc

Nau MM, Brooks BJ, Carney DN, et al. Human small cell lung cancers show amplification and expression of the N-myc gene. Proc Natl Acad Sci USA 83:1092, 1986.

Schwab M. Amplification of N-myc in human neuroblastomas. Trends Genet 1(10):271, 1985.

Seeger RC, Brodeu GM, Sather H, et al. Association of multiple copies of the N-myc oncogene with rapid progression of neuroblastomas. N Engl J Med 313:1111, 1985.

Stanton L, Schwab M, Bishop JM. Nucleotide sequence of the human N-myc gene. Proc Natl Acad Sci USA 83:1772, 1986.

ras

Schwab M, Alitalo K, Varmus HE, et al. A cellular oncogene (c-K-ras) is amplified, overexpressed and located within karotypic abnormalities in mouse adrenocortical tumor cells. Nature 303:497, 1983.

Viola MV, Fromowitz F, Oravez S, et al. Expression of ras oncogene p21 in prostate cancer. N Engl J Med 314:133, 1986.

Others

Gullick WJ, Marsden JJ, Whittle N, et al. Expression of epidermal growth factor receptors on human cervical, ovarian, and vulval carcinomas. Cancer Res 46:285, 1986.

Pelicci PG, Lan Francone L, Brathwaite MD, et al. Amplification of the c-myb oncogene in a case of human acute myelogenous leukemia. Science 224:1117, 1984.

Rovigahi U, Watson DK, Yunis JJ. Amplification and rearrangement of Hu-ets-1 in leukemia and lymphoma with involvement of 11q23. Science 232:398, 1986.

Semba K, Kamata N, Toyoshima K, et al. A v-erb B-related proto-oncogene, c-erb B-2, is distinct from the c-erb B-1/epidermal growth factor-receptor gene and is amplified in a human salivary gland adenocarcinoma. Proc Natl Acad Sci USA 82:6497, 1985.

Slamon DJ, Clark GM, Wong SG, et al. Human breast cancer: correlation of relapse and survival with amplification of the HER-2/neu oncogene. Science 325:117–182, 1987.

Spandidos DA, La Mothe A, Field JK. Multiple transcriptional activation of cellular oncogenes in human head and neck solid tumors. Anticancer Res 5:221, 1985.

Yamamoto T, Kamata N, Kawano H. High incidence of amplification of the epidermal growth factor receptor gene in human squamous carcinoma cell lines. Cancer Res 46:414, 1986.

Deletion of Suppressor Genes

Bodmer WF, Bailey C, Bodmer J, et al. Localization of the gene for familial adenomatous polyposis on chromosome 5. Nature 328:614, 1987.

Klinger HP, Shows TB. Suppression of tumorigenicity in somatic cell hybrids: human chromosome implicated as suppressors of tumorigenicity in hybrids with Chinese hamster ovary cells. J Natl Cancer Inst 71:559, 1983.

Knudson A. Model of hereditary cancers in man. Prog Nuc Acid Res Mol Biol 29:17, 1983.

Mechler B, McGinnis W, Gehring W. Molecular cloning of lethal (2) giant larvae, a recessive oncogene in Drosophila melanogaster. EMBO J 4:1551, 1985.

Naylor SL, Johnson B, Minna JD. Loss of heterozygosity of chromosome 3p markers in small-cell lung cancer. Nature 329:451, 1987.

Oshimura M, Gilmer TM, Barrett J-C. Nonrandom loss of chromosome 15 in Syrian hamster tumors induced by v-H-ras plus v-myc oncogenes. Nature 316:636, 1985.

Sager R, Tanaka K, Lau C-C, et al. Resistance of human cells to tumorigenesis induced by cloned transforming genes. Proc Natl Acad Sci USA 80:7601, 1983.

Solomon E, Voss R, Hall V, et al. Chromosome 5 allele loss in human colorectal carcinomas. Nature 328:616, 1987.

Stanbridge EJ, Der CJ, Doersen C, et al. Human cell hybrids: analysis of transformation and tumorigenicity. Science 215:252, 1982.

Weissman BE, Saxon PJ, Pasquale SR, et al. Introduction of a normal human chromosome 11 into a Wilm's tumor cell line controls its tumorigenic expression. Science 236:175, 1987.

Zbar B, Brauch H, Talmadge C, et al. Loss of alleles on the short arm of chromosome 3 in renal cell carcinoma. Nature 327:721, 1987.

Retinoblastoma

Benedict WF, Murphree AL, Banergee A, et al. Gene for hereditary retinoblastoma assigned to human chromosome 13 by linkage to esterase D. Science 219:971, 1983.

Cavenee WK, Dryja TP, Phillips RA, et al. Expression of alleles by chromosomal mechanisms in retinoblastoma. Nature 305:779, 1983.

Cavenee WK, Hansen MF, Nordenskjold M, et al. Genetic origin of mutations predisposing to retinoblastoma. Science 228:501, 1985.

Cavenee WK, Murphree AL, Shull MM, et al. Prediction of familial predisposition to retinoblastoma. N Engl J Med 314:1201, 1986.

Dryja T, Rapaport J, Joyce J, et al. Molecular detection of deletions involving band q14 of chromosome 13 in retinoblastomas. Proc Natl Acad Sci USA 83:7391, 1986.

Friend SH, Bernards R, Rogelj S, et al. A human DNA segment with properties of the gene that predisposes to retinoblastoma and osteosarcoma. Nature 323:643, 1986.

Fung Y, Murphree A, T'Ang A, et al. Structural evidence for the authenicity of the human retinoblastoma gene. Science 236:1657, 1987.

Knudson AG. Mutations and cancer: statistical study of retinoblastoma. Proc Natl Acad Sci USA 68:820, 1971.

Murphree AL, Benedict WF. Retinoblastoma: clues to human oncogenesis. Science 223:1028, 1984.

Wilms' Tumor

Koufos A, Hansen MF, Copeland NG, et al. Loss of heterozygosity in three embryonal tumors suggests a common pathogenic mechanism. Nature 316:330, 1985.

A Study in Molecular Oncogenesis: Breast Cancer

Brown AMC, Wildin RS, Prendergast TJ, et al. A retrovirus vector expressing the putative mammary oncogene int-1 causes partial transformation of a mammary epithelial cell line. Cell 46:1001, 1986.

Escot C, Theilet C, Lidereau R, et al. Genetic alteration of the c-myc proto-oncogene (myc) in human primary breast carcinomas. Proc Natl Acad Sci USA 83:4834, 1986.

Fasano O, Birnbaum D, Edlund L, et al. New human transforming genes detected by a tumorigenicity assay. Mol Cell Biol 4:1695, 1984.

Kasid A, Lippman ME, Papageorge AG, et al. Transfection of v-ras^{H} DNA into MCF-7 human breast cancer cells bypasses dependents on estrogen for tumor. Science 228:725, 1985.

Kraus MH, Yuasa Y, Aaronson SA. A position 12-activated H-ras oncogene in all HS 578Y mammary carcinosarcoma cells but not normal mammary cells of the same patient. Proc Natl Acad Sci USA 81:5384, 1984.

Lippman ME, Huff K, Jakesz R, et al. Estrogens regulate production of specific growth factors in hormone-dependent human breast cancer. Ann NY Acad Sci 464:11, 1986.

Rochlitz CF, Scott GK, Dodson JM, Liu E, et al. Incidence of activated ras oncogene mutations associated with primary and metastatic breast cancer. Cancer Research, in press, 1989.

Smith HS, Wolman SR, Hackett AJ. The biology of breast cancer at the cellular level. Biochim Biophys Acta 738:103, 1984.

Sukumar S, Notario V, Martin-Zanca M., et al. Induction of mammary carcinomas in rats by nitroso-methylurea involves malignant activation of H-ras-1 locus by single point mutation. Nature 306:658, 1983.

Theillet C, Lidereau R, Escot C, et al. Loss of a c-H-ras-1 allele and aggressive human primary breast carcinoma. Cancer Res 46:4776, 1986.

Zarbl H, Sukumar S, Arthur A, et al. Direct mutagenesis of H-ras-1 oncogenes by N-nitroso-N-methylurea during initiation of mammary carcinogenesis in rats. Nature 315:382, 1985.

7
src and Related Protein Kinases

Overview

The transforming gene of the Rous sarcoma virus, v-*src*, encodes an enzyme that phosphorylates tyrosine residues on proteins. This and related tyrosine kinases comprise a family of regulatory enzymes involved in the control of cellular growth and differentiation. The family includes members closely related to *src*: (1) the *abl* oncogenes that are implicated in the translocation characteristic of chronic myelogenous leukemia, (2) receptors for growth factors, e.g., epidermal growth factor, macrophage colony-stimulating factor, the platelet-derived growth factor, and insulin and insulin-like growth factors (see Chapter 8), and (3) the serine-threonine kinases, e.g., *ral* and *mos*. The activity of many, if not all, of the tyrosine kinases is regulated by phosphorylation.

In the best studied case, *src*, it appears that the transforming gene is generated from the normal cellular proto-oncogene by deletion of a region encoding a tyrosine near the carboxy terminus of the protein. This alteration, also found in many of the growth factor receptor oncogenes, causes these proteins to function abnormally. Although several substrates of the tyrosine kinases have been found, those critical for the initiation and maintenance of the transformed state have not been conclusively identified.

Several oncogenes are formed by fusion of a tyrosine kinase gene with another cellular gene (e.g., v-*fgr*, *met*). This action also may dysregulate protein tyrosine phosphorylations.

Protein phosphorylation has been known to mediate regulation of enzyme function for many years. The most common covalent linkages involve esterification of phosphate to serine and threonine with less frequent phosphorylation of arginine, histidine, aspartic acid, glutamic acid, and cysteine. Tyrosine is an uncommon substrate for phosphorylation.

The hypothesis that tyrosine phosphorylation might be an important mechanism of growth control was first discovered when polyoma middle T antigen (see Chapter 3) was found by Tony Hunter and co-workers to phosphorylate tyrosine. Phosphotyrosine was discovered in Rous sarcoma virus (RSV)-transformed cells in 1979. It soon was recognized that the transforming gene product of RSV, $pp60^{v\text{-}src}$, was a tyrosine protein kinase (Fig. 7.1). A number of other viral oncogenes now are known to belong to the family of tyrosine protein kinases

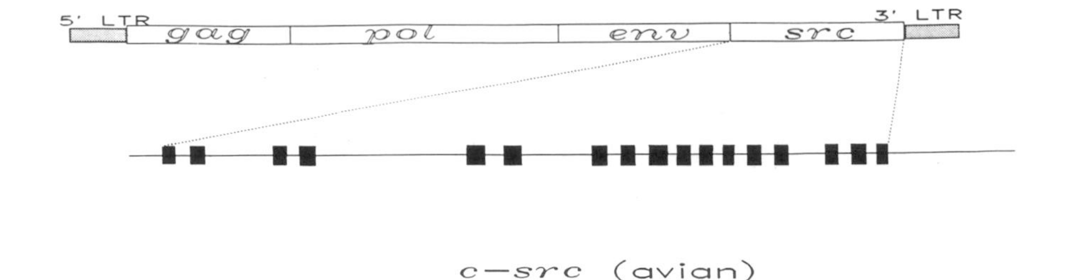

FIGURE 7.1. Structure of the Rous sarcoma virus and the avian c-*src* gene. Because the virus contains complete *gag-pol-env* genes, it is replication-competent.

(Table 7.1); in addition, many growth factor receptors have tyrosine protein kinase activity (Table 7.2).

The fact that genes similar to higher vertebrate tyrosine protein kinases are found among *Drosophila* suggests that tyrosine phosphorylation may play an important role in the control of eukaryotic cellular growth (see below). Protein phosphatases specific for phosphotyrosine have been reported. It is plausible that

TABLE 7.1. *src* and related protein kinases.

Closely related products
c-*src* (v-*src*)
c-*fgr* (v-*fgr*)
c-*yes* (v-*yes*)
c-*syn*
c-*kit* (v-*kit*)
hck, fyn, lyn
Tyrosine kinases lacking transmembrane domains
c-*abl* (v-*abl*)
c-*fes*(fps) (v-*fes*[*fps*])
lsk^{T}/*tck*
Growth factor receptors
Tyrosine kinase with transmembrane domains
c-*erb* B/EGF receptor, v-*erb* B
neu/*erb* B2
c-*fms*/CSF-1 receptor, v-*fms*
c-*ros* (v-*ros*), insulin receptor
PDGF receptor
IGF-1 and IGF-2 receptors
Chimeric src-related products
trk, met, mcf-3, ret, onc D
Serine or threonine protein kinases
c-*mos* (v-*mos*)
c-*raf*/*mil* (v-*raf*/*mil*)
c-*pks*

TABLE 7.2. Tyrosine protein kinases.

Proto-oncogene			Viral oncogene		
Oncogene	Protein	Intracellular location	Oncogene	Protein	Intracellular location
Virus-related					
c-*src*	pp60$^{c\text{-}src}$	Membrane	v-*src*	pp60$^{v\text{-}src}$	Membrane/ cytoskeleton
c-*abl*	p150$^{c\text{-}abl}$	Cytoplasm	v-*abl*	P160$^{gag\text{-}abl}$	Membrane
c-*yes*	pp62$^{c\text{-}yes}$	–	v-*yes*	P90$^{gag\text{-}yes}$	Membrane
c-*fps*	p98$^{c\text{-}fps}$	Soluble	v-*fps*	P140$^{gag\text{-}fps}$	Membrane/ soluble
c-*fes*	p92$^{c\text{-}fes}$	–	v-*fes*	p85$^{gag\text{-}fes}$	Cytoplasm
c-*fgr*	–	–	v-*fgr*	P70$^{gag\text{-}fgr}$	Cytoplasm
c-*ros*	–	–	v-*ros*	P68$^{gag\text{-}ros}$	Membrane
c-*kit*	–	–	v-*kit*	P80$^{gag\text{-}kit}$	–
Growth factor receptors					
c-*erb*-B	EGF receptor	Plasma membrane	v-*erb*-B	gp68/ 74$^{v\text{-}erb\text{-}B}$	Membrane
c-*erb*-B-2(neu)		Plasma membrane	–	gp180	Membrane
–	PDGF receptor	Plasma membrane	–		
c-*fms*	CSF-1 receptor	Plasma membrane	v-*fms*	gp165$^{c\text{-}fms}$	Membrane
–	Insulin receptor	Plasma membrane	–		
–	IGF-1 receptor	Plasma membrane	–		
Others					
–	p56^{lstra}	Membrane	–		
–	p75	Cytoplasm	–		
met	–	Membrane			

just as reversible serine and threonine phosphorylation regulate enzymatic cascades, reversible tyrosine phosphorylation probably controls critical growth pathways. Other interesting findings are those suggesting that these enzymes may mediate phosphorylation of nonprotein substrates such as phosphatidylinositol. Phosphatidylinositol-generated second messengers are known to play an important role in cellular growth control (see Chapter 8).

src Oncogene

STRUCTURE AND FUNCTION

Cellular transformation by RSV is mediated by pp60$^{v\text{-}src}$, the first retroviral tyrosine-specific protein kinase. The c-*src* gene is notable for being the first proto-oncogene identified in the human genome. Bishop, Varmus, and their co-workers at the University of California, San Francisco, used a viral v-*src* probe derived from RSV to find an analogous gene in human DNA. Since these seminal experiments, scientists in many laboratories have used probes from other viral oncogenes to identify other cellular proto-oncogenes.

Although the cellular proto-oncogene c-*src* shares many features with v-*src*, it differs in several important functional aspects. These differences appear to be important for the transforming potential of the v-*src* protein because over-expressed normal c-*src* does not transform cells.

What is known about the structure and function of *src* (and related tyrosine kinase) oncogenes? First, the *src* protein has a 14-carbon fatty acid, myristic acid, bound to the amino terminal glycine (Fig. 7.2). Each pp60^{src} molecule is tightly bound to the inner surface of the cellular plasma membrane by this hydrophobic tail. Mutants of pp60$^{v\text{-}src}$ (e.g., Ala or Glu substitutions for glycine) lacking the myristic acid have levels of kinase activity similar to v-*src* but fail to transform cells. This result is attributed to an inability of the mutant protein, lacking myristate, to associate with the cell membrane. The result also suggests that access to specific substrates is required for *src* to transform cells. The *src* protein has nonenzymatic domains that modulate activity of the tyrosine kinase domain. Phosphorylation appears to play a major role in this regulation.

Regulation of *src* Activity by Phosphorylation

Fibroblasts are not transformed by pp60$^{c\text{-}src}$even when this protein is present at up to 15 times the normal level. The kinase activity of pp60$^{c\text{-}src}$ is lower relative to that of pp60$^{v\text{-}src}$. This difference may be the result of differential regulation of these proteins by autophosphorylation. A major site of tyrosine phosphorylation in chicken pp60$^{c\text{-}src}$ is the tyrosine at position 527, six residues from the carboxy terminus of the protein. In pp60$^{v\text{-}src}$, residues from amino acid 515 to the carboxy terminus are replaced with a different sequence (Fig. 7.2). The increased transforming capacity and kinase activity of the truncated viral oncogene is thought to result from loss of the critical tyrosine at position 527 of c-*src*, which appears to be involved in negative feedback regulation. Similar truncations have likewise altered the function of several growth factor receptors with tyrosine kinase activity (see Chapter 8).

Truncation is not the only way the *src* gene can be activated. Another mechanism involves formation of complexes with unrelated proteins. For example, the DNA tumor virus, polyoma, produces a transforming protein called middle T antigen (see Chapter 3) that forms a tight noncovalent complex with pp60$^{c\text{-}src}$. The bound *src* enzyme has 50-fold higher activity and is phosphorylated in a pattern similar to that observed following addition of the mitogen platelet-derived growth factor (PDGF). (See Chapter 8 for further information on growth factors.)

Can point mutations activate the *src* gene? Most of the v-*src* genes differ from c-*src* at multiple locations in the protein. However, only one mutation is found in

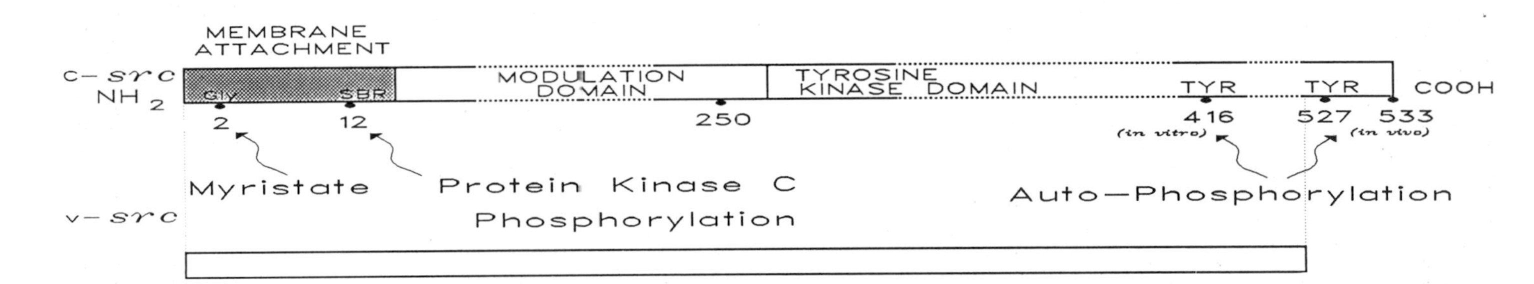

FIGURE 7.2. Comparative structure and function of pp60$^{c\text{-}src}$ and pp60$^{v\text{-}src}$ proteins. Although there are minor amino acid differences between these proteins (not shown), the major difference between the normal protein and the transforming viral protein appears to be the c-terminal truncation of v-*src*. This oncogene lacks a critical tyrosine site (No. 527) for autophosphorylation and autoregulation.

all avian v-*src* genes: threonine 338 located in the kinase domain is changed to isoleucine. In addition several other point mutations known to convert c-*src* to a transforming protein are known. These are arginine to tryptophan at position 95 situated outside the protein kinase domain and glutamic acid 378 to glycine and isoleucine 441 to phenylalanine within the kinase domain. In addition the replacement of tyrosine 527 with phenylalanine which can not be phosphorylated converts c-*src* to a transforming protein. Presumably these changes all function to keep the tyrosine kinase "on" or prevent its autoregulation. To date three-dimensional studies linking structure to function (like those available for the ras genes) have not been completed.

Protein kinase C has been shown to phosphorylate pp60^{src}. The target site, serine 12, is situated among several basic amino acids. This site is similar to the location of threonine 654 of the epidermal growth factor (EGF) receptor, a related tyrosine kinase, also phosphorylated by protein kinase C (see Chapter 8).

Protein kinase C is activated by and represents the major intracellular receptor for tumor promoters such as phorbol esters. Under resting conditions, protein kinase C exists in an inactive soluble form in the cytoplasm. Activation of its kinase activity occurs as a result of binding a molecule of diacylglycerol (DAG), a product of the phosphatidylinositol (second messenger) pathway. The phosphatidylinositol pathway, in turn, is stimulated by mitogenic growth factors, e.g., PDGF. DAG binding to protein kinase C causes association of the kinase with the cell plasma membrane and provides access to membrane substrates such as pp60^{src}. Phorbol esters, which are tumor promoters, compete with DAG for its protein kinase C binding site and cause prolonged activation of the enzyme. Tumor promoters elicit many of the same effects as polypeptide growth factors; the latter stimulate cells through membrane receptors, many of which belong to the tyrosine kinase family of proto-oncogenes.

The activity of pp60$^{c\text{-}src}$ is known to be augmented 10- to 20-fold in the myeloid cell lines HL-60 and U937 when induced to differentiate by phorbol esters. The conserved nature of the protein kinase C phosphorylation site, coupled with rapid occupancy of this site following treatment of cells with tumor promoters, suggests that the site has an important physiological function. What that function is remains to be established.

Thus phosphorylation of pp60^{src} controls its ability to act as a transforming protein and couples it to several other growth factor-induced second messengers.

Role of Tyrosine Phosphorylation in Cell Growth and Differentiation

It is tempting to speculate that the low abundance of phosphotyrosine in cellular proteins and the rapid turnover of phosphate coupled to tyrosine residues

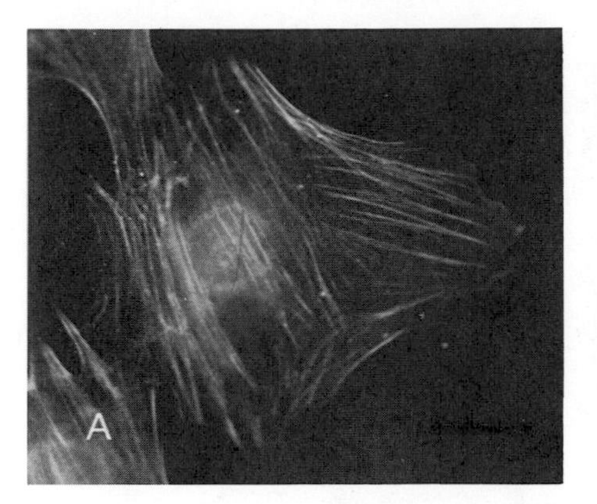

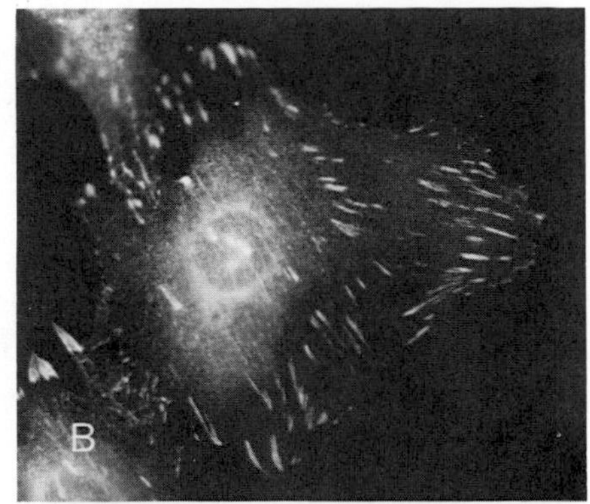

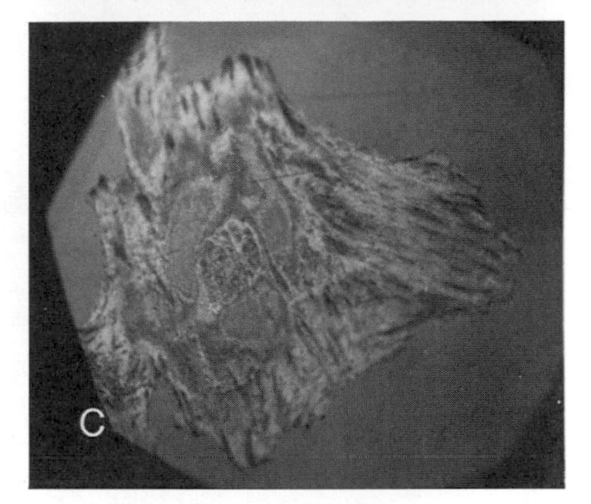

FIGURE 7.3. Localization of viniculin in adhesion plaques of chick embryo cells. (A) Cells stained with anti-actin antibodies. (B) Cells stained with anti-viniculin antibodies. (C) Differential interference contrast micrograph of the same cell. Note location of the adhesion plaques. (Kindly provided by Dr. L. Rohrschneider)

reflects the regulation of rare proteins with functions important for cell growth, development, and differentiation. A large number of proteins containing phosphotyrosine have been detected specifically in cells transformed by RSV. Many of these proteins are phosphorylated in cells transformed by other virus-encoded tyrosine protein kinases. However, the cellular proto-oncogene product, pp60$^{c\text{-}src}$, even when expressed at high levels, does not phosphorylate these substrates. The mechanism by which phosphorylation of such substrates may lead to transformation is unclear. The most common substrates include (1) vinculin (Figs. 7.3 and 7.4); (2) p36 and p81, proteins that are part of the cellular submembranous cortical skeleton; and (3) several soluble proteins including enolase, phosphoglycerate mutase, and lactate dehydrogenase. No functional consequences of these phosphorylations have been shown as yet; however, the substrates so far identified are abundant cellular proteins. Identification of less common proteins with presumed critical regulatory functions modified by tyrosine phosphorylation awaits further study. Nuclear phosphotyrsyl proteins have been identified in *abl*-transformed cells. These proteins bind DNA and may have a regulatory function.

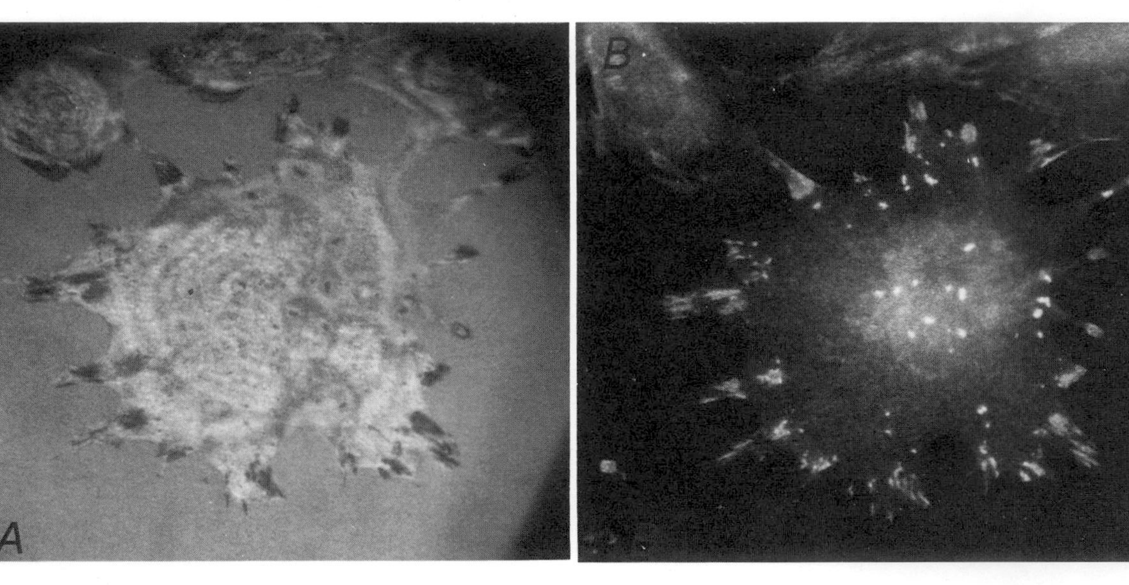

FIGURE 7.4. pp60$^{v\text{-}src}$ localizes to the adhesion plaques of Rous sarcoma virus transformed cells. (A) Differential interference contrast micrograph of a chick embryo cell. Note the location of the adhesion plaques at the periphery of the cell. (B) Staining of the same cell with anti-pp60$^{v\text{-}src}$ antibodies. (Kindly provided by Dr. L. Rohrschneider)

Stimulation of cells with PDGF for as short a time as one minute results in the appearance of an 80–85 kD phosphoprotein with phosphatidylinositol (PI) kinase activity. This protein is also present in cells transformed by polyoma middle T antigen (which complexes to pp60$^{c\text{-}src}$, v-*fms*, and v-*sis*). It is possible that it or related proteins controlling PI turnover are substrates for tyrosine kinases. As described in more detail in Chapter 8 the stimulation of cellular PI turnover with the generation of phosphorylated PI derivatives, PIP and PIP_2 are implicated in cellular responses to a wide variety of growth stimuli. It will be interesting to understand how this tyrosine kinase substrate interacts with protein kinase C and other members of the second messenger cascade.

src in Development and Differentiation

Although *src* and related tyrosine protein kinases were identified first in vertebrates, these enzymes also appear to have conserved structures in protozoans and other simple eukaryotes. D-*src*, a gene homologous to vertebrate c-*src*, has been cloned from *Drosophila*. Nematodes and slime molds possess similar genes. *Drosophila* and vertebrate c-*src* proteins are about 40% identical. The cross-species preservation of critical biochemical landmarks, including the tyrosine kinase domain and the myristylated glycine at the amino terminus, suggests an important, evolutionarily conserved function.

Three classes of *src* transcripts are independently regulated during development in *Drosophila*. Maximal accumulation of message occurs during the initial 15 hours of the embryonic period and again during the pupal period. In situ hybridization shows that the gene is expressed in a time-specific manner. Following the early embryonic period, the gene is expressed primarily in gut smooth muscle, the developing compound eye, the brain, and the ventral ganglia.

The association of the RSV *src* protein with a transformed phenotype suggests that related cellular proteins might be critical for cellular growth control. However, studies of the normal gene during *Drosophila* and chicken development does not support this hypothesis. Maximal *src* expression correlates with differentiation rather than proliferation of the embryonic cells. For example, in chickens the expression of pp60$^{c\text{-}src}$ is elevated in several neural tissues including brain and eye. Immunohistochemical studies show maximal expression in nonproliferating differentiated neural cells.

An in vitro model to investigate the potential function of the *src* gene to promote differentiation has been developed using PC12 pheochromocytoma cells. When infected with recombinant murine retroviruses containing a RSV *src* gene, these cells undergo marked phenotypic conversion. They differentiate into cells resembling sympathetic neurons exhibiting growth of long neurites and synthesis of acetylcholinesterase and neurofilament proteins.

src in Human Tumors

The c-*src* proto-oncogene is found on chromosome 20. Transcripts are commonly 3.9 to 4.0 kilobases (kb) in size. There is at present little information to suggest that genetic rearrangements, gene amplification, or mutations activate this gene in human cancer. However, studies from the National Cancer Institute (NCI) have shown that tissues and cell lines derived from tumors of neuroectodermal origin having a neural phenotype express high levels of c-*src* accompanied by *high* specific kinase activity. These tumors include neuroblastomas, neuroepitheliomas, pheochromocytomas, retinoblastomas, and medulloblastomas. In neuroectodermal tumors that do not express neural characteristics such as glioblastomas and melanomas, there are high protein levels of $pp60^{c\text{-}src}$ with near-normal specific activity. Normal tissues in which most cells are of neuronal origin, e.g., adrenal medulla and brain, also show elevated $pp60^{c\text{-}src}$ levels but, again, normal specific activity.

Although most normal and tumor cell lines and tissues have low levels of $pp60^{c\text{-}src}$ activity, specimens derived from osteogenic sarcoma, Ewing's sarcoma, rhabdomyosarcoma, colon adenocarcinomas, and many breast carcinomas do have elevated kinase activity. The NCI researchers found a specific elevation of $pp60^{c\text{-}src}$ activity in rhabdomyosarcoma tissue compared to normal skeletal muscle and in breast adenocarcinoma tissue compared to normal breast tissue. However, the quantitative amount of *src* protein in the malignant and normal tissues was similar. This result also was observed in human colon carcinoma: elevated $pp60^{c\text{-}src}$ activity with no increase in the quantity of c-*src* protein. Although increased $pp60^{c\text{-}src}$ protein activity has been associated with malignant transformation, the mechanism leading to this increased activity is at present not well understood.

src-Related Oncogenes: c-*yes*, *syn*, lsk^{T}/*tck*, c-*ros*, c-*fps*/*fes*, c-*fgr*, and *kit*

The other members of the *src* proto-oncogene family share a tyrosine kinase function with *src*, however, much less is known about their structures and function(s) in normal cell growth and differentiation or their role in neoplastic transformation. Many growth factor receptors with tyrosine kinase domains have been identified; these molecules are described in Chapter 8.

Several members of the tyrosine protein kinase family are closely related in structure to the c-*src* proto-oncogene (Table 7.2). Many of these genes were identified first in avian sarcoma viruses; they include v-*fps* of the Fujinama sarcoma virus, v-*yes* of the Y73 and Esh sarcoma viruses, and v-*ros* of the UR2 avian sarcoma virus. The v-*fps* oncogene is homologous to the v-*fes* oncogene first isolated from the Gardner-Arnstein (GA) and Snyder-Theilen (ST) strains of feline sarcoma virus. Ten strains of feline sarcoma virus have been characterized transducing six different c-*onc* genes of which 5, v-*fes*, v-*fms*, v-*fgr*, v-*abl*, and v-*kit* encode tyrosine-specific kinases. The predominant malignancies associated with feline sarcoma viruses are fibrosarcomas in cats.

Although the *src* gene of RSV is located 3′ to the envelope gene and this *src* protein is autonomous, the other transforming gene products are all *gag-onc* fusion proteins of 68 to 170 kilodaltons (kD). The viruses carrying them are, in general, replication-defective.

c-*yes*

Antibodies raised against the viral *yes* fusion protein, $p90^{gag\text{-}yes}$, recognize two cellular proteins of 62 and 59 kD, corresponding to two transcripts of 3.9 and 3.7 kb. Although anti-*yes* antibodies do not cross-react with the *src* protein, the overall structures of the c-*yes* proteins are similar to that of *src*. The c-*yes* proteins are phosphorylated on serine and tyrosine. Most importantly, v-*yes* (and v-*fgr*, see below) are known to lack a tyrosine at position 527. It is likely that this lack of tyrosine at the carboxy terminus results in a protein capable of transforming cells via constitutively expressed kinase activity.

The abundance of c-*yes* transcripts is greatest in brain, kidney, retina, and liver and is lower in muscle, bone marrow, spleen, heart, and lung. The molecular layer of the cerebellum has the highest c-*yes* expression in the brain. Despite elucidation of this pattern of expression, little information is currently available on the role of c-*yes* in normal or abnormal cellular proliferation.

The c-*yes* gene maps to human chromosome 18q21, a site near the breakpoint of the translocation t(14:18)(q32:q21) frequently observed in B cell follicular lymphomas.

yes-Related Oncogenes

A *yes*-related proto-oncogene called *syn* has been identified on human chromosome 6 at position q21. The carboxy terminus (amino acid positions 83 to 537 shares 80% homology with *yes*, 77% homology with *src*, and 77% with *fgr*. A 2.8-kb *syn* mRNA is transcribed in placenta, embryonic fibroblasts, lymphoblasts, follicular lymphoma cells, and epidermoid carcinoma cells.

A lymphocyte-specific tyrosine protein kinase gene, lsk^{T}/*tck*, has been cloned. A 2.2-kb message encodes a protein of 509 amino acids (56 kD). The precise regulatory function of these *yes*-related tyrosine protein kinases awaits further study.

c-*ros*

Sequences related to the v-*ros* oncogene are conserved in genomes of humans, mice, fish, and *Drosophila melanogaster*. Although the function of *ros* is unknown, significant homology exists between the tyrosine kinase region of the v-*ros* gene and the human insulin receptor.

The c-*ros* protein possesses homology with the *src* tyrosine kinase. The presence of a membrane-spanning domain suggests that the normal *ros* gene may encode a novel transmembrane receptor, related to the growth factor receptor family described in more detail in Chapter 8.

A related oncogene, *mcf3*, has been detected in NIH 3T3 transformants following transfection with high-molecular-weight human DNA. This gene is the human c-*ros* proto-oncogene. It may have been activated by loss of all but eight amino acids from the putative extracellular domain. The human c-*ros* gene contains at least seven exons spanning more than 20 kb. Transcripts of 3.0 kb have been detected during development of the vertebrate kidney and in muscle tissue in neonates.

c-*fps*/c-*fes*

The c-*fps* *onc*-protein has been identified in human and mouse hematopoietic cells by antisera raised against the viral p140$^{gag\text{-}fps}$ transforming protein. It is a 92-kD phosphoprotein capable of autophosphorylation. The protein shares many features with *src*, including sites for both serine and tyrosine phosphorylation, as well as a membrane location.

The p92$^{c\text{-}fes}$ protein is expressed at high levels in myeloid cells including multipotent murine cells, granulocyte-macrophage progenitors, and differentiated macrophage-like cells, as well as in normal human mononuclear cells. It has been detected in human B lymphoblastoid cell lines but only in a limited number of cell lines outside the hematopoietic lineage. The protein may play a role in normal hematopoietic differentiation.

The human c-*fps*/*fes* proto-oncogene has been assigned to chromosome 15. Although acute promyelocytic leukemia has been associated with t(15:17) translocations, the role of c-*fes* in this malignancy or its normal function(s) is at present unknown.

c-*fgr*

The v-*fgr* oncogene is closely related to v-*src*. It lacks a tyrosine at position 527 and is probably dysregulated by lack of autophosphorylation at this position. Transcripts from the human c-*fgr* proto-oncogene are present in a variety of human lymphoma cell lines, but they are not present in cell lines derived from sarcomas or carcinomas. Mature granulocytes and monocytes express abundant c-*fgr* mRNA, but the message is absent in resting lymphocytes from peripheral blood, thymus, or tonsil. The transcripts are present in low abundance in fetal spleen and lung; no expression is observed in other fetal tissues. The presence and relative stability of c-*fgr* mRNA in mature granulocytes and monocytes suggests that the *fgr* proto-oncogene has a highly specialized (differentiated) function in these hematopoietic cells. The role of this proto-oncogene in malignant transformation is unknown. It is of interest that Epstein-Barr virus (EBV) infection of normal human B lymphocytes increases the steady-state level of the c-*fgr* message. A similar c-*fgr* mRNA increase has been found with EBV-negative Burkitt's lymphoma cell lines as well.

v-*kit*

The v-*kit* oncogene was isolated from another feline sarcoma virus, the Hardy-Zuckerman 4 strain (HZ4-FeSV). The v-*kit* product is an 80-kD *gag-kit* fusion protein with 58% sequence identity to the v-*fms* product (Chapter 8). The predicted amino acid sequence of v-*kit* is compatible with a tyrosine kinase containing both the nucleotide binding domain and the tyrosine phosphorylation site homologous to Tyr 416 of RSV v-*src*. However, attempts to identify a kinase activity associated with the *gag-kit* fusion protein were unsuccessful. The v-*kit* gene is related to but nonidentical with previously described tyrosine kinase genes. Sequences homologous to v-*kit* were detected in normal, uninfected feline, human, and murine DNA. Although v-*kit* can transform NIH 3T3 fibroblasts, its role in in vivo carcinogenesis is unknown.

Functions of the *src*-Like Tyrosine Protein Kinases Remain a Mystery

In summary, a large number of tyrosine protein kinases related to *src* have been identified. The precise function(s) of these proto-oncogenes remain(s) to be discovered, as does their role, if any, in human cancer. Their normal purpose is almost certainly intertwined with growth and differentiation.

abl Oncogene

The formation of a chimeric transforming protein by fusion of the c-*abl* and *bcr* genes in chronic myelogenous leukemia is one of the best examples of a novel *onc*-protein generated by genetic rearrangement in a cancer cell (see Chapter 6).

The Abelson murine leukemia virus (Ab-MuLV) is a replication-defective transforming virus that originally was isolated from a tumor induced in a prednisolone-treated BALB/c mouse inoculated with the Moloney murine leukemia virus. The Moloney virus recombined with the murine c-*abl* gene to form the Abelson virus. The Abelson virus encodes a transforming fusion protein of 160 kD called $p160^{gag\text{-}abl}$. It efficiently transforms immature lymphoid cells and fibroblasts in vitro and induces lymphomas in 80 to 100% of injected neonatal mice after a 3- to 4-week latency period.

The v-*abl* protein shares amino acid sequences with the *src* family of tyrosine kinases. The tyrosine kinase activity of the *abl* gene product is closely associated with the transforming potential of the virus because mutants defective in phosphorylation do not transform either fibroblasts or lymphoid cells. Unlike the *src* protein, the *abl* protein can be found both bound to the plasma membrane and in soluble form.

Cellular homologues of v-*abl* have been identified in a variety of vertebrate cells as well as in *Drosophila*. The mouse and human c-*abl* genes share 71% identity.

Transcripts of c-*abl* are detected in murine fetuses throughout prenatal development with a fivefold drop in expression late in gestation. Expression of c-*abl* coincides with the development of the hematopoietic system; transcripts are readily detectable in spleen and thymus. In mature mice two major mRNA species of 6.5 and 5.3 kb are expressed at comparable levels in most cell types studied; however, the highest level of expression is found in the testes. Four different mRNAs are generated by alternative splicing of the 5′ exons of the two primary c-*abl* transcripts. These differentially spliced messages encode four *abl*-related proteins with different amino termini. The conserved structure of the *abl* gene in evolution and its wide range of tissue expression suggests that the gene plays an essential role in normal cellular metabolism. What that function is remains to be discovered.

Phosphorylation of Ribosomal Protein S6 by *abl*: Possible Importance for Transformation

The critical substrates phosphorylated by the c-*abl* tyrosine protein kinase resulting in cellular transformation have not yet been identified. However, a number of researchers have shown that a cascade of phosphorylation follows transformation by both the Rous sarcoma virus and the Abelson murine leukemia virus. This effect leads to an increase in protein-bound serine and protein-bound tyrosine. Among the phosphoseryl proteins, ribosomal protein S6 is of particular interest. S6 phosphorylation is correlated with growth-promoting stimuli in a wide variety of systems including (1) serum and growth factor-induced cell proliferation; (2) *Xenopus* oocyte fertilization and maturation; (3) regenerating rat liver; and (4) viral transformation. The precise function of S6 is not known; however, it may play a role in binding mRNAs to ribosomes, thereby controlling recruitment of specific mRNAs and hence the synthesis of certain proteins.

In normal Swiss mouse 3T3 fibroblasts, phosphorylation of S6 depends on growth-promoting serum stimulation. Studies from the laboratory of Nobelist David Baltimore at MIT have shown that in Abelson virus-transformed NIH 3T3 fibroblasts S6 is constitutively phosphorylated. This result correlates with reduced serum dependence for these cells. These scientists have shown that microinjection of the v-*abl* protein into *Xenopus* oocytes induces a 7- to 15-fold increase in serine phosphorylation of S6. How the *abl* tyrosine kinase modulates activity of cellular serine kinases is not known. Tyrosine phosphorylation may influence activity of a serine kinase directly, or it might inactivate a serine phosphatase. Clearly, this area is exciting for future research.

abl and Human Leukemias

Chronic myelogenous leukemia (CML) is a clonal hematological malignancy characterized in more than 95% of cases by a reciprocal translocation between chromosomes 9 and 22 (see Chapter 6). The cellular proto-oncogenes c-*abl* and c-*sis* have been mapped to human chromosomes 9q34 and 22q11, respectively. The translocation results in displacement of the c-*abl* proto-oncogene from chromosome 9 to a "breakpoint cluster region" (bcr) on chromosome $22q^-$. This translocation generates an aberrant 8.7-kb *bcr-abl* hybrid transcript that may well be involved in the pathogenesis of CML. The displaced c-*sis* gene apparently is not rearranged by the CML translocation. However, it is transcriptionally active in a subset of cells from more advanced CML patients.

The normal product of the c-*abl* gene is a protein of 145 kD; *bcr* codes for two related phosphoproteins of 160 and 190 kD, respectively. A protein of 210 kD with elevated tyrosine kinase activity is coded for by the hybrid gene. The *bcr* gene encodes the amino terminus, whereas the *abl* gene encodes the carboxy terminus of this protein. Transcription from the normal *bcr* gene results in mRNAs of 7.0 and 4.5 kb. These transcripts are expressed in cells from multiple lineages, including hematopoietic and nonhematopoietic tissue and cell lines. The *bcr* mRNAs are present in resting as well as active cells and in both induced and uninduced cell lines. The gene is evolutionarily conserved in vertebrates. At present the function of *bcr* is unknown.

It should be noted that the c-*abl* gene is not always activated by translocation. For example, acute nonlymphocytic leukemia (ANLL) involves a reciprocal translocation between the short arm of chromosome 6 and the long arm of chromosome 9 [t(6;9)(p23;q34)]. CML and ANLL share a common myeloid lineage and an identical breakpoint on chromosome 9. Clinically, they are associated with bone marrow basophilia and are resistant to chemotherapy in blast crisis. However, the c-*abl* gene is not activated in ANLL as it is in CML. Other as yet unidentified genes must be involved in ANLL translocations to account for the unique clinical presentation of this leukemia.

In summary, the *abl* oncogene was first identified as a fusion gene in the Abelson leukemia virus. It subsequently was shown to be activated by translocation to *bcr* on chromosome 22 in CML. This displacement not only generated a chimeric 210-kD transforming protein but identified another novel gene, *bcr*, with an important (as yet unknown) function in normal cellular metabolism.

met Oncogene

The *met* oncogene is another example of a transforming gene generated by fusion of two chromosomal loci. The oncogene originally was activated by treatment of human osteosarcoma cell line (HOS) in vitro with *N*-methyl-*N'*-nitronitrosoguanidine (MNNG) and was isolated from *met*-transformed NIH 3T3 cells. Activation resulted from a DNA rearrangement joining two separate loci: *tpr*, mapping to chromosome 1, and c-*met*, mapping to chromosome 7q21-31, by in situ hybridization. Nucleotide sequence analysis of the 3′ end of *met* revealed sequence identity with tyrosine kinases. Close identity (about 70%) in the kinase domain was demonstrated with the human insulin receptor and the murine v-*abl* product.

The c-*met* gene generates a 9.0-kb message in human fibroblast and epithelial cell lines. The truncated, chimeric *met* oncogene is expressed as a novel 5.0-kb transcript consisting of 5′ sequences derived from the *tpr* locus and 3′ sequences from the *met* locus. The size of the normal *met* protein is 140 kD, whereas the size of the novel polypeptide encoded by the *met* oncogene is 65 kD. The latter protein is phosphorylated on tyrosine and serine residues and has tyrosine-specific autophosphorylation activity. Presumably, genetic rearrangement and truncation of the *met* gene results in its abnormal regulation. In the c-*src* protein, the domain modulating kinase activity lies 5′ to the tyrosine kinase domain. The *tpr* gene has been substituted for this modulating region in the activated *met* oncogene.

The long arm of chromosome 7 often is associated with nonrandom chromosome deletions in ANLL. It is interesting that the *met* oncogene was first discovered in a cell line treated with a chemical carcinogen because ANLL is commonly found in patients subjected to cytotoxic drug therapy for a primary malignancy and in patients exposed to environmental carcinogens. The role, if any, of this gene in human malignancy remains to be determined.

Using restriction fragment length polymorphisms (RFLPs) (see Chapter 13), White and his colleagues from the University of Utah have shown that the *met* gene maps near the as yet unidentified gene(s) for cystic fibrosis (CF). CF is the most common genetic disease in Caucasian populations, with a carrier frequency of approximately 1 in 20. The biochemical basis of the disease is unknown, although its association with abnormal sweat chloride concentrations suggests an ion channel defect. This disorder is inherited in an autosomal recessive fashion. Although tyrosine kinase-associated growth factor receptors are linked in some instances to sodium ion-proton transport, the provocative association of the *met* gene with the CF locus is at present unclear.

Other Chimeric Oncogenes

Most of the genes identified by transfection of NIH 3T3 cells with tumor cell genomic DNA have been members of the *ras* gene family (see Chapters 6 and 9). However, several novel genes identified by this method include *onc* D or *trk* (from a carcinoma of the ascending colon), *met* (from MNNG-HOS cells), *dbl* (from a diffuse B cell lymphoma), and a transforming allele of c-*raf*-1 (from a carcinoma of the stomach). The *mcf*-3 and *ret* proto-oncogenes also become activated during transfection of NIH 3T3 cells with human DNA.

Sequence analysis of a cDNA clone of *onc* D revealed that it was formed by fusion of a truncated nonmuscle tropomyosin gene and a previously unknown tyrosine kinase gene. The extracellular domain of a transmembrane tyrosine kinase was replaced by the first 221 amino acids of the nonmuscle tropomyosin molecule. This chimeric protein is tumor-specific; no rearrangement of DNA in normal colon tissue adjacent to the tumor specimen was observed. Another example of a chimeric oncogene is v-*fgr* derived from the Gardner-Rasheed strain of feline sarcoma virus. It encodes a tyrosine kinase attached to a 128-amino-acid domain of α-actin.

How do these chimeric oncogenes transform cells? The answer is not known, but conceivably the tropomyosin or actin domains direct the protein kinase to aberrant locations within the cell where phosphorylation of illegitimate substrates occurs. It is also possible that tropomycin or actin sequences replace an allosteric control region of the kinase or, even more simply, that they stabilize the chimeric mRNA, thereby increasing levels of the kinase protein product.

Serine-Threonine Kinases

raf/mil Oncogene

The *raf* gene was isolated originally from a murine sarcoma virus, MSV 3611. The viral oncogene product is a 75 kD *gag-raf* fusion protein. The *mil* (*mht*) gene was transduced from chicken DNA by the avian retrovirus, MH2, a virus that also carries the c-*myc* gene. The cloned v-*mil* gene is unable to induce transformation of macrophages directly but appears to induce production of a macrophage growth factor that greatly enhances the capacity of v-*myc* to induce monocytic neoplasms in vivo. The viral mil product again is a fusion protein, p100 *gag-mil* (*mht*). Sequence analysis of the murine and avian cDNA clones revealed *raf* and *mil* to be homologous. The v-*mil* gene contains sequences from two additional 5′ exons compared to the v-*raf* gene accounting for the different sizes of their respective proteins. These proteins are associated with a serine-threonine kinase activity; the deduced amino acid sequence shows a distant relationship to tyrosine-specific kinases.

Four raf-related genes have been identified in the human genome. The human homolog of the murine and avian virally transduced gene is c-*raf*-1 located on chromosome 3p25. The human c-*raf*-1 gene contains 17 exons and spans approximately 48 kb of DNA. The overall sequence identity between human c-*raf*-1 and v-*raf* is 90%. The human c-*raf* messenger RNA is 3.4 kb. The predicted amino acid sequence of c-*raf*-1 codes for a protein of 73 kD (648 amino acids). The viral genes v-*raf* and v-*mil* contain gag sequences that replace the 5′ end of the cellular gene: v-*raf* and v-*mil* begin within exons 9 and 7 of the mouse and chicken c-*raf* genes, respectively. There is evidence that truncation of the c-*raf* gene at its 5′ end may lead to activation: five transforming variants with such a structural rearrangement have been described including retroviral LTR insertion between exons 5 and 6, and two transforming *raf* genes recovered from human tumors.

Another human *raf* gene, c-*raf*-2, was localized to chromosome 4. Sequence analysis revealed this gene to possess no intron sequences and numerous small deletions and insertions compared to c-*raf*-1 resulting in several termination codons in each of the reading frames. Thus c-*raf*-2 is a pseudogene thought to have been generated by reverse transcription of spliced mRNA.

Using a c-*raf*-1 probe, Huebner et al. isolated a *ras*-related gene from murine c-DNA, termed A-*raf*. With the murine A-*raf* gene as probe, two human A-*raf* genes were isolated. Human A-*raf*-1 was localized to the X chromosome whereas A-*raf*-2 was mapped to chromosome 7. Sequence identity between c-*raf*-1 and

A-*raf*-1 is on the order of 70–76%. Human A-*raf*-1 mRNA was found in myeloid and T cell lines; however, transcripts corresponding to A-*raf*-2 were not detected in any cell line tested, suggesting that A-*raf*-2, like c-*raf*-2, may be a pseudogene. Sequence analysis of A-*raf*-1 DNA predicts a protein closely related to the c-*raf*-1 serine-threonine kinase. Therefore the A-*raf*-1 product also is predicted to be a serine-threonine kinase.

The location of human c-*raf*-1 on chromosome 3p25 is a site frequently altered in neoplasia, particularly small cell lung carcinoma. Transforming c-*raf*-1 genes also have been identified in conjunction with human stomach cancer and glioblastomas. In human lung carcinoma cell lines, the expression of c-*raf*-1 appears to be amplified compared to virally transformed NIH 3T3 cells.

Scientists at the NIH have developed a mouse model of lung carcinogenesis. More than 90% of newborn mice treated transplacentally with ethylnitrosourea (ENU) and "promoted" with injections of butylated hydroxytoluene at birth developed lung tumors and lymphomas by 5 to 14 weeks of age. These tumors and the cell lines derived from them express high levels of c-*raf*-1 RNA. Moreover, DNA prepared from these cells transforms NIH 3T3 cells. Acquisition of transforming activity by the c-*raf* proto-oncogene also may be due to recombination within noncoding regions of the gene. Such a change probably led to activation of rat c-*raf* during transfection of DNA into the target cells. DNA from (1) a rat hepatocellular carcinoma induced by a food carcinogen, 1-amino-3-methylimidazo (4,5-*f*) quinoline, and (2) a human hepatocellular carcinoma (the DNA of which had originally been transfected into NIH 3T3 cells) had similar rearrangements around exons 6 to 7 of the c-*raf* gene.

pks Oncogene

Using the murine v-*raf* gene as a probe, Mark et al. at the National Cancer Institute identified another *raf*-related gene in a human fetal liver cDNA library. This gene shares 71% nucleotide sequence identity with c-*raf*-1. The predicted protein product shares 75% amino acid sequence identity with c-*raf*-1 product. The pks gene was localized to the human X chromosome (Xp11). Southern blots of human genomic DNA revealed another *pks*-related gene by hybridization with a *pks* probe. It is likely that pks and A-*raf*-1 are identical and that the second pks sequence corresponds to A-*raf*-2.

A *pks* transcript of 2.7 kb was elevated in the leukocytes of two patients with angioimmunoblastic lymphadenopathy with dysproteinemia and autoimmunity. There was no similar elevation of c-*raf*-1, *fos*, *abl*, or *myc* transcripts in these patients. It was speculated that pks could conceivably play a role in X-linked immunodeficiency diseases.

mos Oncogene

The product of the *mos* oncogene has also been described as a serine-threonine kinase. This gene is discussed in more detail in Chapter 8.

Bibliography

src and Related Tyrosine Kinases

General: Tyrosine Protein Kinases

Hunter, T. A thousand and one protein kinases. Cell 50:823, 1987.

Hunter T, Cooper JA. Protein-tyrosine kinases. Annu Rev Biochem 54:897, 1985.

Nishimura J, Huang JS, Deuel TF. Platelet-derived growth factor stimulates tyrosine-specific protein kinase activity in Swiss mouse 3T3 cell membranes. Proc Natl Acad Sci USA 79:4303, 1982.

Petruzzelli L, Herrera R, Rosen OM. Insulin receptor is an insulin-dependent tyrosine protein kinase: co-purification of insulin-binding activity and protein kinase activity to homogeneity from human placenta. Proc Natl Acad Sci USA 81:3327, 1984.

src

Alema S, Casalbore P, Agostini E, et al. Differentiation of PC 12 pheochromocytoma cells induced by v-src oncogene. Nature 316:557, 1985.

Bolen JB, Rosen N, Israel MA. Increased pp60^{c-src} tyrosyl kinase activity in human neuroblastomas is associated with amino-terminal tyrosine phosphorylation of the src gene product. Proc Natl Acad Sci USA 82:7275, 1985.

Brugge JS, Cotton PC, Queral AE, et al. Neurones express high levels of a structurally modified activated form of pp60^{c-src}. Nature 316:554, 1985.

Cartwright C, Eckhart W, Simon S, et al. Cell transformation by pp60^{c-src} mutated in the carboxy-terminal regulatory domain. Cell 49:83, 1987.

Cooper JA, Gould KL, Cartwright CA, et al. Tyr^{527} is phosphorylated in pp60^{c-src}: implications for regulation. Science 231:1431, 1986.

Cotton PC, Brugge JS. Neural tissues express high levels of the cellular src gene product pp60^{c-src}. Mol Cell Biol 3:1157, 1983.

Courtneidge S, Ralston R, Alitalo K, et al. Subcellular location of an abundant substrate (p36) for tyrosine-specific protein kinases. Mol Cell Biol 3:340, 1983.

Gilmore TD, Radkey K, Martin GS. Tyrosine phosphorylation of a 50 K cellular polypep tide associated with the Rous sarcoma virus transforming protein pp60^{src}. Mol Cell Biol 2:199, 1982.

Hunter T. Transforming gene product of Rous sarcoma virus phosphorylates tyrosine. Proc Natl Acad Sci USA 77:1311, 1980.

Hunter T. A tail of two src's: mutatis mutandis. Cell 49:1, 1987.

Iba H, Takeya T, Cross FR, et al. Rous sarcoma virus variants that carry the cellular src gene instead of the viral src gene cannot transform chicken embryo fibroblasts. Proc Natl Acad Sci USA 81:4424, 1984.

Jakobovitz EB, Majors JE, Varmus HE. Hormonal regulation of the Rous sarcoma src gene via a heterologous promoter defines a threshold dose for cellular transformation. Cell 38:757, 1984.

Johnson PJ, Coussens PM, Danko AV, et al. Overexpressed pp60^{c-src} can induce focus formation without complete transformation of NIH 3T3 cells. Mol Cell Biol 5:1073, 1985.

Kampes MM, Buss JE, Sefton BM. Mutation of NH2-terminal glycine of p60^{src} prevents both myristoylation and morphological transformation. Proc Natl Acad Sci USA 82:4625, 1985.

Kaplan DR, Whitman M, Schaffhausen B, Pallas DC, White M, Cantley L, Roberts TM. Common elements in growth factor stimulation and oncogenic transformation: 85 kD phosphoprotein and phosphatidylinositol kinase activity. Cell 50:1021, 1987.

Kmiecik T, Shalloway D. Activation and suppression of $pp60^{src}$ transforming ability by mutation of its primary sites of tyrosine phosphorylation. Cell 49:65, 1987.

Le Beau MM, Westbrook CA, Diaz MO, et al. Evidence for two distinct c-src loci on human chromosomes 1 and 20. Nature 312:70, 1984.

Maher PA, Pasquale EB, Wang JYJ, et al. Phosphotyrosine-containing proteins are concentrated in focal adhesions and intercellular junctions in normal cells. Proc Natl Acad Sci USA 82:6576, 1985.

Piwnica-Worms H, Saunders K, Roberts T, et al. Tyrosine phosphorylation regulates the biochemical and biological properties of $pp60^{c\text{-}src}$. Cell 49:75, 1987.

Oskarsson M, McClements WL, Blair DG, et al. Properties of a normal mouse cell DNA sequence (sarc) homologous to the src sequence of Moloney sarcoma virus. Science 207:1222, 1980.

Parker RC, Varmus HE, Bishop JM. Expression of v-src and c-src in rat cells demonstrates qualitative differences between $pp60^{v\text{-}src}$ and $pp60^{c\text{-}src}$. Cell 37:131, 1984.

Rohrschneider LR. Adhesion plaques of Rous sarcoma virus-transformed cells contain the src gene product. Proc Natl Acad Sci USA 77:3514, 1980.

Sakaguchi AY, Naylor SL, Shows TB. A sequence homologous to Rous sarcoma virus v-src is on human chromosome 20. Prog Nucleic Acid Res Mol Biol 29:279, 1983.

Sefton BM, Hunter T, Ball EH, et al. Vinculin: a cytoskeletal target of the transforming protein of Rous sarcoma virus. Cell 24:165, 1981.

Shalloway D, Coussens PM, Yaciuk P. Overexpression of the c-src protein does not induce transformation of NIH 3T3 cells. Proc Natl Acad Sci USA 81:7071, 1984.

Simon MA, Drees B, Kornberg T, et al. The nucleotide sequence and the tissue-specific expression of Drosophila c-src. Cell 42:831, 1985.

Sorge LK, Levy BT, Maness PF. $pp60^{c\text{-}src}$ is developmentally regulated in the neural retina. Cell 36:249, 1984.

Sugimoto Y, Whitman M, Cantley LC, et al. Evidence that the Rous sarcoma virus transforming gene product phosphorylates phosphatidylinositol and diacyglycerol. Proc Natl Acad Sci USA 81:2117, 1984.

Swanstrom R, Parker RC, Varmus HE, et al. Transduction of a cellular oncogene—the genesis of Rous sarcoma virus. Proc Natl Acad Sci USA 80:2519, 1983.

Takeya T, Hanafusa H. Structure and sequence of the cellular gene homologous to the RSV src gene and the mechanism for generating the transforming virus. Cell 32:881, 1983.

Willingham MC, Pastan I, Shih TY, et al. Localization of the src gene product of the Harvey strain of MSV to the plasma membrane of transformed cells by electron microscopic immunocytochemistry. Cell 19:1005, 1980.

Yonemoto W, Jarvis-Morar M, Brugge JS, et al. Tyrosine phosphorylation within the aminoterminal domain of $pp60^{c\text{-}src}$ molecules associated with polyoma virus middle-sized tumor antigen. Proc Natl Acad Sci USA 82:4568, 1985.

yes

Marth JD, Peet R, Krebs EG, et al. A lymphocyte-specific protein-tyrosine kinase gene is rearranged and overexpressed in the murine T cell lymphoma LSTRA. Cell 43:393, 1985.

Semba K, Nishizawa M, Miyajma N, et al. yes-Related proto-oncogene, syn, belongs to the protein-tyrosine kinase family. Proc Natl Acad Sci USA 83:5459, 1986.

Shibuya M, Hanafusa H, Balduzzi PC. Cellular sequences related to three new onc genes of avian sarcoma virus (fps, yes, and ros) and their expression in normal and transformed cells. J Virol 42:143, 1982.

Yoshida M, Sasaki M, Misc K, et al. Regional mapping of the human proto-oncogene c-yes to chromosome 18 at band g21.3. Jpn J Cancer Res 76:559, 1985.

ros

Birchmeier C, Birnbaum D, Waitches G. Characterization of an activated human ros gene. Mol Cell Biol 6:3109, 1986.

Petruzzelli L, Herrera R, Rosen OM. Insulin receptor is an insulin-dependent tyrosine protein kinase: copurification of insulin-binding activity and protein kinase activity to homogeneity from human placenta. Proc Natl Acad Sci USA 81:3327, 1984.

Ullrich A, Bell JR, Chen EY, et al. Human insulin receptor and its relationship to the tyrosine kinase family of oncogenes. Nature 313:756, 1985.

fps/fes

Foster DA, Hanafusa H. A fps gene without gag gene sequences transforms cell into culture and induces tumors in chickens. J Virol 48:744, 1983.

Foster DA, Shibuya M, Hanafusa H. Activation of the transformation potential of the cellular fps gene. Cell 42:105, 1985.

Huang GC, Hammond C, Bishop JM. Nucleotide sequence and topography of chicken c-fps: genesis of a retroviral oncogene encoding a tyrosine-specific protein kinase. J Mol Biol 181:175, 1985.

Samarut J, Mathey-Prevot B, Hanafusa H. Preferential expression of the c-fps protein in chicken macrophages and granulocytic cells. Mol Cell Biol 5:1067, 1985.

Shibuya M, Hanafusa H. Nucleotide sequence of Fujinami sarcoma virus: evolutionary relationship of its transforming gene with transforming genes of other sarcoma viruses. Cell 30:787, 1982.

Weinmaster G, Zoller MJ, Smith M, et al. Mutagenesis of Fujinami sarcoma virus: evidence that tyrosine phosphorylation of $P130^{gag\text{-}fps}$ modulates its biological activity. Cell 37:559, 1984.

fgr

Naharro G, Dunn CY, Robbins KC. Analysis of the primary translational product and integrated DNA of a new feline sarcoma virus, GR-FeSV. Virology 125:502, 1983.

Schultz A, Oroszlan S. Myristylation of gag-onc fusion proteins in mammalian transforming retroviruses. Virology 133:431, 1984.

kit

Besmer P, Murphy J, George P, et al. A new acute transforming feline retrovirus and relationship of its oncogene v-kit with the protein kinase gene family. Nature 320:415, 1986.

abl

Bell JC, Mahadevan LC, Colledge WH, et al. Abelson-transformed fibroblasts contain nuclear phosphotyrosyl proteins which preferentially bind to murine DNA. Nature 325:552, 1987.

Collins SJ, Groudine M. Rearrangement and amplification of c-abl sequences in the human chronic myelogenous leukemia cell line K-562. Proc Natl Acad Sci USA 80:4813, 1983.

De Klein A, Geurts van Kessel A, Grosveld G, et al. A cellular oncogene is translocated to the Philadelphia chromosome in chronic myelocytic leukemia. Nature 300:765, 1982.

Groffen J, Stephenson JR, Heisterkamp N, et al. Philadelphia chromosomal breakpoints are clustered within a limited region, bcr, on chromosome 22. Cell 36:93, 1984.

Heisterkamp N, Stam K, Groffen J, et al. Structural organization of the bcr gene and its role in the Ph^1 translocation. Nature 315:758, 1985.

Konopka JB, Watanabe SM, Witte ON. An alteration of the human c-abl protein in K562 leukemia cells unmasks associated tyrosine kinase activity. Cell 37:1035, 1984.

Maller JL, Foulkes JG, Erickson E, et al. Phosphorylation of ribosomal protein S6 on serine after microinjection of the Abelson murine leukemia virus tyrosine-specific protein kinase into Xenopus oocytes. Proc Natl Acad Sci USA 82:272, 1985.

Prywes R, Hoag J, Rosenberg N, et al. Protein stabilization explains the gag requirement for transformation of lymphoid cells by Abelson murine leukemia virus. J Virol 53:123, 1985.

Stam K, Heisterkamp N, Grosveld G, et al. Evidence of a new chimeric bcr/c-abl mRNA in patients with chronic myelocytic leukemia and the Philadelphia chromosome. N Engl J Med 313:1429, 1985.

met

Dean M, Park M, Le Beau MM, et al. The human met oncogene is related to the tyrosine kinase oncogenes. Nature 318:385, 1985.

Park M, Dean M, Cooper CS, et al. Mechanism of met oncogene activation. Cell 45:895, 1986.

White R, Woodward S, Leppert M, et al. A clearly linked genetic marker for cystic fibrosis. Nature 318:382, 1985.

raf and *pks*

Bonner T, Opperman H, Seeburg P, et al. The complete coding sequence of the human raf oncogene and the corresponding structure of the c-raf-1 gene. Nuc Acid Res 14:1009, 1986.

Huebner K, ar-Rushdi A, Griffin CA, et al. Actively transcribed genes in the raf oncogene group, located on the X chromosome in mouse and human. Proc Natl Acad Sci USA 83:3934, 1986.

Mark GE, Seely TW, Snows TB, et al. pks, a raf-related sequence in humans. Proc Natl Acad Sci USA 83:6312, 1986.

Other Tyrosine Kinase Genes

Flordellis C, Kan NC, Lautenberger K. Analysis of the cellular proto-oncogenes mht/raf . . . Virology 141:267, 1985.

Graf T, Weizaecker F, Grieser S, et al. V-mil induces autocrine growth and enhanced tumorigenicity in V-myc transformed avian macrophages. Cell 45:357, 1986.

Marth JD, Peet R, Krebs EG, et al. A lymphocyte-specific protein-tyrosine kinase gene is rearranged and overexpressed in the murine T-cell lymphoma LSTRA. Cell 43:393, 1985.

Martin-Zanca D, Hughes SH, Barbacid M. A human oncogene formed by the fusion of truncated tropomycin and protein-tyrosine kinase sequences. Nature 319:743, 1986.

Moelling KB, Heimann, Beimling P, et al. Serine and threonine-specific protein kinase activities of purified gag-mil and gag-raf proteins. Nature 312:558, 1984.

Reinach F, MacLeod AR. Tissue specific expression of the human tropomycin gene involved in the generation of the trk oncogene. Nature 322:648, 1986.

Srivastava S, Wheelock R, Aaronson SA, et al. Identification of the protein encoded by the human diffuse β cell lymphoma (dbl) oncogene. Proc Natl Acad Sci USA 83:8868, 1986.

Voronova AF, Sefton BM. Expression of a new tyrosine protein kinase is stimulated by retrovirus promoter insertion. Nature 319:682, 1986.

8
Growth Factors and Receptors

Overview

Growth factors and growth factor receptors may be involved in the abnormal growth characteristic of cancer, and both entities are encoded by known proto-oncogenes. Growth factors may act as self-stimulants (*autocrine control*), stimulants of nearby cells (*paracrine control*), or stimulants of distant tissue (*hormonal control*). Growth factor receptors often are cell surface molecules that trigger internal cascades of events when stimulated by their particular growth factors. "Second messengers" of several varieties transduce or transmit signals from the cell surface to the nucleus, resulting in gene transcription. The *sis* oncogene is related to the platelet-derived growth factor (PDGF) and is the only known example of a secreted oncogene product. Other oncogenes are related to growth factor receptors containing a tyrosine protein kinase function: *erb* B is homologous to the epidermal growth factor (EGF) receptor; *erb* B-2 (*neu*) is a related oncogene; the *fms* gene encodes a product similar to the macrophage colony-stimulating factor-1 receptor; the *ros* oncogene product shows some homology with the insulin receptor. In these instances the normal cellular genes have been rendered transforming by point mutations and/or deletions. Proepidermal growth factor (a precursor of the epidermal growth factor, EGF) shares similarities with the EGF receptor. In addition, it has sequence homology with the *mos*-oncogene and with the low density lipoprotein receptor gene. The *erb* A gene shares sequences with the thyroid hormone receptor, a steroid hormone nuclear receptor. Still other growth factors and their receptors appear to play a role in oncogenesis but are not yet associated with known oncogenes and their products. Such proteins include transforming growth factors α and β and the T cell growth factor interleukin-2.

Growth Factors

Some growth factors are polypeptides that stimulate cell proliferation by binding to specific high affinity plasma membrane receptors (Table 8.1). These receptors exert their effects intracellularly by generating various second messengers (see below). Evidence for the involvement of growth factors in cell proliferation dates

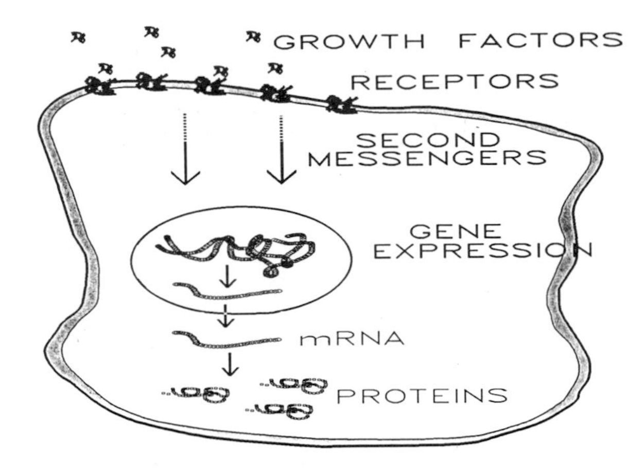

FIGURE 8.1. Growth factor-stimulated pathway of cellular proliferation. Elements in the pathway include the growth factor, growth factor receptor, and second messenger cascade. Ultimately, binding of a growth factor leads to modulation of gene expression in the nucleus. Examples of neoplastic transformation caused by alteration of proto-oncogenes corresponding to each step in the pathway are described in the text.

from the early days of tissue culture when it was shown that supplementation with serum of cultured primary cells and cell lines in vitro promoted the growth of those cultures. Early investigators also found that neoplastically transformed cells require less serum. With the development of serum-free media by Gordon Sato and others, it became clear that cancer cells would proliferate in the absence, or in much reduced amounts, of specific growth factors. Loss of the requirement for specific growth factors could be mediated by the autologous synthesis of a growth factor (autocrine growth), by synthesis of altered growth factor receptors, or by direct activation of various elements of the second messenger cascades (Fig. 8.1).

Autocrine/Paracrine Growth Factors

The multiplicity of growth factors, their receptors, the timing of their interactions with cells, and tissue receptor heterogenicity provide for the coordinate growth of normal cells during development and maturation. Often several growth factors are required to elicit a maximal cell proliferative response, in part because their mitogenic stimuli act at different points in the cell cycle. Autostimulation (*autocrine*) and stimulation of nearby cells (*paracrine*) by local release of growth factors contribute to the finely tuned coordinate growth of cells within a tissue (Fig. 8.2). These types of growth factor differ from the classical *endocrine* growth factors that typically are produced in one tissue and stimulate a distant tissue. It is reasonable to expect that the uncontrolled production of an autologous growth factor by a cell requiring this factor might result in a growth

TABLE 8.1. Growth factors and their receptors.

Name(s)	Primary translation product	Mature factor size	Cell source	Target cell	Receptor
PDGF	241 aa (B chain); A chain unknown; B chain encoded in c-*sis* proto-oncogene	32 kD (16 kD B chain; 14–18 kD A chain), + CHO	Blood platelets, endothelial cells, placenta	Mesenchymal cells smooth muscle, placental tropho-blast	185 kD tyrosine kinase
EGF	1168 or 1217 aa	6 kD (53 aa)	Submaxillary gland, Brunner's gland, possibly parietal cells	Wide variety of epithelial and mesenchymal cells	c-*erb*B gene; 170 kD tyrosine kinase
TGFα	160 aa	5.6 kD (50 aa)	Transformed cells, placenta, embryos	Same as EGF	Binds to EGF receptor, may have separate one
CSF-1	252 aa	70 kD (2 × 35 kD); 60% CHO	Mouse L-cells	Macrophage pro-genitors	c-*fms* proto-oncogene; 170 kD tyrosine kinase
CSF-2 (granulocyte-macrophage CSF)	144 aa	15–28 kD (127 aa) (1–50% CHO)	Endotoxin-induced lung; placenta	Macrophage and granulocyte pro-genitors	Unknown
Multi-CSF (IL-3)	144 aa	28 kD (134 aa) (50% CHO)	T-lymphocytes	Eosinophil, mast cell, granulocyte, macrophage pro-genitors, T-lymphocytes	Unknown

IGF-1	130 aa	7 kD (70 aa)	Adult liver and (somatomedin C)	Epithelial, mesen-chymal	450 kD complex (2 α chains of 130 kD; 2 β chains of 85 kD)
IGF-11	180 aa	7 kD (67 aa)	Fetal liver, placenta (somatomedin A)	Epithelial, mesen-chymal	Single polypeptide chain of 260 kD
IL-1			Mononuclear phagocytes	Wide variety; fibroblasts, T-cells, endothelial cells	
IL-2	169 aa (mouse) 153 aa (human)	15 kD (133 aa)	T-helper cells Some CHO	Cytotoxic T-lymphocytes	55 kD (33 kD protein + 22 kD CHO)
FGF	Unknown	14–18 kD (basic FGF is 146 aa)	Brain, pituitary chondrosarcoma	Endothelial, fibroblasts	Unknown
β-NGF	307 aa	26 kG (1 × 118 aa)	Submaxillary gland	Sympathetic and sensory neurons	130 kD (possibly kinase)

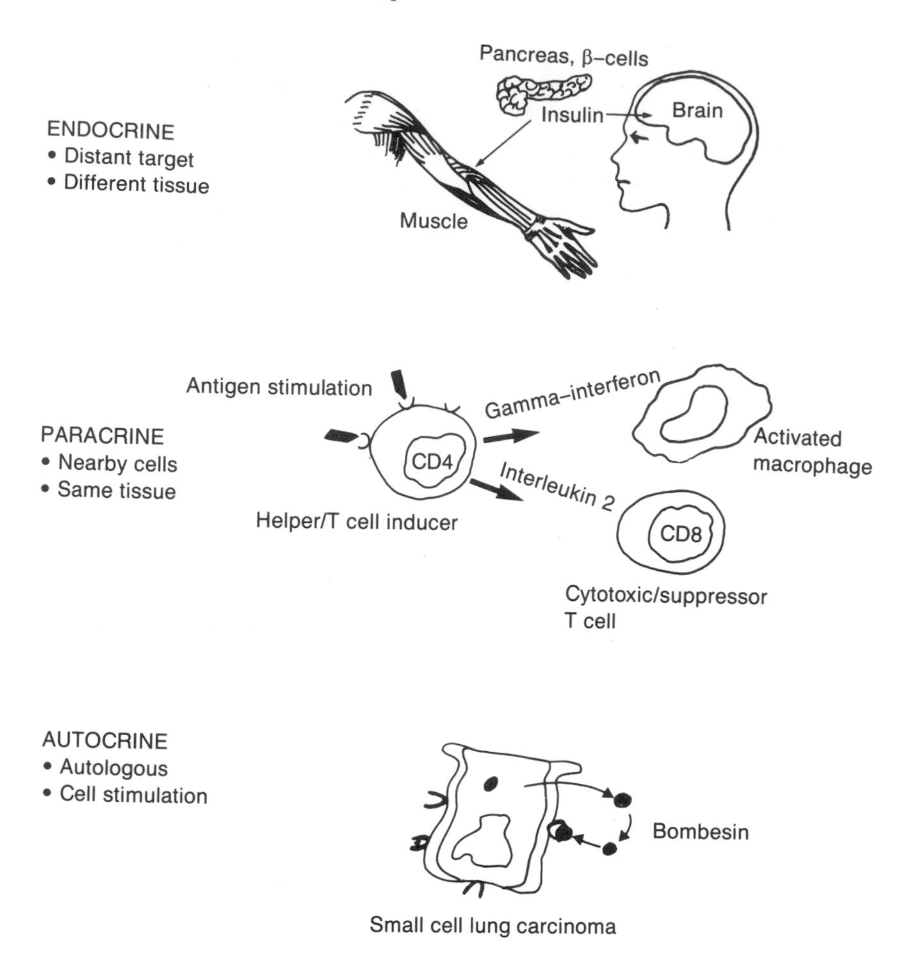

FIGURE 8.2. Autocrine/paracrine and endocrine growth control. An autocrine growth factor stimulates the cell that produces it. A paracrine growth factor stimulates neighboring cells. An endocrine growth factor stimulates distant cells, usually of a different tissue type. Autocrine and paracrine growth factors are also known as cytokines.

advantage for the clone derived from this cell. Examples of autocrine-stimulated cells include smooth muscle cells [insulin growth factor-1 (IGF-1)], chemically transformed mouse fibroblasts [transforming growth factor beta (TGFβ)], human osteosarcoma cells U-20S (PDGF), T cell leukemia cells [interleukin-2 (IL-2)], and small cell lung carcinoma cells (bombesin). In all of these cases antibodies that bind to the autocrine growth factors specifically inhibit the responsive cells, and in one case (IL-2) a monoclonal antibody binding to the IL-2 receptor also blocks proliferation of the T cells. In the latter case, the gibbon ape leukemia cell line MLA-144 both produces and responds to IL-2.

It is likely that local production of growth factor contributes not only to the growth potential of tumor cells but also to the proliferation of stromal cells necessary to support the tumor cell mass. For example, several tumor angiogenesis factors produced by tumor cells or infiltrating inflammatory cells have been characterized. These factors stimulate the local proliferation of vascular cells that support the growing mass of tumor cells. The complex interplay of paracrine growth factors, the host's stromal response, and the autocrine factors acting directly on a nascent tumor cell may ultimately determine how rapidly the tumor grows and destroys the host.

Transmembrane Signaling

To understand better the possible mechanisms of transformation by growth factors and their receptors, it is necessary to describe in simple terms what is known about transmembrane signaling. Many ligands that act as cellular signals bind to a cell surface receptor. The binding generates a second messenger in the cytoplasm. Figure 8.3 shows the best-studied signaling pathways.

Many ligands produce an increase or a decrease in the cyclic nucleotides cyclic adenosine monophosphate (cAMP) and cyclic guanosine monophosphate (cGMP). Adenyl or guanyl cyclase enzymes generate these compounds. These "second messengers" then modulate the function of many cytoplasmic enzymes, among them the cAMP-dependent protein kinases. The link between the cell surface receptor and the cyclase is provided in some cases by any one of a number of so-called G proteins. Comprising three subunits, these proteins can mediate stimulatory or inhibitory effects on the cyclases (Fig. 8.3).

Another second messenger generated by surface receptor binding is inositol-3-phosphate (IP3). This phosphorylated sugar is derived from the phosphoinositol cycle (Fig. 8.3). A membrane-associated phosphodiesterase linked to a surface receptor by one or more G protein-like molecules generates IP3 and diacylglycerol from phosphatidylinositol 4,5 biphosphate (PIP_2). Diacylglycerol stimulates the activity of protein kinase C. Sphingosine and related substances appear to inhibit this enzyme. IP3 acts on the endoplasmic reticulum and possibly other sites to release calcium. The released cytoplasmic calcium also performs a second messenger role: It binds to calmodulin, modulating the activity of the calmodulin calcium-dependent protein kinase. Calcium also acts on protein kinase C. Minute changes in cytosolic-free calcium result as well from changes in the flux of calcium through cell membrane calcium channels. The activity of these and other ion pores is regulated by membrane potential (voltage difference across the membrane) as well as by interactions with membrane receptors and phospholipids. Calcium ions are known to directly affect many intracellular metabolic pathways.

Thus cyclic nucleotides, calcium, IP3, and a number of protein kinases mediate the intracellular effects of occupied cell surface receptors. Undoubtedly, other ion channels (e.g., Na^+/K^+) and other second messengers will be described. For

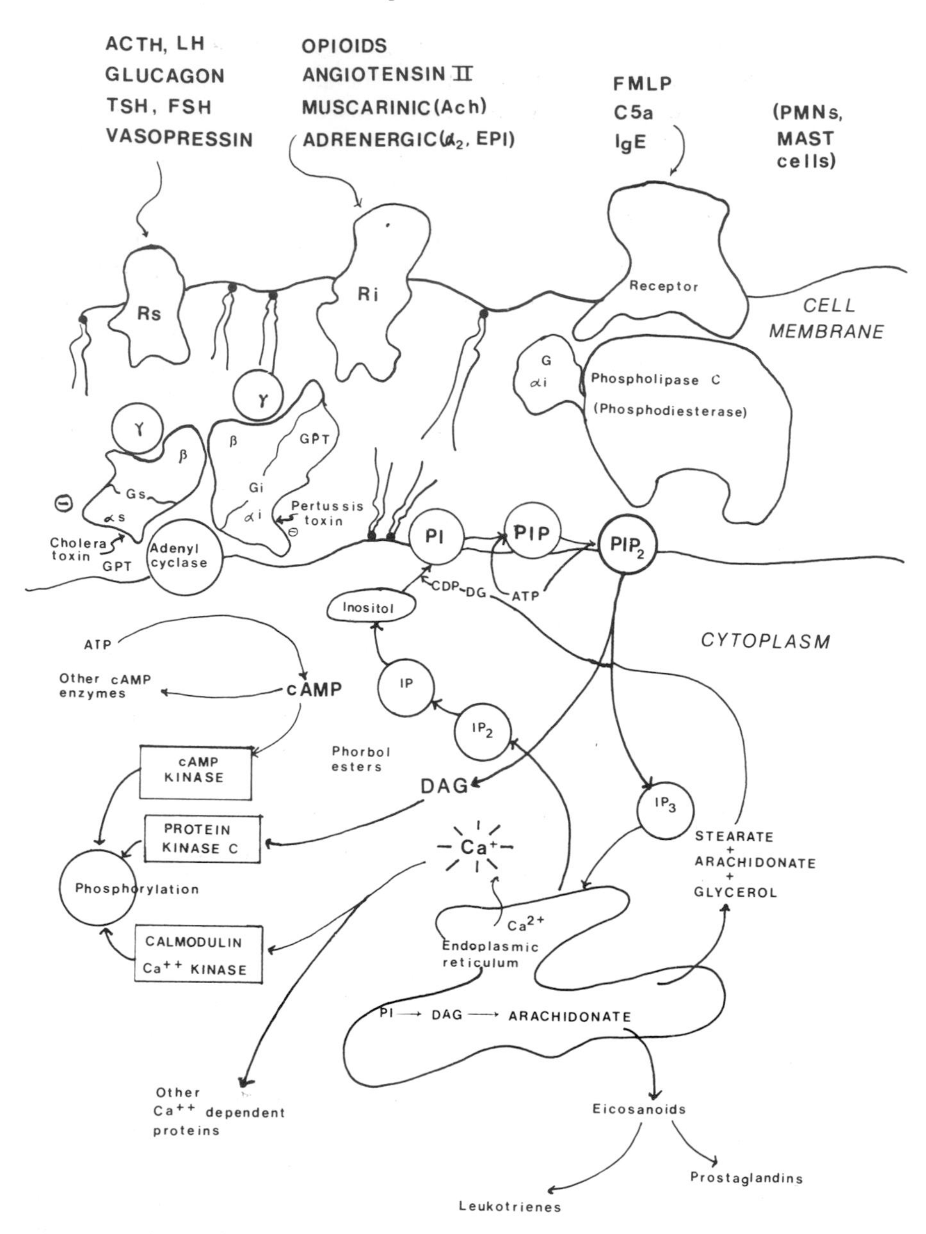

FIGURE 8.3. Transmembrane signaling. Growth factors and other ligands trigger cells by binding to cell surface receptors. The receptors are coupled to the second messengers generating enzymes by G proteins. The second messengers IP3, DAG, and cyclic nucleotides control the release of intracellular calcium and the activity of various kinase (phosphorylation) enzymes. The intracellular calcium concentration, the activity of the kinases, and the release of arachidonate metabolites are important in the control of cellular functions, including growth and differentiation (see text for more details).

our purposes here, it is important to realize that a number of the controlling enzymes in these modulating cascades may be encoded by proto-oncogenes. When disrupted, such pathways can lead to misregulated cell growth and neoplastic transformation.

v-*sis* ONCOGENE AND PDGF

The *sis* oncogene is the transforming gene of simian sarcoma virus (SSV). The protein product was first identified in 1982 and has been cloned and expressed in *Escherichia coli*. Despite the availability of cloned protein, the mode of action of this oncogene product was unclear until more recently. It did not show sequence homology to the tyrosine kinase family of oncogenes or to the nuclear oncogenes of the *myc* family. In 1983 sequence analysis of the human PDGF gene by two independent groups revealed shared sequences with the *sis* oncogene. This discovery was the first in a series that increased our understanding of the relation between growth factors and cancer. Currently the c-*sis* product is the only known example of an oncogene-encoded growth factor.

Platelet-derived growth factor, released by platelets during blood clotting, is one of the major growth factors present in serum. It is a potent mitogen and chemotactic factor for various mesenchymal cells including fibroblasts and smooth muscle cells. PDGF is normally released from the α-granules of platelets during hemostasis at the site of a wound, and it is thought to promote wound repair by stimulating growth of fibroblasts. Fibroblasts do not themselves produce PDGF, but they do have PDGF receptors. Many other mesenchymal cells produce PDGF or a related PDGF-like molecule(s).

Purified human PDGF contains two distinct but related polypeptide chains: PDGF-A 14–18 kD and PDGF-B (16 kD). When disulfide-linked as heterodimers, these subunits yield proteins of about 30 kD. The amino acid sequence of the human PDGF-B chain was found to be similar to the amino acid sequence predicted for the v-*sis* product from the DNA sequence of the cloned simian viral gene (Fig. 8.4). The differences between the two proteins could be ascribed to the different species from which they were isolated. Thus v-*sis* is thought to encode a form of PDGF-B.

The v-*sis* gene of SSV is inserted into the 3′ end of the viral envelope gene, and SSV-transformed cells produce a chimeric protein containing 38 amino acids derived from the viral "*env*" gene and 220 amino acids encoded by v-*sis* sequences. Products of the v-*sis* gene are recognized by anti-PDGF antibodies. This finding is no surprise given their amino acid sequence homology. When the v-*sis* gene encoding the simian PDGF-B protein is introduced into a fibroblast by infection with SSV, the infected cell begins to produce its own growth factor (autocrine control), which enables the cell to proliferate in an uncontrolled fashion. Several lines of evidence support the autocrine growth hypothesis for PDGF/*sis*.

1. Anti-PDGF antibodies inhibit acute transformation by SSV, and anti-PDGF antibodies block the growth of some but not all SSV-transformed cell lines.

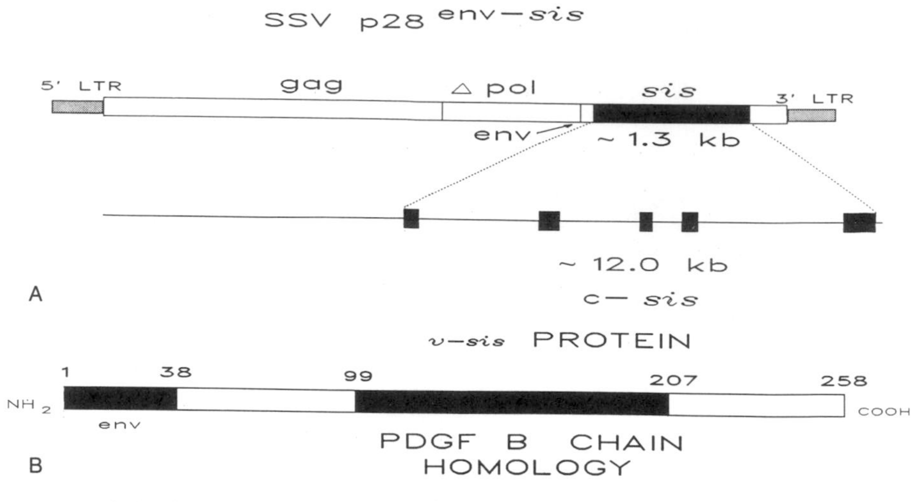

FIGURE 8.4. The v-*sis* gene and PDGF. The v-*sis* gene encodes a protein of 258 amino acids with an amino terminal fusion of 38 amino acids from the envelope gene of simian sarcoma virus. The c-*sis* gene encodes a protein of 241 amino acids; 99 to 207 are homologous with the B chain of PDGF. (A) Gene structures of v-*sis* and c-*sis*. (B) Homology of v-*sis* for PDGF B chain.

2. Culture medium of most SSV-transformed cells contains a PDGF-like mitogenic factor.
3. The v-*sis* protein product competes with PDGF for binding to PDGF receptors.
4. Only cells with PDGF receptors (e.g., fibroblasts, glial cells, myoblasts) are transformed by SSV.
5. Studies with deletion mutants have shown that the v-*sis* protein must be secreted and bind to the PDGF receptor to promote transformation.

The similarity of the v-*sis* protein and PDGF, presumably coded by the c-*sis* gene, raises the question of why one of these proteins is associated with transformation and the other is not, and, more importantly, if "dysregulated" autocrine production of c-*sis*/PDGF is associated with human cancer. Among normal cells, placental cytotrophoblast cells and embryonic smooth muscle cells have PDGF receptors and produce PDGF-like factors. Cell lines derived from placentas expressing cell surface receptors for PDGF respond to exogenous PDGF with activation of the c-*myc* gene and stimulation of DNA synthesis. Because these cells are among the fastest growing and most invasive normal cells, the expression of PDGF receptors in this tissue may help account for their "pseudomalignant" behavior.

Platelet-derived growth factor receptors also are found on a variety of mesenchymal cells. Fibroblasts have 400,000 receptors per cell with a dissociation constant of 10 to 1000 picomolar (pM). Most epithelial cells do not have

PDGF receptors. Binding of PDGF induces tyrosine autophosphorylation of its 185-kD receptor.

The c-*sis* gene has been assigned to human chromosome 22 and is involved in the reciprocal 9:22 translocation observed in chronic myelogenous leukemia (CML). In this case, however, the c-*abl* gene is activated and the c-*sis* gene remains transcriptionally silent (see Chapter 7).

A 4.2-kb c-*sis* transcript has been observed in five of six sarcomas and three of five glioblastomas, but it is normally not expressed in fibroblasts or connective tissue. Molecules with PDGF-like properties have been detected in a variety of transformed cells including a human osteosarcoma cell line, U20S, glioma U-343, bladder carcinoma T24, and hepatoma HepG2. What role these PDGF-like molecules played in the transformation of these cells, some of which lack PDGF receptors, is unclear. Presumably mutations or chromosomal rearrangements have occurred in the tumor cells such that the endogenous PDGF genes are no longer repressed. These genetic lesions may be in the PDGF genes themselves or in genes controlling PDGF expression.

Many tumors secrete growth factors that can confer a reversible transformed phenotype on normal established cell lines [so-called transforming growth factors (TGFs)]. One of these factors, TGFα, is related to EGF, whereas another TGFβ is not (see below). PDGF and PDGF-like growth factors are known to potentiate the effects of TGFα and TGFβ when stimulating growth of normal cells. However, no oncogene products other than the v-*sis* protein are known to be secreted.

Growth Factor Receptors

Tyrosine Kinase Growth Factor Receptors

Several of the best characterized plasma membrane receptors for growth factors are members of the tyrosine kinase family of proto-oncogenes. This group comprises the EGF receptor (v-*erb* B), the macrophage colony-stimulating factor (CSF-1) receptor (v-*fms*), and the insulin receptor, which shows 30 to 40% identity with the product of the *ros* oncogene. The PDGF receptor appears to be another member based on nucleotide sequence homology, but to date it has not been associated with a known oncogene product. Figure 8.5 compares this group of receptors, their related viral oncogene products, and several other plasma membrane molecules described below. The viral oncogenes of this group share an important feature with v-*src*: All have truncated carboxyl termini compared to their proto-oncogene homologues.

As discussed in Chapter 7, the truncation of v-*src* relative to c-*src* removes tyrosine 527 and produces a transforming protein, $pp60^{v\text{-}src}$, with elevated kinase activity. It is likely that tyrosine residues located in the carboxyl termini of members of the tyrosine kinase receptor group also may be important sites of phosphorylation that regulate kinase activity.

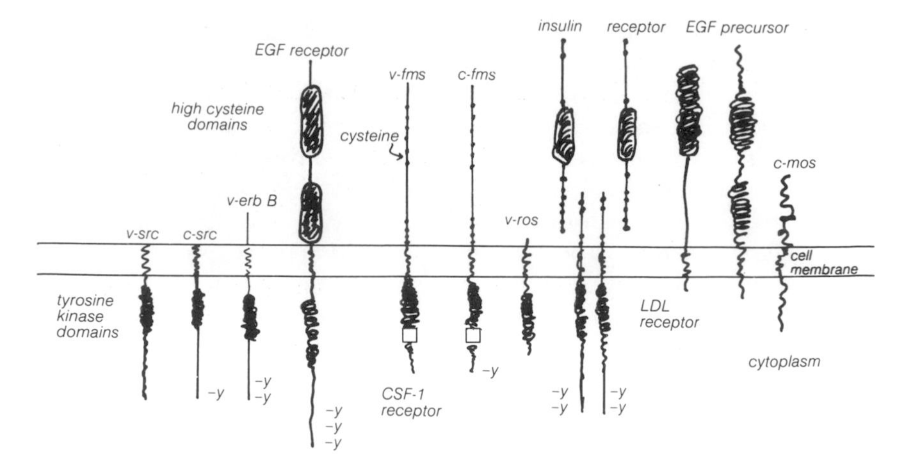

FIGURE 8.5. Family of cell surface receptors having tyrosine kinase activity. The position of tyrosines (y) thought to be important substrates for phosphorylation are shown. Domains sharing similar function/sequence include the cytoplasmic tyrosine kinase domains and the extracellular "high cysteine" domains. The location of cysteins is also noted (•).

Several members of this related group of receptors share other features. For example, the EGF receptor and the insulin receptor belong to a family of related transmembrane proteins that possess cysteine-rich sequence repeats in their extracellular ligand-binding domains. Two proteins distantly related to this family are the low-density lipoprotein (LDL) receptor, which is not a tyrosine kinase, and the EGF precursor (pro-EGF), which in addition to being the precursor of EGF may also be a receptor for a yet to be identified ligand. No viral oncogenes have been associated with the LDL receptor; however, the *mos* oncogene of the Moloney murine sarcoma virus is distantly related to pro-EGF.

Both v-*erb* B and v-*ros* lack most of the extracellular ligand binding domains found in their respective proto-oncogenes, including the cysteine-rich repeats. In contrast, the v-*fms* gene retains the entire extracellular ligand-binding domain it shares with c-*fms* (the CSF-1 receptor itself lacks the cysteine-rich repeats) and is truncated only at its carboxyl terminus.

v-*erb* B Oncogene and EGF Receptor

Epidermal growth factor was first recognized for its ability to stimulate precocious eyelid opening and tooth eruption in newborn mice. It subsequently was purified from male mouse submaxillary glands. Only later did investigators realize that it was a potent mitogen for fibroblasts in vitro. The active molecule is a 53-amino-acid polypeptide chain derived from a larger precursor. In situ hybridization analyses of sections of whole newborn mice indicate that EGF message is expressed in a large variety of tissues.

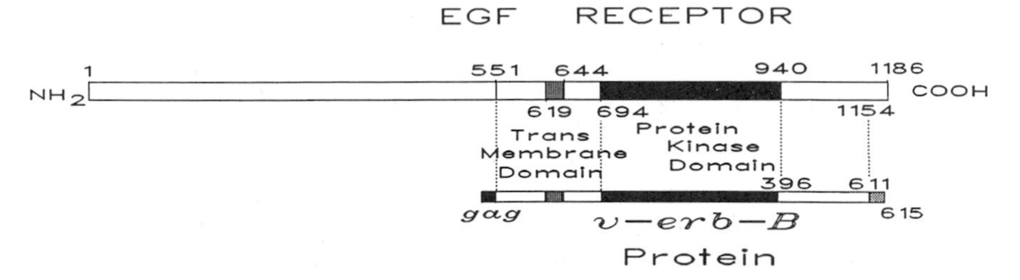

FIGURE 8.6. The v-*erb* B protein and EGF receptor. The v-*erb* B protein is a truncated version of the EGF receptor that lacks most of the extracellular ligand binding domain. It shares most homology in the tyrosine protein kinase domain.

The human protein was first purified from urine as urogastrone. Distal tubules of the kidney express high levels of an unprocessed higher-molecular-weight form of EGF, thought to be the precursor for urogastrone. EGF is strongly mitogenic for many cell types, and its growth effects are strongly potentiated by insulin and PDGF. To date no tumors are known to synthesize EGF, although many secrete a closely related factor, tumor growth factor alpha (TGFα).

The EGF receptor provided the first direct support for a relation between growth factor receptors and oncogenes. Investigators first showed that tryptic peptides derived from purified human EGF receptor were almost identical to the amino acid sequence predicted by the DNA sequence of the v-*erb* B oncogene from avian erythroblastosis virus (AEV). When the human cDNA for EGF was cloned, this preliminary work was confirmed: The AEV v-*erb* B gene must have been derived from the chicken EGF receptor gene (Fig. 8.6).

The EGF receptor is a 175-kD glycoprotein with tyrosine kinase activity. Its autophosphorylation is stimulated by the binding of EGF. Although it seems likely that phosphorylation of specific tyrosine residues on proteins associated with growth regulation is somehow involved in the mitogenic response, the details of this cascade have not been worked out. The single polypeptide chain of the EGF receptor is comprised of two domains: one inside and the other outside the cell. The internal carboxyl half of the protein shares striking homology with the catalytic domain of the *src* tyrosine-protein kinase (see below and Chapter 7). It is within this region that the human EGF receptor shares more than 90% homology with the chicken v-*erb* B gene product. The v-*erb* B protein is a truncated, predominantly cytoplasmic analogue of the EGF receptor (Figs. 8.5 and 8.6).

How does the v-*erb* B gene transform cells? It is known that the glycosylated 70 to 80-kD v-*erb* protein must be localized to the cell surface membrane for infected cells to exhibit a transformed phenotype. However, does this mean that the v-*erb* B protein mimics an occupied constitutively activated EGF receptor?

The answer is "maybe." In many cells transformed by viruses encoding tyrosine protein kinases, there are obvious elevations in total cellular phosphotyrosine, and a number of phosphotyrosine proteins have been identified. In AEV-transformed cells, however, there is only slight stimulation of phosphotyrosine content. Until the critical mitogenic substrates of the v-*erb* B and other *src*-like tyrosine kinases are identified, it is difficult to know if the lower activity found in AEV-transformed cells is sufficient to generate the transformed phenotype. Another confusing issue is the finding that erythroblasts, the usual target for AEV, do not normally contain EGF receptors. Perhaps other growth factors for this lineage of cells (e.g., interleukin-3 or erythropoietin) have tyrosine kinase activity that is replaced by the v-*erb* protein.

The level of c-*erb* B expression is low in most tissues, and c-*erb* B transcripts are not found in many tumors despite the fact that EGF receptors are found on many tissue types. However, an occasional tumor expresses high levels of EGF receptors. For example, some squamous cell carcinoma cell lines from head and neck tumors have between 1×10^6 and 15×10^6 receptors per cell, and the epidermoid carcinoma cell line used to clone the EGF receptor A431 has 3×10^6 to 5×10^6 receptors per cell compared to 1.5×10^5 per normal epidermal cell expressing the receptor. Glioblastomas also have been shown to have large numbers of EGF receptors.

At least two mechanisms can account for elevated EGF receptors: (1) chromosome translocation and (2) gene amplification. Cytogenetic studies of A431 cells show translocation of chromosome 7. The c-*erb* B gene is located on human chromosome 7, region 7p11-p13, and it is plausible that it is activated by the translocation. In addition, elevated expression of EGF receptors with 5- to 40-fold gene amplification has been found in A431 and in HN10 and HN5 squamous carcinoma cell lines.

erb B-Related Oncogenes: *neu* and *erb* B-2

Rat neuroglioblastomas induced by exposure in utero to ethylnitrosourea frequently carry an oncogene detectable upon transfection into NIH 3T3 cells. This oncogene, a tyrosine kinase called *neu*, was found to be more related to c-*erb* B than to other members of the tyrosine kinase class of oncogenes. The *neu* oncogene encodes a tumor antigen, p185, which is serologically related to the EGF receptor and is present on the surface of induced rat neuroblastomas. More than 50% of the amino acids coding for *neu* are shared with the EGF receptor, and there is 80% homology in the tyrosine kinase domain. The *neu* oncogene, but not its normal counterpart, yields transformed foci when transfected into NIH 3T3 cells.

Structural comparison of the normal and transforming *neu* genes revealed no gross rearrangements, and comparable levels of p185 are expressed in nontransformed and transformed cell lines. Thus the alteration responsible for activation of *neu* is thought to result in a single amino acid substitution.

Construction of in vitro recombinants between the normal and transforming genes resulted in discovery of a unique mutation responsible for activation of the *neu* proto-oncogene. DNA from four cell lines containing independently activated *neu* oncogenes had a single point mutation in the transmembrane domain of the molecule. The mutation was predicted to change valine to glutamic acid at this site.

The *neu* oncogene also can be activated by exposure of animals to another (related) alkylating agent, methylnitrosourea. Studies of Baf/N rats exposed to methylnitrosourea have shown that all of nearly 100 tumors contained H-*ras* oncogenes activated by identical G to A transitions.[1] The *neu* gene is activated by T to A transversions. Although carcinogens cause widespread damage to cellular DNA, the specificity of the changes leading to activation of these oncogenes is remarkable. Because the ultimate appearance of a transformed cell is a multistep process, additional not yet understood factors must contribute to tumor formation.

The human *neu* gene, called c-*erb* B-2, maps to chromosome 17p11-q21 (the map position of the human c-*erb* A-1 locus as well). This chromosome band is often involved in a nonrandom reciprocal translocation, t(15:17)(q22:q21), in acute promyelocytic leukemia. The human c-*erb* B-2 gene encodes a protein of 138 kD with a transmembrane topology similar to that of the EGF receptor. Thus it is likely that the gene encodes a yet to be discovered growth factor receptor.

In one study amplification of the c-*erb* B-2 gene was found in three human adenocarcinomas: one of salivary origin, one mammary, and one gastric cancer cell line (MKN-7). In another study of 101 fresh samples of human malignant tumors representing 21 tumor types, the c-*erb* B-2 gene was amplified in 5 of 63 adenocarcinomas and in none of the 38 other tumors tested. Amplification of *erb* B-2 in human breast carcinomas is correlated with more aggressive tumors and a worse prognosis (see Chapter 6). The c-*erb* B-2 gene was amplified in one of eight squamous cell carcinomas.

Other work suggests that antibodies recognizing *neu*/*erb* B-2 can reverse the in vitro phenotype of cells carrying this oncogene. If this gene is expressed in human cancer, it would be an ideal target for immunotoxins (see Chapter 14).

c-*fms* Proto-oncogene: Macrophage CSF-1 Receptor

Colony-stimulating factor-1 signals hematopoietic precursor cells to form colonies containing predominantly mononuclear phagocytes. CSF-1 is a lineage-specific growth factor, in contrast to the granulocyte/macrophage colony-stimulating factor (GM-CSF) and interleukin-3 (multilineage CSF), that also directly induces mononuclear phagocyte proliferation. It is an acidic molecule composed of two disulfide-bonded, approximately 14-kD polypeptide chains that are heavily glycosylated. It binds specifically to mononuclear phagocytes and their precursors irrespective of their tissue of origin or state of maturation.

The v-*fms* gene is the transforming gene of the McDonough strain of feline sarcoma virus (SM-FeSV) and belongs to the family of *src*-related oncogenes that

[1]Transitions change one purine base to the other purine base (e.g., G→A) or one pyrimidine base to the other pyrimidine base (e.g., T→C). Transversions change purine to pyrimidine or pyrimidine to purine (e.g., T→A).

have tyrosine-specific kinase activity. The transforming protein of this virus is a *gag-onc* fusion protein, gp180$^{gag\text{-}fms}$ (Fig. 8.7). v-*fms* itself encodes a transmembrane glycoprotein of 142 kD. Expression of this protein at the cell surface is required for cellular transformation.

The feline c-*fms* proto-oncogene product is a 170-kD glycoprotein with associated tyrosine kinase activity. This glycoprotein is expressed on mature cat macrophages from peritoneal inflammatory exudates and splenic tissue. The receptor for the murine colony-stimulating factor (CSF-1) is also expressed on mononuclear phagocytes and is a 165-kD glycoprotein with associated tyrosine kinase activity. Rabbit antisera to a recombinant v-*fms*-coded polypeptide precipitates the feline c-*fms* product and specifically cross-reacts with the 165-kD glycoprotein from mouse macrophages. This result suggests that c-*fms* encodes the gene for the normal CSF-1 receptor.

The complete nucleotide sequence of the human c-*fms* gene has now been determined. The 972-amino-acid c-*fms* protein has an extracellular domain, a membrane-spanning region, and a cytoplasmic tyrosine kinase domain. The c-*fms* protein from macrophages is phosphorylated at tyrosine residues after addition of CSF-1. The tyrosine residue at position 969 of c-*fms*, similar to the tyrosine at position 1,173 of the EGF receptor, is the major site for this ligand-induced autophosphorylation. The c-*fms* gene is activated by truncation of the amino terminus of its protein in a manner similar to the c-*erb* B gene; however, the truncated protein retains the ability to bind CSF-1.

The CSF-1 receptor gene has been localized to human chromosome 5q. Although no direct associations with malignancy have been made, this region of 5q is associated with several hematological neoplasms. In particular, the 5q$^-$ syndrome, in which the long arm of chromosome 5 is deleted, is characterized by myeloid dysplasia and a tendency to develop myeloid leukemia.

v-*ros* Oncogene and the Human Insulin Receptor

The metabolic effects of insulin are mediated by its high affinity binding to specific cell surface receptors. The insulin receptor is an integral membrane glycoprotein composed of two α subunits (M_r ~120,000–130,000) and two β subunits (M_r ~90,000) linked by disulfide bands. Based on photoaffinity labeling of the receptor, insulin is thought to bind to the α subunit. The α subunit contains a region rich in cysteine residues similar to that present in the ligand-binding domains of the EGF receptor and the LDL receptor. Insulin binding to the receptor stimulates autophosphorylation of the β subunit on serine and tyrosine residues. The β subunit shares structural homology with the EGF receptor and the *src* family of tyrosine kinases (Fig. 8.5).

There is one insulin receptor gene in the haploid human genome that is highly conserved in vertebrate and invertebrate cells. However, multiple RNA transcripts are generated in various cells, the significance of which is not known. It is clear that the α and β subunits of the receptor are derived from a single polypeptide precursor of ~190,000 daltons, which is cleaved by a proteolytic processing enzyme prior to insertion into the membrane.

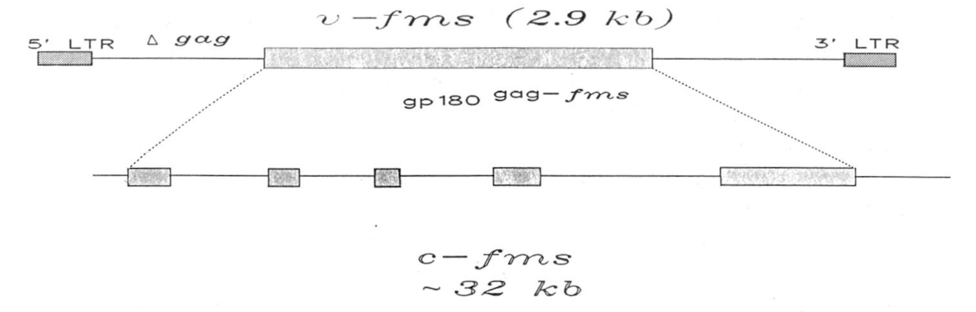

FIGURE 8.7. The v-*fms* gene and its cellular homologue c-*fms*, the CSF-1 receptor. Structure of the SM-FeSV proviral DNA and the c-*fms*-specific exon sequences found in human DNA are shown.

It is probable that the oncogene v-*ros* carried by the UR2 retrovirus is a truncated avian insulin receptor. This oncogene shares as much as 73% identity with the human insulin receptor in the tyrosine protein kinase domain.

The insulin receptor shares structural homology with the insulin-like growth factor (IGF-1) receptor but is more distantly related to other transmembrane receptors with tyrosine kinase activity (Fig. 8.5). It remains to be seen how insulin receptors signal cells to produce the pleiotropic effects that occur with insulin binding.

mos AND PROEPIDERMAL GROWTH FACTOR

The *mos* oncogene is the transforming gene of Moloney murine sarcoma virus (Mo-MSV), a defective virus that arose from a rhabdomyosarcoma in a mouse that had been infected with Moloney murine leukemia virus (Fig. 8.8). Analysis of the v-*mos* gene from MSV revealed that it differs from its normal c-*mos* homologue at many positions (25 base differences of 1157 nucleotides). This finding might at first suggest that activation of c-*mos* to an oncogene requires specific structural changes in the gene product, such as the point mutations that activate c-*ras*. However, insertional activation of c-*mos* expression has been observed in mouse cells, indicating that changes in gene expression are sufficient to account for its transforming potential.

Analysis of the *mos* locus in a number of mouse plasmacytomas has revealed a novel method of oncogene activation. In these cells an endogenous retrovirus-like element referred to as an intracisternal A particle (IAP) has become integrated within the coding sequence of c-*mos*, resulting in a slight truncation of c-*mos* coding sequences at the amino terminus and presumably in altered regulation of the remaining gene sequences. A search for like events in similar human cells failed to reveal any alterations in the human c-*mos* locus. Furthermore, human c-*mos*, unlike its murine counterpart, cannot be activated to become a transforming gene by in vitro manipulation of sequences governing its

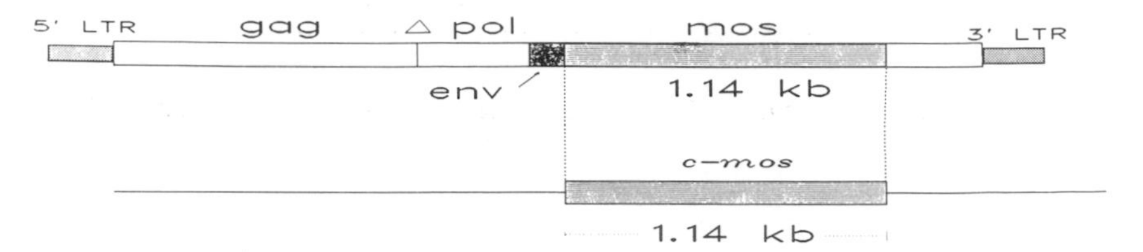

FIGURE 8.8. The *mos* oncogene of the Moloney murine sarcoma virus.

expression. In fact, it is not yet known if this oncogene has any role at all in human cancer. However, human cells are susceptible to transformation by murine v-*mos* in appropriate expression vectors.

Moloney murine sarcoma virus induces fibrosarcomas in vivo and transforms NIH 3T3 cells in vitro. The *mos* oncogene codes for a protein of 37 kD that is present in transformed cells in low copy number. Evidence suggests that the *mos* oncogene shares sequences with pro-EGF. EGF is synthesized as part of a large precursor that spans the cell membrane. The pro-EGF molecule shares many similarities with the EGF receptor. Both molecules are membrane-bound glycoproteins of approximately 1200 amino acids with extensive duplications in their extracellular domains (Fig. 8.5).

The LDL receptor shares sequences (38% identity) with the pro-EGF molecule but not with c-*mos*. Conversely, the cAMP-dependent protein kinase and the yeast cell division control gene share 21% and 22% identity, respectively, with the c-*mos* gene and none with the pro-EGF molecule. Thus a number of cell surface receptor and regulatory molecules appear distantly related to one another and probably share common ancestors.

It appears that the *mos* oncogene is not a tyrosine kinase. Most of the *mos* sequences shared with pro-EGF and of pro-EGF with the EGF receptor (a known tyrosine kinase) are in the external domains. Furthermore, cells infected with the Moloney sarcoma virus do not contain elevated levels of phosphotyrosine. The *mos* protein more likely phosphorylates serines and threonines, like the cAMP-dependent kinase with which it shares sequences.

In support of this idea, $p37^{mos}$ was shown to have serine kinase activity. The target of $p37^{mos}$ may be a nuclear protein, p55. Minimal phosphorylation of p55 may result in cell transformation, and extensive phosphorylation apparently leads to cell death.

The c-*mos* gene is expressed at relatively high levels in murine gonads and appears to be developmentally regulated during germ cell differentiation. Low levels of the 1.7-kb mouse c-*mos* transcript are detectable during the 3 weeks following birth, but thereafter levels of the c-*mos* transcript increase 10- to 100-fold

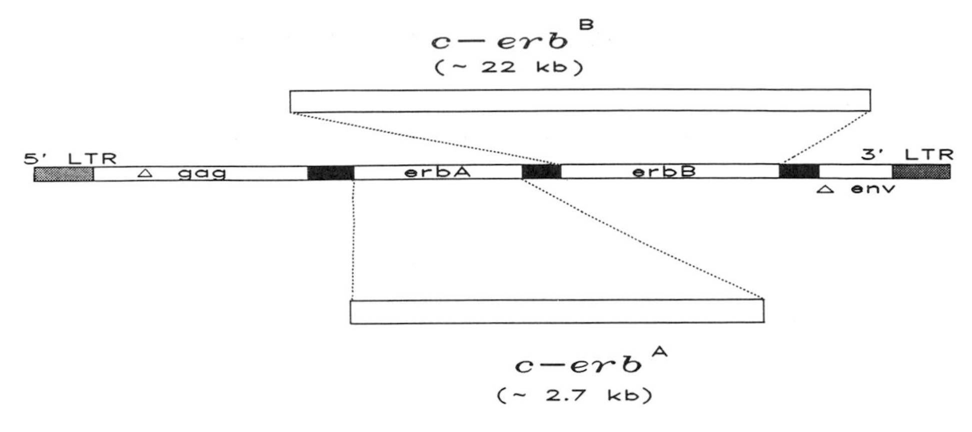

FIGURE 8.9. The avian erythroblastosis virus contains two oncogenes. The v-*erb* A gene product is translated as a 75-kD *gag-erb* A fusion protein. The v-*erb* B gene encodes a 68-kD-phosphorylated glycoprotein. The corresponding cellular proto-oncogenes span 2.7 kb (c-*erb* A) and 22 kb (c-*erb* B), respectively. (Not drawn to scale)

to reach adult values by 5 weeks of age. It is of interest that a 4.7-kb *abl* transcript follows similar kinetics in testis.

The *mos* transcripts are found predominantly in pre- and post-meiotic germ cells. Transcripts of different size are found in mouse ovary (1.4 kb) and mouse embryo (1.3 kb). Human c-*mos* transcripts of about 1.0 kb have been detected in human testicular tissue.

The human c-*mos* proto-oncogene is located on chromosome 8 at band q22. The t(8:21)(q22:q2) chromosome rearrangement is associated with acute myeloblastic leukemia subgroup M2. Interestingly, the c-*mos* gene remains on the 8q chromosome; and the c-*myc* gene, which maps to the 8q24 arm, is translocated to the 21q2 chromosome. No rearrangement of the c-*mos* gene was detected by the Rowley group using Southern blot analysis.

c-*erb* A and Steroid Hormone Receptors

Avian erythroblastosis virus contains two viral oncogenes, each of which has a normal cellular counterpart (Fig. 8.9). The c-*erb* B gene is related to the cellular proto-oncogene encoding the EGF receptor, as already discussed.

The human c-*erb* A gene encodes a protein of 456 amino acids, which specifically binds thyroid hormones with high affinity. Like the estrogen and glucocorticoid receptors, this gene comprises a series of functional domains. The carboxy ends of these molecules share about 17% identity and are responsible for hormone binding. The middle portion of the molecules is most conserved (about 50% identity) and includes a cysteine-rich region with structural similarity to various transcriptional regulatory proteins. This region, with the amino terminus, mediates specific DNA binding and transcriptional activation of hormone-responsive genes.

The v-*erb* A protein is located in the nucleus and binds DNA. The viral oncogene comprises a truncated 5′ erb A sequence fused to the viral *gag* gene. The p75$^{gag\text{-}v\text{-}erbA}$ fusion protein has no detectable hormone-binding capacity and probably acts as a constitutively active thyroid hormone. Like thyroid hormones themselves, which play a modulatory role in human and animal oncogenesis, the v-*erb* A gene apparently enhances transformation. It has a poor capacity to transform cells alone, yet functions synergistically with v-*erb* B and other oncogenes.

Because the v-*erb* A protein lacks the transcriptional regulatory portion of the c-*erb* A gene, it is unlikely that it acts by stimulating runaway transcription. Rather, it may act in a manner analogous to either of two known mutant glucocorticoid receptors. If c-*erb* A functions to promote differentiation of erythroid cells, a defective v-*erb* A protein might compete for DNA binding sites and disturb the differentiating signals transduced by c-*erb* A.

Other Growth Factors Possibly Related to Oncogenesis

Transforming Growth Factor Alpha

Tumor growth factor alpha is a 5.6-kD single-chain polypeptide of 50 amino acids. It was first identified as sarcoma growth factor but later shown to share considerable identityy (35% shared amino acids with conservation of all cysteine residues) with EGF and to bind to the EGF receptor on an equimolar basis with purified EGF. TGFα has been found in a variety of virally transformed cells, and in human placenta, and rodent embryos. It is of note that vaccinia virus encodes an EGF/TGFα-like molecule that may play a role in the benign hyperplasia induced by this pox virus. TGFα has been postulated to be an embryonic form of EGF because it is expressed in transformed cells and fetal tissues. This factor may play a role in neoplastic pathogenesis by autocrine stimulation of cells with EGF receptors.

Research suggests that TGFα is a potent stimulant of angiogenesis, which is important in wound healing, luteinization, neovascularization in diabetes, and outgrowth of solid tumors. Some agents, e.g., heparin, copper, and E-type prostaglandins, promote migration of inflammatory and endothelial cells. Other factors such as the heparin-binding peptides, fibroblast growth factors, endothelial growth factors, angiogenesis factors, and TGFα, presumably have a direct mitogenic effect on cells of the capillary bed.

Transforming Growth Factor Beta

Unlike TGFα, which shares many properties with EGF, TGFβ differs in molecular characteristics from known growth factors. Depending on the cell type, TGFβ can have stimulatory or inhibitory effects. Furthermore, its actions often depend on other mitogenic factors present in tissue culture medium, suggesting that it is a permissive factor that determines how cells respond to other growth factors.

Transforming growth factor beta is a 25-kD protein composed of two identical 112-amino-acid polypeptide chains linked by disulfide bonds. These polypeptides are cleaved from a larger precursor of 391 amino acids that is translated from a 2.5-kb mRNA. Using Northern blot analysis, this transcript has been found in a wide variety of normal and transformed cells, and the protein has been detected in normal liver, lung, heart, brain, and kidney, as well as embryos and placenta. Mitogen stimulation of human peripheral blood lymphocytes results in an increase in TGFβ release. A number of cells in culture both produce and respond to TGFβ. These cells do not proliferate out of control, in part at least because the growth factor is released in an inactive form.

The TGFβ moiety is mitogenic for a variety of fibroblastic cell types in tissue culture. In a portion of these cells its actions are mediated through PDGF, and binding of TGFβ may lead to expression of the c-*sis* gene.

In some instances TGFβ can exhibit an inhibitory effect on cell growth. For example, a variety of neoplastically transformed epithelial cells are inhibited by TGFβ. In serum-free media, normal human prokeratinocytes are inhibited by TGFβ, whereas certain squamous carcinoma cell lines grown under identical conditions are not. It is tantalizing to postulate that loss of the inhibitory actions of TGFβ or similar growth-inhibitory molecules might contribute to transformation of some cell types. This situation might be caused by mutation or dysfunction of the genes encoding TGFβ or its receptor.

The TGFβ receptor has not been as well studied as the EGF or PDGF receptors. Low numbers (10,000–40,000/cell) of high affinity (25–140 pM) receptors are found on a variety of epithelial and mesenchymal cell types. The human TGFβ receptor is a glycoprotein dimer with an unreduced M_r of 615 kD. Despite the provocative suggestions made above linking TGFβ to cellular growth control, to date no specific correlations have been made between this growth factor or its receptor and known oncogenes.

Interleukin-2: T Cell Growth Factor

Treatment of human peripheral blood T cells with a number of mitogens, including plant lectins such as concanavalin A and phytohemagglutinin A, causes these cells to proliferate and release a variety of "interleukins," molecules that provide signals for amplification or inhibition of the immunoinflammatory response. One of these factors, interleukin 2 (IL-2) was originally called T cell growth factor because it supported the long-term in vitro growth of normal cytotoxic T lymphocytes.

The secreted IL-2 glycoprotein has 133 residues with one internal disulfide bond. Although IL-2 originally was thought to be a lineage-specific growth factor for T cells, it is now known that its receptors are also found on B cells, macrophages, and endothelial cells. This factor may be a means for T cells to communicate with several of the other cells important in the regulation of the immunoinflammatory network. The activation of T and B cells involves a hierarchy of signals. IL-2 receptors are not found on resting T cells. The expression of functional receptors requires the prior binding of antigen or mitogen to cells. The most specific signal, the antigen, triggers only members of a clonal population.

The process of cellular activation leads to expression of other receptors for both general growth factors and lineage-specific growth factors. An example of the former is transferrin receptors. An example of the latter are factors that promote the final differentiation of B cells into plasma cells, including B cell growth and differentiation factors. IL-2 can function as an autocrine or a paracrine growth factor for T cells (Fig. 8.2). The fact that normal T cells can be made to proliferate for long periods of time in the presence of IL-2 and antigens suggests that mutations in the IL-2 gene or its receptor might be associated with T cell malignancies. The *tat* gene from human T cell leukemia virus I promotes the increased expression of IL-2 and its receptor. Presumably, this expression gives the infected cells a growth advantage. It has been hypothesized that such autonomously growing cells may undergo further somatic genomic changes that could eventually lead to a truly transformed phenotype of adult T cell lymphoma/leukemia. Although cell surface IL-2 receptors from normal and transformed lymphocytes are slightly different in size (55 and 60 kD, respectively), the meaning of this difference is unclear. Although several IL-2 receptor transcripts have been identified, no clear association of unusually processed messages has been associated with lymphomas. Thus despite the potential role of IL-2 and its receptor for involvement in lymphocyte malignancy, to date no clear-cut genetic rearrangements or other changes in these genes have been found.

Additional Factors

Novel connections between growth factors and oncogenes have been and will continue to be made. A number of examples have been discussed. Another example includes bombesin, a tetradecapeptide related to gastrin-releasing peptide. This entity is a growth factor for small-cell carcinoma of the lung as well as for NIH 3T3 cells. Polyoma middle T antigen, the transforming protein of this DNA tumor virus shares amino acid sequences with gastrin. The meaning of this interesting finding is unclear, but it does suggest a novel connection between a DNA tumor virus and a peptide growth factor gene.

Bibliography

Growth Factors

Barnes D, Sato G. Serum-free cell culture: a unifying approach. Cell 22:649, 1980.

Goustin AS, Leof EB, Shipley GD, et al. Growth factors and cancer. Cancer Res 46:1015, 1986.

Heldin CH, Westermark B. Growth factors: mechanism of action and relation to oncogenes. Cell 37:9, 1984.

Kaplan PL, Anderson M, Ozanne B. Transforming growth factor production enables cells to grow in the absence of serum: an autocrine system. Proc Natl Acad Sci USA 79:485, 1982.

Sporn MB, Roberts AB. Autocrine growth factors and cancer. Nature 313:745, 1985.

Transmembrane Signaling

Berridge MJ, Irvine RF. Inositol triphosphate, a novel second messenger in cellular signal transduction. Nature 312:315, 1984.

Rozengurt E. Early signals in the mitogenic response. Science 234:161, 1986.

Sibley DR, Benovic JL, Caron MG, et al. Regulation of transmembrane signaling by receptor phosphorylation. Cell 48:913, 1987.

erb A

Debuire B, Henry C, Benaissa M, et al. Sequencing the erb A gene of avian erythroblastosis virus reveals a new type of oncogene. Science 224:1456, 1984.

Graf T, Beug H. Role of the v-erb A and v-erb B oncogenes of avian erythroblastosis virus in erythroid cell transformation. Cell 34:7, 1983.

Kahn P, Frykberg L, Brady C, et al. v-erb A cooperates with sarcoma oncogenes in leukemic cell transformation. Cell 45:349, 1986.

Riedel H, Dall TJ, Schlessinger J, et al. A chimaeric receptor allows insulin to stimulate tyrosine kinase activity of epidermal growth factor receptor. Nature 324:68, 1986.

Weinberger C, Hollenberg SM, Rosenfeld MG, et al. Domain structure of human glucocorticoid receptor and its relationship to the v-erb A oncogene product. Nature 318:670, 1985.

Weinberger C, Thompson CC, Ong ES, et al. The c-erb A gene encodes a thyroid hormone receptor. Nature 324:641, 1986.

v-*sis*/PDGF

Bowen-Pope DF, Vogel A, Ross R. Production of platelet-derived growth factor-like molecules and reduced expression of platelet-derived growth factor receptors accompany transformation by a wide spectrum of agents. Proc Natl Acad Sci USA 81:2396, 1984.

Chiu I-M, Reddy EP, Givol D, et al. Nucleotide sequence analysis identified the human c-sis proto-oncogene as a structural gene for platelet-derived growth factor. Cell 37:123, 1984.

Cochran BH, Reffel AC, Stiles CD. Molecular cloning of gene sequences regulated by platelet-derived growth factor. Cell 33:939, 1983.

Dalla-Favera R, Gallo RC, Giallongo A, et al. Chromosomal localization of the human homologue (c-sis) of the simian sarcoma virus onc gene. Science 218:686, 1982.

Deuel TF, Huang JS, Huang SS, et al. Expression of a platelet-derived growth factor-like protein in simian sarcoma virus transformed cells. Science 221:1348, 1983.

Doolittle RF, Hunkapillar MW, Hood LE, et al. Simian sarcoma virus onc gene, v-sis, is derived from the gene (or genes) encoding a platelet-derived growth factor. Science 221:275, 1983.

Groffen J, Heisterkamp N, Stephenson JR, et al. c-sis is translocated from chromosome 22 to chromosome 9 in chronic myelocytic leukemia. J Exp Med 158:9, 1983.

Josephs SF, Guo C, Ratner L, et al. Human proto-oncogene nucleotide sequences corresponding to the transforming region of simian sarcoma virus. Science 223:486, 1984.
Josephs SF, Ratner L, Clarke MF, et al. Transforming potential of human c-sis nucleotide sequences encoding platelet-derived growth factor. Science 225:636, 1984.
Niman HL. Antisera to a synthetic peptide of the sis viral oncogene product recognize human platelet-derived growth factor. Nature 307:180, 1984.
Rao CD, Igarashi H, Chiu I-M, et al. Structure and sequence of the human c-sis/platelet-derived growth factor 2 (sis/PDGF-2) transcriptional unit. Proc Natl Acad Sci USA 83:2392, 1986.
Robbins KD, Devare SG, Reddy EP, et al. In vivo identification of the transforming gene product of simian sarcoma virus. Science 218:1131, 1982.
Swan DC, McBride OW, Robbins KC, et al. Chromosomal mapping of the simian sarcoma virus onc gene analogue in human cells. Proc Natl Acad Sci USA 79:4691, 1982.
Waterfield MD, Scrace GT, Whittle N, et al. Platelet-derived growth factor is structurally related to the putative transforming protein $p28^{sis}$ of simian sarcoma virus. Nature 304:35, 1983.

v-*erb* B/EGF Receptor

Brown JP, Twardzik DR, Marquardt H, et al. Vaccinia virus encodes a polypeptide homologous to epidermal growth factor and transforming growth factor. Nature 313:491, 1985.
Cohen S, Urisho H, Stoscheck C, et al. A native 170,000 epidermal growth factor receptor-kinase in A431 cell membrane vesicles. J Biol Chem 257:1523, 1982.
Cowley G, Smith JA, Gusterson B, et al. The amount of EGF receptor is elevated on squamous cell carcinoma. Cancer Cells 1:5, 1984.
Downward J, Yarden Y, Mayes E, et al. Close similarity of epidermal growth factor receptor and v-erb B oncogene protein sequence. Nature 307:521, 1984.
Fung Y-KT, Lewis WG, Crittenden LB, et al. Activation of the cellular oncogene c-erb B by LTR insertion: molecular basis for induction of erythroblastosis by avian leukosis virus. Cell 33:357, 1983.
Gilmore T, DeClue JE, Martin GF. Protein phosphorylation at tyrosine is induced by the v-erb B gene product in vivo and in vitro. Cell 40:609, 1985.
Hayman M, Beug H. Identification of a form of the avian erythroblastosis virus erb B gene product at the cell surface. Nature 309:460, 1984.
Liberman TA, Nusbaum HR, Razos N, et al. Amplification, enhanced expression and possible rearrangement of EGF receptor gene in primary human brain tumors of glial origin. Nature 313:144, 1985.
Lin CR, Chen WS, Kruijer W, et al. Expression cloning of human EGF receptor-complementary DNA: gene amplification and three related messenger RNA production A431 cells. Science 224:843, 1984.
Meyers MB, Merluzzi VJ, Spengler BA, et al. Epidermal growth factor receptor is increased in multidrug-resistant Chinese hamster and mouse tumor cells. Proc Natl Acad Sci USA 83:5521, 1986.
Shimizu N, Kondo I, Gamou S, et al. Genetic analysis of hyperproduction of epidermal growth factor receptors in human epidermoid carcinoma A431 cells. Somatic Cell Mol Genet 10:45, 1984.
Spurr NK, Solomon E, Jansson M, et al. Chromosomal localization of the human homologue to the oncogenes erb A and B. EMBO J 3:159, 1984.

Ullrich A, Coussens L, Hayflick JS, et al. Human epidermal growth factor receptor cDNA sequence and aberrant expression in A431 epidermoid carcinoma cells. Nature 309:418, 1984.
Ushiro H, Cohen SJ. Identification of phosphotyrosine as a product of epidermal growth factor-activated protein kinase in A431 cell membranes. J Biol Chem 255:8363, 1980.
Yamamoto T, Ikawa S, Akiyama T, et al. Similarity of protein encoded by human c-erb B-2 gene to epidermal growth factor receptor. Nature 319:230, 1986.
Yokota J, Yamamoto T, Toyoshima K, et al. Amplification of c-erb B-2 oncogene in human adenocarcinomas in vivo. Lancet 1:765, 1986.

FGF

Bohlen P, Baird A, Esch F, et al. Isolation and partial molecular characterization of pituitary fibroblast growth factor. Proc Natl Acad Sci USA 81:5364, 1984.
Gospodarowicz D, Massoglia S, Chen J, et al. Isolation and pituitary fibroblast growth factor by fast protein liquid chromatography (FPLC); partial chemical and biological characterization. J Cell Physiol 122:323, 1985.
Thomas KA, Rios-Candelore M, Fitzpatrick S. Purification and characterization of acidic fibroblast growth factor from bovine brain. Proc Natl Acad Sci USA 81:357, 1984.

neu

Akiyama T, Sudo C, Ogawara H, et al. The product of the human c-erb B-2 gene: a 185-kilodalton glycoprotein with tyrosine kinase activity. Science 232:1644, 1986.
Bargmann C, Hung M-C, Weinberg R. The neu oncogene encodes an epidermal growth factor-related protein. Nature 319:226, 1986.
Bargmann CI, Hung M-C, Weinberg RA. Multiple independent activations of the neu oncogene by a point mutation altering the transmembrane domain of p185. Cell 45:649, 1986.
Schecter AL, Stern DF, Vaidayanathan L, et al. The neu oncogene: an erb-B-related gene coding a 185,000-M_r tumor antigen. Nature 312:513, 1984.

v-*fms*/CSF-1 Receptor

Anderson SJ, Gonda MA, Rettenbeier CW, et al. Subcellular localization of glycoproteins encoded by the viral oncogene v-fms. J Virol 51:730, 1984.
Coussens L, Van Beveren C, Smith D, et al. Structural alteration of viral homologue of receptor proto-oncogene fms at carboxyl terminus. Nature 320:222, 1986.
Kawasaki ES, Ladner MB, Wang AM, et al. Molecular cloning of a complementary DNA encoding human macrophage-specific colony-stimulating factor (CSF-1). Science 230:291, 1985.
Manger R, Najita L, Nichols EJ, et al. Cell surface expression of the McDonough strain of feline sarcoma virus (fms) gene product (gp140fms). Cell 39:327, 1984.
Nienhuis AW, Bunn HF, Turner PH, et al. Expression of the human c-fms proto-oncogene in hematopoietic cells and its deletion in the 5q-syndrome. Cell 42:421, 1985.
Rettenmier CW, Chen JH, Roussel MF, et al. The product of the c-fms proto-oncogene: a glycoprotein with associated tyrosine kinase activity. Science 228:320, 1985.
Roussel MF, Dull TJ, Rettenmeier CW, et al. Transforming potential of the c-fms proto-oncogene (CSF-1 receptor). Nature 325:549, 1987.

Sherr CJ, Rettenmier CW, Sacca R, et al. The c-fms proto-oncogene product is related to the receptor for the mononuclear phagocyte growth factor, CSF-1. Cell 41:665, 1985.

GM-CSF

Cantrell MW, Anderson D, Cerretti DP, et al. Cloning, sequence, and expression of a human granulocyte/macrophage colony-stimulating factor. Proc Natl Acad Sci USA 82:6250, 1985.

Metcalf D. The granulocyte-macrophage colony-stimulating factors. Science 229:16, 1985.

v-*ros*/Insulin Receptor

Kasuga M, Van Obberghen E, Nissley SP, et al. Demonstration of two subtypes of insulin-like growth factor receptors by affinity cross-linking. J Biol Chem 257:53, 1981.

Massagué J, Czech MP. The subunit structures of two distinct receptors for insulin-like growth factors I and II and their relationship to the insulin receptor. J Biol Chem 257:5028, 1982.

Petruzzelli L, Herrera R, Rosen OM. Insulin receptor is an insulin-dependent tyrosine protein kinase: copurification of insulin-binding activity and protein kinase activity to homogeneity from human placenta. Proc Natl Acad Sci USA 81:3327, 1984.

Ullrich A, Bell JR, Chen EY, et al. Human insulin receptor and its relationship to the tyrosine kinase family of oncogenes. Nature 313:756, 1985.

White MF, Maron R, Kahn CR. Insulin rapidly stimulates tyrosine phosphorylation of an M_r-185,000 protein in intact cells. Nature 318:183, 1985.

mos

Baldwin GS. Epidermal growth factor precursor is related to the translation product of the Moloney sarcoma virus oncogene mos. Proc Natl Acad Sci USA 82:1921, 1985.

TGF-α/β

Anzano MA, Roberts AB, Smith JM, et al. Sarcoma growth factor from conditioned medium of virally transformed cells is composed of both type α and type β transforming growth factors. Proc Natl Acad Sci USA 80:6264, 1983.

Assoian RK, Komoriya A, Meyers CA, et al. Transforming growth factor-β in human platelets: identification of a major storage site, purification and characterization. J Biol Chem 258:7155, 1983.

Bringman T, Lindguist PB, Derynck R. Different transforming growth factor-α species are derived from a glycosylated and palmitoylated transmembrane precursor. Cell 48:429, 1987.

Cheifetz S, Weatherbee J, Tsang ML-S, et al. The transforming growth factor β system; a complex pattern of cross reactive ligands and receptors. Cell 48:409, 1987.

Derynck R, Jarrett JA, Chen EY, et al. Human transforming growth factor-β cDNA sequence and expression in tumor cell lines. Nature 316:701, 1985.

Derynck R, Roberts AB, Winkler ME, et al. Human transforming growth factor-α: precursor structure and expression in E. coli. Cell 38:287, 1984.

Roberts AB, Anzano MA, Wakefield LM, et al. Type β-transforming growth factor: a bifunctional regulator of cellular growth. Proc Natl Acad Sci USA 82:119, 1985.

Shipley GD, Tucker RF, Moses HL. Type β-transforming growth factor/growth inhibitor stimulates entry of monolayer cultures of AKR-2B cells into S phase after a prolonged prereplicative interval. Proc Natl Acad Sci USA 82:4147, 1985.

Sporn MB, Roberts AB, Wakefield LM, et al. Transforming growth factor-β: biological function and chemical structure. Science 233:532, 1986.

Schreiber AB, Winkler ME, Derynck R. Transforming growth factor-α: a more potent angiogenic mediator than epidermal growth factor. Science 232:1250, 1986.

IL-2

Knabbe C, Lippman ME, Wakefield LM. Evidence that transforming growth factor-β is a hormonally regulated negative growth factor in human breast cancer cells. Cell 48:417, 1987.

Leonard WJ, Depper JM, Crabtree GR, et al. Molecular cloning and expression of cDNAs for the human interleukin-2 receptor. Nature 311:636, 1985.

Leonard WJ, Depper JM, Kanehisa M, et al. Structure of the human interleukin-2 receptor gene. Science 230:633, 1985.

Morgan DA, Ruscetti FW, Gallo R. Selective in vitro growth of T lymphocytes from normal human bone marrows. Science 193:1007, 1976.

Nikaido T, Shimizu A, Ishida N, et al. Molecular cloning of cDNA encoding human interleukin-2 receptor. Nature 311:631, 1984.

Smith KA. Interleukin 2. Annu Rev Immunol 2:319, 1984.

Smith KA, Cantrell DA. Interleukin-2 regulates its own receptors. Proc Natl Acad Sci USA 82:864, 1985.

Taniguchi T, Matsui H, Takashi F, et al. Structure and expression of a cloned cDNA for human interleukin-2. Nature 302:305, 1983.

IL-3

Ihle JN, Keller J, Henderson L, et al. Biological properties of homogeneous interleukin-3: demonstration of WEH1-3 growth factor activity, mast cell growth factor activity, P-cell stimulating activity, colony stimulating factor activity, and histamine producing cell stimulating factor activity. J Immunol 131:282, 1983.

Other Growth Factors

Bauknecht T, Kiechle M, Bauer G, et al. Characterization of growth factors in human ovarian carcinomas. Cancer Res 46:2614, 1986.

Cuttita F, Carney DN, Mulshine J, et al. Bombesin-like peptides can function as autocrine growth factors in human small cell lung cancer. Nature 316:823, 1985.

Huff KK, Kaufmann D, Gabby HH, et al. Secretion of an insulin-like growth factor-I-related protein by human breast cancer cells. Cancer Res 46:4613, 1986.

Klagsbrun M, Sasse J, Sullivan R, et al. Human tumor cells synthesize an endothelial cell growth factor that is structurally related to basic fibroblast growth factor. Proc Natl Acad Sci USA 83:2448, 1986.

Liotta LA, Mandler R, Murano G, et al. Tumor cell autocrine motility factor. Proc Natl Acad Sci USA 83:3302, 1986.

9
ras Family of Oncogenes

Overview

The ras genes are of three varieties, each with a similar exon–intron structure and each encoding a protein of 21 kilodaltons (kD): p21. The Harvey *ras* oncogene was the first to be directly implicated in human neoplasia. An H-*ras* gene from a bladder carcinoma cell line was isolated and characterized by its ability to transform the murine cell line NIH 3T3.

Single-base-pair differences at specific codons distinguish normal proto-*ras* genes from their oncogenic counterparts. These changes translate to single amino acid substitutions in the mutant p21 products. In addition, elevated levels of normal p21, corresponding in some cases to amplification of the *ras* gene, have been associated with neoplasia.

The p21 protein appears to be similar to the G proteins that act as second messengers, possibly in the phosphatidyl inositol (PI) system, although the latter has not yet been conclusively proved. Normal p21 has GTPase activity that converts the protein from an active to an inactive form. Mutant p21s may have reduced or ineffectual GTPase activity leading to constitutive activation of the protein. Activated protein may transduce a signal(s) important for cell growth; thus constitutive activation may lead to misregulated growth.

The *ras* genes are widely conserved among animal species and have been found in yeast as well. Yeast *ras* genes bear some homology to vertebrate *ras* genes and have been shown to be of importance to yeast growth and cell cycle.

Mutant *ras* genes and proteins can be experimentally distinguished from their normal counterparts by specific DNA probes and antibodies. These and similar reagents may eventually prove to be of importance in cancer diagnostics and therapeutics.

The *ras* genes originally were isolated from Harvey and Kirsten murine sarcoma viruses. The *ras* sequences in these viruses were derived from rat DNA that had been transduced by the murine retroviruses. Subsequently, sequences corresponding to both Harvey (H-) and Kirsten (K-) *ras* genes were detected in human, avian, murine, and nonvertebrate genomes. A third member of the *ras* family, N-*ras*, was detected by its expression in human neuroblastoma and sarcoma cell lines.

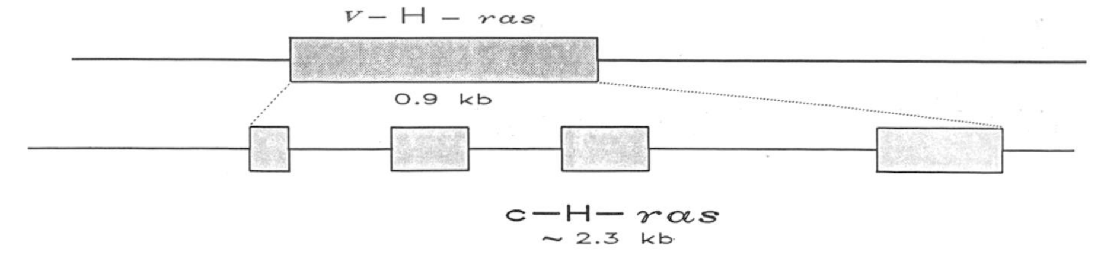

FIGURE 9.1. Structure of Harvey-*ras* gene.

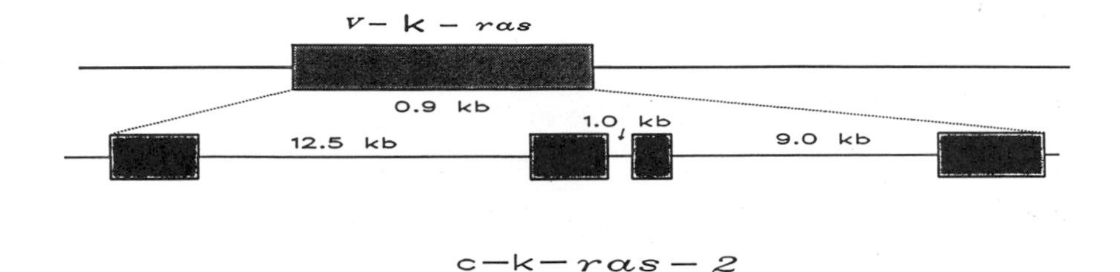

FIGURE 9.2. Structure of Kirsten-*ras* gene.

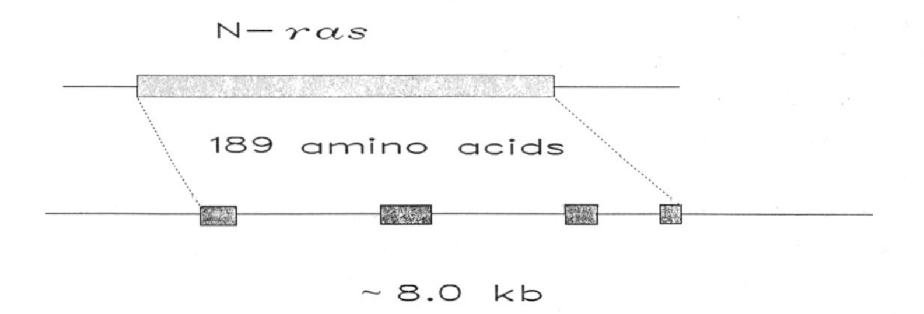

FIGURE 9.3. Structure of N-*ras* gene.

All vertebrate *ras* genes that have been molecularly analyzed have a similar structure (Figs. 9.1, 9.2, and 9.3). There are four exons spanning up to 30 kilobases (kb) of DNA. The K-*ras*-2 gene differs slightly in having two alternative fourth coding exons (IVA and IVB). The fact that *ras* sequences are expressed ubiquitously suggests an important physiological function for members of this proto-oncogene family. The protein product of each vertebrate *ras* gene is approximately 21 kD in size, containing 188 or 189 amino acids. The activity of p21 proteins is described in some detail below.

Isolation of *ras* Transforming Genes

The *ras* oncogenes (Tables 6.1, 10.1) were the first oncogenes to be implicated in human cancer and are the best characterized oncogenes at the genetic and structural level, although their function remains obscure. Their presence in DNA from human tumors was discovered by a series of experiments conducted in several laboratories, including those of M. Barbacid, G. Cooper, R. Weinberg, and M. Wigler. The crux of their experiments was that DNA from human tumor-derived cell lines could transform mouse NIH 3T3 cells in vitro, whereas normal human DNA did not affect these recipient cells.

The NIH 3T3 transfection assay has been described (see Fig. 2.16 and 6.1). Cells from foci transfected by human tumor DNA were found to contain human DNA, proving that the human DNA was somehow involved in the transformation process. In addition, the presence of human DNA provided an assay for isolation of the effector gene using the *alu* probe specific for human DNA.

Researchers isolated DNA from the original transformants and used it to transfect new NIH 3T3 cells. In this secondary transfection experiment, foci of transformation appeared at a higher frequency than previously. Analysis of the transformed cells using the *alu* probe showed sequences equivalent to 0.1% of the human genome. The secondary DNA was used to transfect a third set of NIH 3T3 cells, again enriching for the transforming sequences. At this point, the human gene responsible was identified by Southern blotting as a member of the Harvey-*ras* family of oncogenes.

Characterization of *ras* Transforming Genes

Because normal human DNA contains H-*ras* genes but does not transform NIH 3T3 cells, it was clear that the *ras* oncogene from human tumor DNA must differ in some way from its proto-oncogene counterpart so as to activate its transforming potential. It might be thought that merely sequencing the two genes would reveal the crucial difference. However, the genes contain approximately 2000 base pairs; and because there was no normal tissue corresponding to the tumor cell line DNA, sequences of the two genes might be expected to contain several

TABLE 9.1. The *ras* family of oncogenes.

ras oncogenes
H-*ras*
K-*ras*
N-*ras*
ras-related oncogenes
ral
R-*ras*
mel

	Exons	Transforming Capacity
Normal ras gene	□ □ □ □	-
Transforming ras gene	■ ■ ■ ■	+
Plasmid 1	□ □ ■ ■	-
Plasmid 2	■ ■ □ □	+
Plasmid 3	□ ■ ■ ■	-
Plasmid 4	■ □ □ □	+

THE DEDUCTION FROM THIS EXPERIMENT IS THAT A MUTATION HAD OCCURRED IN EXON 1, WHICH CONVERTED THE RAS GENE INTO A TRANSFORMING GENE.

1. RESTRICTION SITES LOCATED BETWEEN THE EXONS ARE USED TO CONSTRUCT THESE CHIMAERIC GENES.
2. OPEN BOXES REPRESENT EXONS FROM THE NORMAL RAS ALLELE, BLACK BOXES REPRESENT EXONS FROM THE TRANSFORMING RAS ALLELE.

FIGURE 9.4. Principle of mix and match experiments.

nucleotide differences as a result of random allelic variation. To get around this problem, the functional differences between the normal and transforming genes were exploited by "mix and match" experiments (Fig. 9.4). It was found that the activated region occurred in the first exon of the gene. Sequence analysis of this smaller region, to which the transforming mutation mapped, revealed a single base pair difference resulting in a predicted substitution of a valine for the normal glycine at the twelfth amino acid (Table 6.1).

The original experiments were performed with human bladder carcinoma EJ/T24 cells. Similar mutations were quickly found in other activated *ras* genes. Normal rat *ras* proteins also contain glycine at amino acid 12, whereas Harvey sarcoma virus *ras* encoded arginine and Kirsten sarcoma virus *ras* encoded serine at the twelfth position. A *ras* transforming gene isolated from the Calu-1 human lung carcinoma encoded cysteine at position 12. These and other results are summarized in Table 6.2. It is now clear that amino acid 12 is not the only position mutated in activated *ras* proteins. Mutations at position 61 and, more recently, at position 13 have been detected in human tumors or cell lines derived from these tumors. Furthermore, mutations are found not only in the H-*ras* gene but in K-*ras* and N-*ras* genes as well. Experiments suggest that activated *ras* genes may be present in at least 15 to 20% of all human tumors.

The experiments to isolate and characterize *ras* transforming genes are in many ways landmarks of molecular oncology. For the first time a functional assay was used to detect a difference between DNA from normal and cancer cells. Even more remarkable was the fact that these DNAs differed by a single base pair change. A major question raised by these results is the following: If a single base change is sufficient to activate a *ras* gene, why is the frequency of tumor initiation much lower than the expected mutation rate? It might be argued that because tumor cells are known to have higher mutation rates than normal the mutations are an epiphenomenon of the tumor rather than its cause. On the other hand, it might be argued that such *ras* mutations contribute to tumor progression rather than initiation of malignancy (see Chapter 6). In addition, it has been noted that *ras* genes alone are unable to transform primary cells in culture (see Chapter 5) and that at least one additional gene expressing "immortalization function" (e.g., *myc*, adenovirus ElA or polyoma large T antigen) is required.

ras Genes and Cancer

The most frequently identified transformation-associated genes found in human solid tumors are members of the *ras* family. Single-base substitutions represent one means of gene activation. Mutations of this type appear to code for an altered p21 product with increased transforming ability (see below). Approximately 15% of solid tumors examined contain single-base alterations in one or more *ras* genes.

Quantitative differences in normal *ras* gene expression also have been found to be associated with transformation: When cloned normal *ras* genes transcribed from strong promoters were introduced into NIH 3T3 cells, analysis of the small number of resulting transformants showed them to express high levels of p21 protein. Researchers in Barbacid's group analyzed many DNA samples from human tumors and demonstrated amplification of at least one *ras* gene in some of them.

Elevated levels of *ras*-specific mRNA and p21 have been found in bladder, colon, and prostate tumors. Benign lesions, on the other hand, were consistently negative for p21. In certain cases the degree of p21 expression has been correlated with histological tumor grade. For example, in one study of prostate cancer, the level of p21 provided a better correlation with tumor grade than classical tumor markers such as carcinoembryonic antigen (CEA) or prostatic antigen.

Activated *ras* genes have been detected in premalignant lesions as well, suggesting a potential role in tumor initiation. For example, benign papillomas induced by carcinogens or tumor promoters in experimental animals demonstrate a high frequency of c-H-*ras* activation. Similar papillomas have been induced in mice by infection with v-H-*ras* followed by application of phorbol esters, suggesting that the presence of an activated *ras* gene alone is sufficient to initiate neoplasia.

ras Product p21

The biochemical role of *ras* genes in normal and tumor cells remains unknown at the present time. However, several biochemical properties of the protein products are known and give some clues as to their functions. First, p21 protein extracted from mammalian cells is covalently bound to the fatty acid palmitic acid. This binding probably serves to anchor the protein in the inner surface of the plasma membrane. Second, Scolnick and co-workers discovered that p21 is a guanine nucleotide binding protein capable of binding GTP or GDP with high affinity but incapable of binding other nucleotides, e.g., GMP or ATP (see Gibbs et al. 1984b). Further biochemical analysis was facilitated by the availability of large quantities of normal p21 through expression of the human protein in bacterial cells. This p21 was discovered to have GTPase activity, catalyzing the slow hydrolysis of GTP to GDP, which remains bound to the p21 protein. This result confirmed an earlier suggestion by Scolnick and co-workers that p21 proteins are related to the G protein family (see below). Furthermore, it was found that an oncogenic mutant of p21 (the valine-12 mutant of human H-*ras*) was deficient in GTPase activity. It was the first demonstration of a biochemical difference between normal and mutant p21, and it has led to an interesting model that goes some way to explaining how mutations activate the protein (Fig. 9.5).

The model was developed by analogy with the G proteins that act as second messengers in transmitting signals from activated receptors to regulatory enzymes (see Chapter 8). For example, stimulation of the beta-adrenergic receptor leads to binding of GTP to the alpha subunit. This subunit (the *ras* analogue) then dissociates from other components of the receptor and activates adenylate cyclase, the enzyme catalyzing cAMP synthesis. Because the G protein has endogenous GTPase activity, it converts bound GTP to GDP; the GDP form is inactive and returns to the receptor to await a new round of stimulation.

Evidence suggests that $p21^{N\text{-}ras}$ might be the G protein that transduces growth factor receptor signals to the phosphatidyl inositol second messenger cascade. Although this work needs to be substantiated, it might be imagined how mutations of p21 that reduce GTPase activity could bring about transformation. Because the mutant p21 converts bound GTP to GDP more slowly, it might remain in the active state longer and signal the cell to undergo inappropriate cell proliferation. This model also accounts for the ability of normal p21 to transform cells when expressed at high levels: These levels simply provide more of the GTP-bound forms of p21. If only the GTP-bound form were active, a critical level of this form might be achieved and again cause inappropriate cell growth.

Comparison of the primary sequence of p21 with a GDP/GTP binding protein, whose three-dimensional structure is partly known (elongation factor EF-Tu) (Fig. 9.6), has led to a model for the tertiary structure of p21 (Fig. 9.5). This model predicts that p21 adopts different conformations according to whether GDP or GTP is bound. this conformational change probably results in differential binding of an unidentified cellular component. The p21 protein then can be seen as a kind of biochemical switch in which the "on" position may be the GTP-bound form, and the "off" position is the GDP-bound form. Turning the protein "on" is accomplished when an external signal is received at the cell surface. The switch turns itself "off" through its intrinsic GTPase activity, which converts bound GTP to GDP. The oncogenic form may be thought of as a defective switch in which the "off" function is impaired.

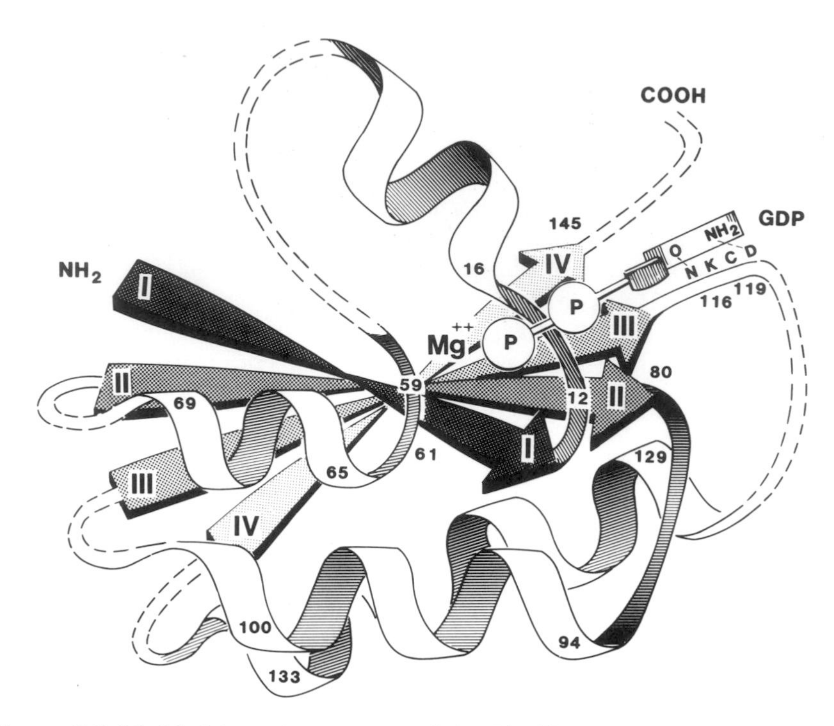

FIGURE 9.5. Model of the tertiary structure of p21. Significant amino acids and β-strands are numbered, the latter with Roman numerals. Broken lines represent regions with limited homology to EF-Tu. Loops I to IV comprise the guanine nucleotide binding site. The phosphoryl binding region is made up of sequences from 10 to 16 and from 57 to 63. The latter sequence may contain a site for magnesium binding. Amino acids defining guanine specificity are asparagine 116 and aspartic acid 119 as well as sequences around amino acid 145. Mutations at position 12 that activate the transforming potential of p21 change the configuration of the loop, which would affect the interaction of the loop with bound nucleotide or with magnesium and water molecules in the nucleotide binding site. Activating mutations at positions 59 and 61 would affect GTP hydrolysis. (Kindly provided by Frank McCormick)

Evidence is accumulating that some amino acid substitutions that activate p21 to the oncogenic form may work by causing direct conformational changes in the protein, such that it is constitutively in an activated form. With these mutants GTPase activity is not substantially reduced from wild-type levels, but this activity is not effective in turning off the protein's function (Fig. 9.7).

Scientists at Genentech and Merck have used oligonucleotide-directed mutagenesis to alter human *ras* genes at various positions. Substitutions at residues 16, 116, or 119 of the H-*ras* protein significantly reduced the ability of both normal and activated forms of p21 to bind and hydrolyze GTP. However, the mutations did not diminish the ability of the activated proteins to transform Rat-1

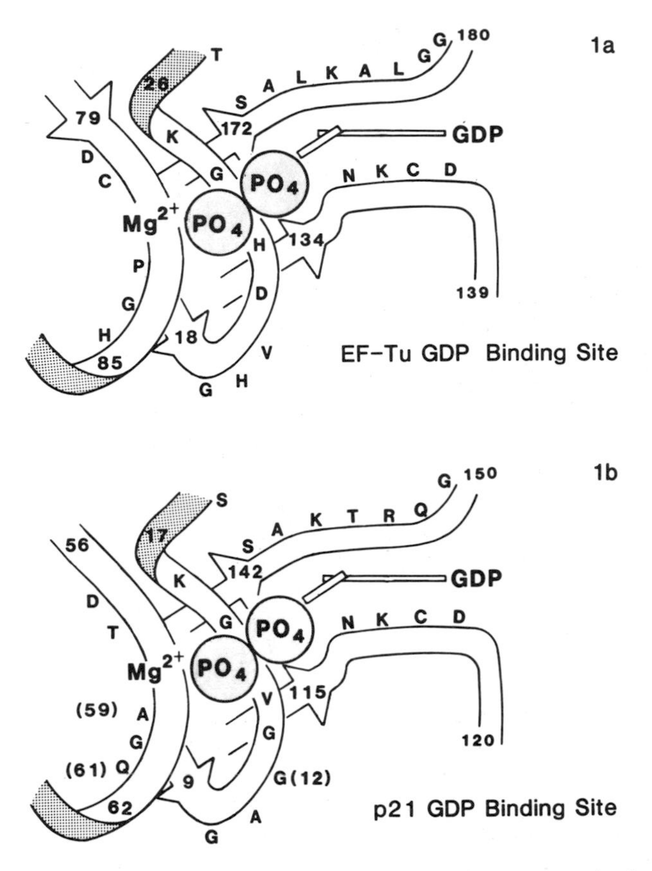

FIGURE 9.6. EF-Tu and p21 GDP binding sites. Amino acid sequences in the loops proposed to constitute the GDP binding sites are indicated by single letter codes: A = alanine. C = cysteine. D = aspartate. E = glutamate. G = glycine. H = histidine. K = lysine. L = leucine. N = asparagine. P = proline. Q = glutamine. R = arginine. S = serine. T = threonine. V = valine. (Kindly provided by Frank McCormick)

cells. Thus although p21 is hypothesized to interact with cellular components as part of a signal-transducing system, the proposed models are stiil incomplete.

ras in Yeast

The yeast *Saccharomyces cerevisiae* contains two *ras*-related genes: *RAS* 1 and *RAS* 2. (By convention, wild-type yeast genes are designated by capital letters.) These genes code for proteins of 309 and 322 residues. They are about 90% identical to the human p21 protein in their first 90 amino terminal amino acids and 60% identical over the next 80 amino acids. The carboxy terminal 150

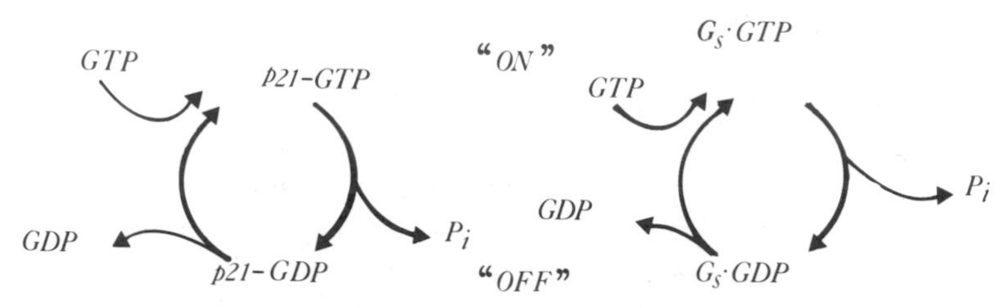

FIGURE 9.7. Model of p21 as a regulatory protein analogous to the G proteins. By analogy to the G alpha subunit (G_S) that modulates the function of adenyl cyclase, there are two states: "on" and "off." The binding of β-adrenergic agonists (for G_S) or a growth factor (e.g., bombesin) to its receptor (for p21) promotes the exchange of GTP for bound GDP. The G_S·GTP complex stimulates adenyl cyclase until the GTP is hydrolyzed. Continued occupation and stimulation of the receptor by its ligand promotes the continuous hydrolysis of GTP. By analogy, p21 is hypothesized to interact with a cellular target, possibly the inositol phosphate cycle signal-transducing system. Mutations that inhibit the GTP hydrolytic activity would keep the protein in a constant "on" state. Conversely, mutations that affected the capacity of p21 to stimulate its cellular target would also produce an *onc*-protein. (Kindly provided by Frank McCormick)

amino acids are dissimilar to vertebrate *ras* genes and to each other. Either *RAS* 1 or *RAS* 2 is necessary for yeast cell viability, but disruption of both is lethal. Wigler and co-workers have shown that yeast cells are still viable if a human H-*ras* gene replaces *RAS* 1 and *RAS* 2 (see e.g. Kataoka et al. 1985). The conserved amino terminal region of yeast *RAS* 1 binds guanine nucleotides and has GTP hydrolytic activity similar to that of mammalian p21.

What happens to yeast when mutations equivalent to the mutations of human p21 proteins are introduced into yeast *ras* genes? When glycine at position 19 is replaced by valine in *RAS* 2 (equivalent to changes at position 12 of mammalian p21) the resulting mutant strain had elevated adenyl cyclase activity producing high levels of cAMP with resulting increases in cAMP-dependent protein phosphorylation. Cyclic AMP has been implicated in growth and cell cycle control in some mammalian cells, and it is possible that cAMP-dependent protein phosphorylations may be involved in the transformation of mammalian cells. However, it does not appear that human *ras* genes are directly involved in this pathway because no consistent perturbation of cAMP metabolism has been detected in mammalian *ras*-transformed cells.

Interestingly, when exposed to conditions where nutrients were limited, mutant *RAS* 2-containing yeast cells exited the cell cycle without accumulating carbohydrate reserves, and the cells eventually failed to sporulate and died. Although no direct connection can be made between altered protein phosphorylation and cell cycle control in yeast and mammalian cells, the yeast *ras* studies show that oncogenes do have important conserved roles in eukaryotic cellular growth control.

Mutant p21 as a Tumor Marker

Activated forms of p21 appear to contribute directly to the development of cancer. These forms have not been detected in normal tissue. They therefore can be considered to be highly specific cancer markers that may someday be useful for detecting or monitoring neoplasia (see Chapter 13). Antibodies that can distinguish some oncogenic forms of p21 from the normal protein are already available and are being evaluated for their diagnostic potential. DNA probes capable of detecting mutant *ras* genes also are being evaluated.

Inactivation of Mutant p21: Reversion of Transformed Cells to a Normal Phenotype

Many researchers believe that when the mechanism of *ras* protein function and the way in which oncogenic p21 differs from normal p21 is understood we will be in a position to devise new therapeutic agents that inactivate the *onc*-protein specifically. As a demonstration of this potential, Feramisco and co-workers microinjected an antibody capable of binding to oncogenic p21 into living cells transformed by the *ras* oncogene. The result (Fig. 9.8) was reversion of the transformed cells to a normal phenotype. Because this antibody does not bind normal p21, the revertant cells continued to grow normally. Eventually, these cells returned to their transformed state as the microinjected antibody was degraded. This experiment demonstrated that an agent capable of specifically inactivating an oncogene product is able to reverse the transformed morphology and offers hope that drugs with these properties eventually can be developed.

Other *ras* Genes

The *ral* Gene

Using a 20 bp oligonucleotide probe, Chardin and Taritian isolated a *ras*-related gene termed *ral* from a cDNA library derived from a simian B-lymphocyte cell line. The probe corresponded to amino-acids 57-63 of the K-*ras* protein, an area highly conserved among all *ras* genes. A *ral* probe prepared from the cloned gene hybridized weakly under low stringency conditions to human K-*ras* and N-*ras* DNA and also to yeast *RAS*-1 and *RAS*-2 DNA. Hybridization was stronger to H-*ras* DNA.

The *ral* gene is expressed in murine cell lines (transcript sizes 2.8 kb and 1.3 kb). The protein predicted from the *ral* gene sequence comprises 206 amino acids (M^r = 23.5 kD) with 52% sequence identity with human H-*ras*, K-*ras*, or N-*ras*. The *ral* protein has an additional 11 N-terminal amino acids and an additional 6 C-terminal amino acids. Regions in the GTP binding domain are highly conserved between *ral* and the previously described *ras* genes as is a C-terminal

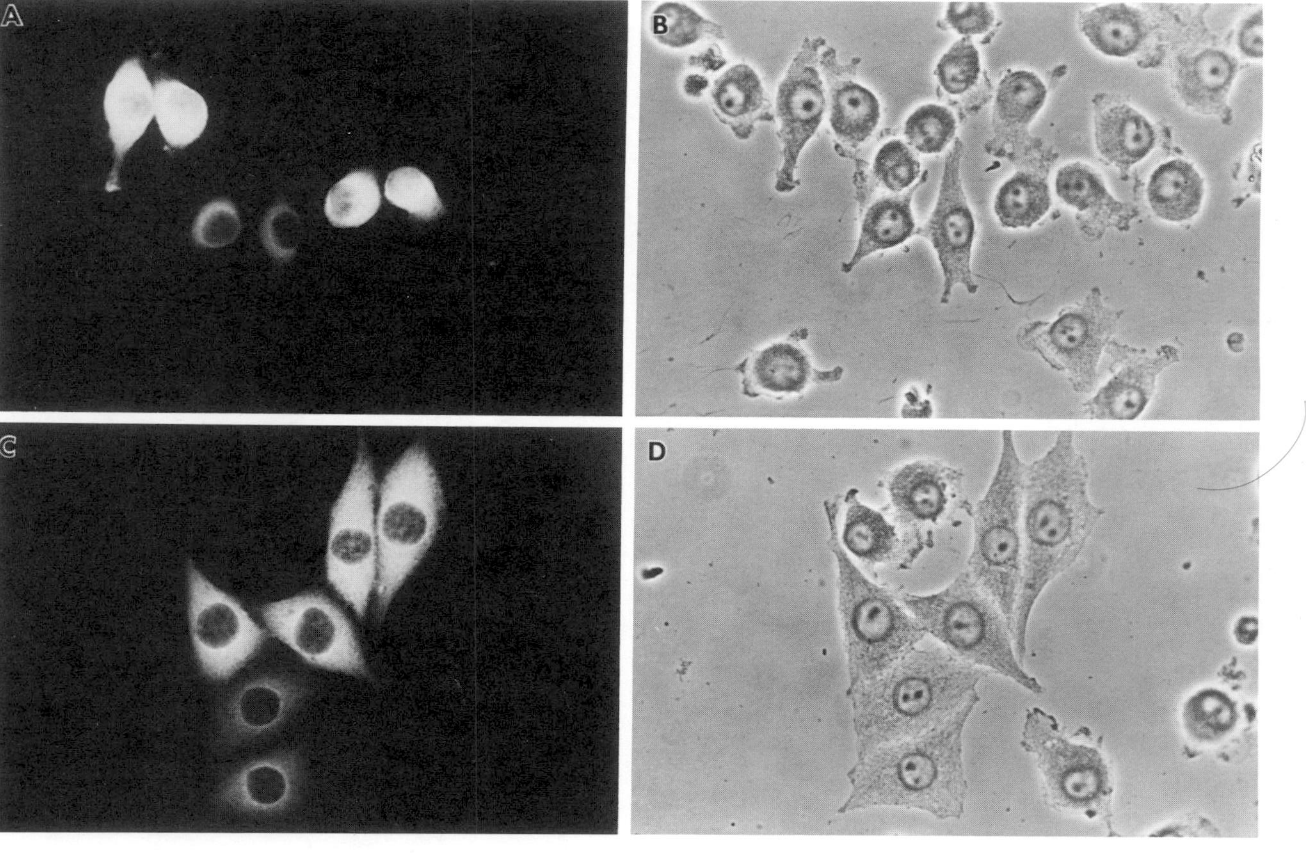

FIGURE 9.8. Anti-p21 antibodies cause reversion of Kirsten-*ras* transformed cells. Correspondence of the flattened (normal) appearance of K-NRK cells with microinjection of anti-p21-Ser antibodies. K-NRK cells (growing on glass coverslips) were injected with nonimmune immunoglobulin G (IgG) (a,b) or anti-p21-Ser (c,d) and fixed 24 hours after injection. The cells were treated with fluorescein isothiocyanate-labeled goat anti-rabbit IgG, which stained only the injected cells. The field of cells were photographed by epifluorescence (A,B) and phase-contrast (B,D) microscopy. (A,B) The same field of cells injected with control IgG. (C,D) The same field of cells injected with anti-p21-Ser. (Kindly provided by Frank McCormick.)

sequence required for posttranslational lipid binding and membrane anchoring. This similarity suggests the *ral* product is a GTP binding protein with membrane localization similar to the better characterized 21-kD proteins of the H-*ras*, K-*ras*, and N-*ras* genes.

THE R-*ras* GENE

A human *ras*-related gene was isolated from genomic DNA by hybridization at low stringency with a v-H-*ras* probe. The cloned and sequenced gene spanned approximately 6.7 kb; it comprised 6 exons in contrast to the four-exon structure observed for the H-*ras*, K-*ras*, and N-*ras* genes. The gene was localized to human chromosome 19. A similar murine R-*ras* gene was mapped to murine chromosome 7.

Expression of the R-*ras* gene is widely variable in human cell lines: A 1-kb mRNA transcript was observed at highest concentration in diploid foreskin fibroblasts and fibrosarcoma cells whereas moderate to low levels of expression were observed in prostate and renal adenocarcinoma cells. Low to undetectable levels were observed in a papillary adenocarcinoma, a mammary carcinoma, and diploid lung fibroblasts.

The predicted amino acid sequence of the R-*ras* product comprises 218 residues with an extra 26 N-terminal amino acids compared to H-*ras* p21. The two proteins share 55% amino acid identity including conservation of the GTP binding site and residues at positions comparable to H-*ras* p21: 12, 13, 59, 61, and 63, i.e. those residues implicated in activation of H-*ras* proto oncogenes. Therefore R-*ras* product is predicted to be a GTP-binding and GTPase protein similar to H-*ras* p21.

The relationship of R-*ras* to *ral* is unknown.

THE *mel* GENE

A transforming gene with weak homology to the H-*ras* gene was isolated by transfection of high-molecular-weight DNA derived from a melanoma cell line into NIH 3T3 cells. Designated *mel*, it could be detected in three of eight melanoma cell lines. Sequences homologous to a *mel* probe were detected in human, dog, mouse, and hamster DNA suggesting a conserved, ubiquitous gene. The human *mel* gene has been mapped to chromosome 19 p13.2–q13.2, a locus not corresponding to any previously described oncogene. The relationship of *mel* to R-*ras* which was also isolated with a H-*ras* probe at low stringency and also maps to chromosome 19 is currently unknown.

BIBLIOGRAPHY

ras: Structure and Function

Campisi J, Gray HE, Parder AB, et al. Cell-cycle control of c-myc but not c-ras expression is lost following chemical transformation. Cell 36:241, 1984.

Capon DJ, Seeburg PH, McGrath JP, et al. Activation of K-ras-2 gene in human colon and lung carcinomas by two different point mutations. Nature 304:507, 1983.

Chang EH, Furth ME, Scolnick EM, et al. Tumorigenic transformation of mammalian cells induced by a normal human gene homologous to the oncogene of Harvey murine sarcoma virus. Nature 297:429, 1982.

Chang EH, Gonda MA, Ellis RW, et al. Human genome contains four genes homologous to transforming genes of Harvey and Kirsten murine sarcoma viruses. Proc Natl Acad Sci USA 79:4848, 1982.

Davis M, Malcolm S, Hall A, et al. Localization of the human N-ras oncogene to chromosome lcen-p21 by in situ hybridization. EMBO J 2:2281, 1983.

Feramisco JR, Gross M, Kamata T, et al. Microinjection of the oncogene form of the human H-ras [T-24] protein results in rapid proliferation of quiescent cells. Cell 38:109, 1984.

Gibbs JB, Ellis RW, Scolnick EM. Autophosphorylation of v-H-ras p21 is modulated by amino acid residue 12. Proc Natl Acad Sci USA 81:2674, 1984a.

Gibbs JB, Sigal IS, Poe M, et al. Intrinsic GTPase activity distinguishes normal and oncogenic ras p21 molecules. Proc Natl Acad Sci USA 81:5704, 1984b.

Gilman AG. G proteins and dual control of adenylate cyclase. Cell 36:577, 1984.

Jurnak F. Structure of the GDP domain of E-Tu and location of the amino acids homologous to ras oncogene proteins. Science 230:32, 1985.

Kamata T, Feramisco JR. Epidermal growth factor stimulates guanine nucleotide binding activity and phosphorylation of ras oncogene proteins. Nature 310:147, 1984.

Lochrie MA, Hurley JB, Simon MI. Sequence of the alpha subunit of photoreceptor G protein: homologies between transducin, ras, and elongation factors. Science 228:96, 1985.

Manne V, Bekesi E, Kung H-F. H-ras proteins exhibit GTPase activity: point mutations that activate H-ras gene products result in decreased GTPase activity. Proc Natl Acad Sci USA 82:376, 1985.

McCormick F, Clark BFC, la Cour TFM, et al. A model for the tertiary structure of p21, the product of the ras oncogene. Science 230:78, 1985.

McGrath JP, Capon DJ, Goeddel DV, et al. Comparative biochemical properties of normal and activated human ras p21 protein. Nature 310:644, 1984.

Mulcahy LS, Smith MR, Stacey DW. Requirement for ras proto-oncogene function during serum-stimulated growth of NIH 3T3 cells. Nature 313:241, 1985.

Newbold RF, Overell RW. Fibroblast immortality is a prerequisite for transformation by EJ c-H-ras oncogene. Nature 304:648, 1983.

Pincus MR, van Renswoude J, Harford JB, et al. Prediction of the three-dimensional structure of the transforming region of the EJ/T24 human bladder oncogene product and its normal cellular homologue. Proc Natl Acad Sci USA 80:5253, 1983.

Reddy EP, Reynolds RK, Santos E, et al. A point mutation is responsible for the acquisition of transforming properties by the T24 human bladder carcinoma oncogene. Nature 300:149, 1982.

Seeburg PH, Colby WW, Capon DJ, et al. Biological properties of human c-H-ras-1 genes mutated at codon 12. Nature 312:71, 1984.

Shih TY, Stokes PE, Smythers GW, et al. Characterization of the phosphorylation sites and the surrounding amino acid sequences of the p21 transforming proteins coded for by the Harvey and Kirsten strains of murine sarcoma viruses. J Biol Chem 257:11767, 1982.

Shimizu K, Goldfarb M, Suard Y, et al. Three human transforming genes are related to the viral ras oncogenes. Proc Natl Acad Sci USA 80:2112, 1983.

Sigal IS, Gibbs JB, D'Alonzo J, et al. Mutant ras-encoded proteins with altered nucleotide binding exert domination biological effects. Proc Natl Acad Sci USA 83:952, 1986.
Stacy DW, Kung H-F. Transformation of NIH 3T3 cells by microinjection of H-ras p21 protein. Nature 310:508, 1984.
Sukumar S, Pulciani S, Doniger J, et al. A transforming ras gene in tumorigenic guinea pig cell lines initiated by diverse chemical carcinogens. Science 223:1197, 1984.
Sweet RW, Yokoyama S, Kamata T, et al. The product of ras is a GTPase and the T24 oncogenic mutant is deficient in this activity. Nature 311:273, 1984.
Tabin CJ, Bradley SM, Bargmann CI, et al. Mechanism of activation of a human oncogene. Nature 300:143, 1982.
Taparowsky E, Suard Y, Fasano O, et al. Activation of T24 bladder carcinoma transforming gene is linked to a single amino acid change. Nature 300:762, 1982.
Walselam MJO, Davies SA, Houslay MD, et al. Normal $p21^{N\text{-}ras}$ couples bombesin and other factor receptors to inositol phosphate production. Nature 323:173, 1986.
Walter M, Clark SG, Levinson AD. The oncogenic activation of human $p21^{ras}$ by a novel mechanism. Science 233:649, 1986.
Zarbl H, Sukumar S, Arthur AV, et al. Direct mutagenesis of H-ras-1 oncogenes by N-nitroso-N-methylurea during initiation of mammary carcinogenesis in rats. Nature 315:382, 1985.

ras: N-*ras*

Brown R, Marshall CJ, Pennie SG, et al. Mechanism of activation of an N-ras gene in the human fibrosarcoma cell line HT1080. EMBO J 3:1321, 1984.
Taparowsky E, Shimizu K, Goldfarb M, et al. Structure and activation of the human N-ras gene. Cell 34:581, 1983.
Yuasa Y, Gol RA, Chang A, et al. Mechanism of activation of N-ras oncogene of SW-1271 human lung carcinoma cells. Proc Natl Acad Sci USA 81:3670, 1984.

ras: In Yeast

Birchmeier C, Broek D, Wigler M. RAS proteins can induce meiosis in Xenopus oocytes. Cell 43:615, 1985.
Broek D, Samiy N, Fasano, et al. Differential activation of yeast adenylate cyclase by wild type and mutant RAS proteins. Cell 41:763, 1985.
Broek D, Toda T, Michael T. The S. cerevisiae CDC25 gene product regulates the RAS/Adenylate cyclase pathway. Cell 48:789, 1987.
De Feo-Jones D, Scolnick EM, Koller R, et al. *ras*-related gene sequences identified and isolated from Saccharomyces cerevisiae. Nature 306:707, 1983.
Kataoka T, Powers S, McGill C, et al. Genetic analysis of yeast RASl and RAS2 genes. Cell 37:437, 1984.
Kataoka T, Powers S, Cameron S, et al. Functional homology of mammalian and yeast RAS genes. Cell 40:19, 1985.
Powers S, Kataoka T, Fasano O, et al. Genes in S. cerevisiae encoding proteins with domains homologous to the mammalian ras proteins. Cell 36:607, 1984.
Tatchell K, Chalriff DT, DeFeo-Jones D, et al. Requirement of either of a pair of ras-related genes of Saccharomyces cerevisiae for spore viability. Nature 309:523, 1984.
Toda T, Uno I, Ishikawa T, et al. In yeast, ras proteins are controlling elements of adenylate cyclase. Cell 40:27, 1985.

ras: Clinical Relevance

Albino AP, Le Strange R, Oliff AI, et al. Transforming ras genes from human melanoma: a manifestation of tumor heterogeneity. Nature 308:69, 1984.

Balmain A, Pragnell IB. Mouse skin carcinomas induced in vivo by chemical carcinogens have a transforming Harvey-ras oncogene. Nature 303:72, 1983.

Balmain A, Ramsden M, Bowden GT, et al. Activation of the mouse cellular Harvey-ras gene in chemically induced benign skin papillomas. Nature 307:658, 1984.

Der CJ, Cooper GM. Altered gene products are associated with activation of cellular ras genes in human lung and colon carcinomas. Cell 32:201, 1983.

Eva A, Tronick SR, Gol RA, et al. Transforming genes of human hematopoietic tumors: frequent detection of ras-related oncogenes whose activation appears to be independent of tumor phenotype. Proc Natl Acad Sci USA 80:4026, 1983.

Feig LA, Bast RC Jr, Knapp RC, et al. Somatic activation of ras gene in a human ovarian carcinoma. Science 223:698, 1984.

Feinberg AP, Vogelstein B, Droller MJ, et al. Mutation affecting the 12th amino acid of the c-H-ras oncogene product occurs infrequently in human cancer. Science 220:1175, 1983.

Fujita J, Srivastana S, Kraus M, et al. Frequency of molecular alterations affecting ras proto-oncogenes in human urinary tract tumors. Proc Natl Acad Sci USA 82:3849, 1985.

Hand PH, Thor A, Wunderlich D, et al. Monoclonal antibodies of predefined specificity detect activated ras gene expression in human mammary and colon carcinomas. Proc Natl Acad Sci USA 81:5227, 1984.

Jhanwar SC, Neel BG, Hayward WS, et al. Localization of c-ras oncogene family on human germ line chromosomes. Proc Natl Acad Sci USA 80:4704, 1983.

Kraus MH, Yuasa Y, Aaronson SA. A position 12-activated H-ras oncogene in all HS578Y mammary carcinosarcoma cells but not normal mammary cells of the same patient. Proc Natl Acad Sci USA 81:5384, 1984.

McBride OW, Swann DC, Santos E, et al. Localization of the normal allele of T24 human bladder carcinoma oncogene to chromosome 11. Nature 300:773, 1982.

O'Brien SJ, Nash WG, Goodwin JL, et al. Dispersion of the ras family of transforming genes to four different chromosomes in man. Nature 301:839, 1983.

Pulciani S, Santos E, Lauver AV, et al. Oncogenes in solid human tumors. Nature 300:539, 1982.

Santos E, Martin-Zanca D, Reddy EP, et al. Malignant activation of a K-ras oncogene in lung carcinoma but not in normal tissue of the same patient. Science 223:661, 1984.

Viola MV, Fromowitz F, Oravez S, et al. ras oncogene p21 expression is increased in premalignant lesions and high-grade bladder carcinoma. J Exp Med 161:1213, 1985.

Viola MV, Fromowitz F, Oravez S, et al. Expression of ras oncogene p21 in prostate cancer. N Engl J Med 314:133, 1986.

Weinberg R. ras-oncogenes and the molecular mechanisms of carcinogenesis. Blood 64:1143, 1984.

ras-Related Genes

Chardin P, Taritian A. The ral gene: a new ras related gene isolated by use of a synthetic probe. EMBO J 5:2203, 1986.

Lowe DC, Capon DJ, Delwart E, et al. Structure of the human and murine R-ras genes, novel genes closely related to ras proto-oncogenes. Cell 48:137, 1987.

The *mel* gene

Padua R, Barrass N, Currie G. A novel transforming gene in a human malignant melanoma cell line. Nature 311:671, 1984.

Spurr N, Hughes D, Goodfellow P, et al. Chromosomal assignment of c-mel, a human transforming oncogene to chromosome 19p13.2–q13.2. Som Cell Mol Genet 12:637, 1986.

10
myc and Other Nuclear Oncogenes

Overview

The nuclear oncogenes include the *myc* family—v-*myc*; c-*myc*; N-*myc*; L-*myc* —and additionally *myb*, *fos*, and *ski* transduced originally by retroviruses; a cellular-derived gene, p53, and several DNA tumor virus genes: the T large antigens of SV40 and polyoma; adenovirus E1A; and papilloma virus E6. A good deal of phenomenology has been accumulated with regard to these genes; but their protein products are ill-defined, and their role in normal and neoplastic cellular activities is obscure.

The *myc* family genes have been studied most extensively. v-*myc* originally was isolated from an avian myelocytomatosis virus. c-*myc* is the cellular homologue of this gene. The gene structure has been characterized in detail and expression of the gene in normal and neoplastic tissue defined. Regulation of the expression of c-*myc* appears to reside in 5′ untranslated sequences. Expression is related to the cell cycle and may be developmentally associated. The *myc* protein is known to have a nuclear location and may be associated with small nuclear ribonuclear proteins. Other *myc* genes include N-*myc* from neuroblastoma, L-*myc* from small-cell carcinoma of the lung, and R-*myc* from rhabdomyosarcoma.

The *myb* gene first was characterized as an avian retroviral oncogene. The c-*myb* transcripts are detected predominantly in hematopoietic cells. The mRNA has a short half-life and appears to be degraded by a labile protein. Both *myc* and *myb* genes are regulated, at least in part, by posttranscriptional mechanisms. The *myb* protein is even less well characterized than the *myc* products.

The *fos* transcripts are among the first detected after quiescent cells are stimulated to divide. The pattern of expression is complex, at least as so far deduced. Expression in vivo is highest among differentiating cells, whereas in vitro cell lines show higher expression during undifferentiated growth.

The *ski* gene, which is poorly characterized, was isolated originally from avian embryo cells infected with transformation-defective avian leukosis virus.

The p53 gene was noted as a normal cellular protein complexed with various viral oncogenes in transformed cells. Studies with mutants suggest that it is needed for manifestation of a fully transformed phenotype.

The DNA tumor virus oncogenes have been studied in some detail with regard to viral life cycles. No cellular homologues appear to exist. Each of these genes can complement *ras*, as does *myc*, in the primary rat embryo cell transformation

TABLE 10.1. Nuclear oncogenes.

c-*myc* Family (v-*myc*)
c-*myc*
N-*myc*
L-*myc*
R-*myc*
c-*myb* (v-*myb*)
c-*fos* (v-*fos*)
c-*ski* (v-*ski*)
p53
SV40 large T antigen
Polyoma large T antigen
Adenovirus E1A protein
Papilloma virus E6

assay. Their importance lies in the independent evolution of additional genes in this category, underlining the importance of such genes to living organisms.

Regulation of growth ultimately resides in the nucleus; therefore it is not surprising that some products of oncogenes are located in the nucleus. Included among the nuclear oncogenes (Table 10.1) are the *myc*-related genes (c-, N-, L-, and v-*myc*), *myb, fos, ski*, adenovirus E1A, papilloma virus E6, SV40 and polyoma large T genes, and the cellular p53 gene. In addition to having a common nuclear location for their protein product, some of these genes (N-*myc*, c-*myc*, v-*myc*, p53, adenovirus E1A, large T) have a similar ability to cooperate with an activated *ras* oncogene to transform primary rat embryo cells (Table 10.2) (see Chapter 5). Furthermore, some of these oncogenes appear activated not by mutations within the coding region but by mechanisms that augment expression or abrogate the normal regulatory controls for gene expression (c-*myc*, N-*myc*). Because of their importance in a variety of neoplasms in many species, we first discuss the *myc* genes as prototypes of nuclear oncogenes.

TABLE 10.2. Characteristics of the protein products of viral nuclear oncogenes.

Oncogene	Capacity to immortalize	Transforms established cell lines	Binds to double stranded DNA	Binds to p53
E1A (12S)	+	+	0	0
E1A (13S)	+	+	0	0
E1B (21 kD)	0	0	0	0
E1B (55 kD)	0	0	0	+
SV40-large T	+	+	+	+
Polyoma-large T	+	0	+	0
v-*myc*	+	+	+	0
v-*myb*	+	+	+	0
v-*fos*	?	+	?	0
v-*ski*	?	+	?	0

myc Family Proto-oncogenes

myc Is a Ubiquitous Oncogene

The *myc* oncogene first was identified in the myelocytomatosis virus 29 (MC29), a replication-defective avian retrovirus. This virus induces acute transformation of a wide variety of cells in vivo and in vitro, including fibroblasts, epithelial cells, and myeloid cells. Subsequently, transduction of *myc* by several other replication-defective avian retroviruses, e.g., MH2, CMII, and OK10, and by feline leukemia viruses was demonstrated.

The v-*myc* gene from MC29 was isolated as a *gag-myc* fusion product consisting of 1358 base pairs of *gag* sequence linked to 1568 base pairs of the *myc* gene. Initially, this finding raised the question of whether the fusion between *gag* and *myc* was necessary for transformation. The identification of a spliced subgenomic mRNA encoding only the v-*myc* gene in OK10 and the transformation of primary avian fibroblasts in culture by avian retroviral constructions expressing only v-*myc*, chicken c-*myc* equivalents, or the human *myc* gene argue that *myc* alone can transform avian cells.

Augmented *myc* Expression Is Associated with Tumorigenesis

Subsequent to the discovery of *myc* genes in these acutely transforming retroviruses, the *myc* gene was found also to be involved in the pathogenesis of a slow transforming virus, the avian leukosis virus (ALV). ALV is a replication-competent retrovirus that does not contain a transforming oncogene. Chickens infected by this virus develop B cell lymphomas within 6 to 8 months. In most of these cases ALV infection results in provirus integration within the first intron of c-*myc*, thereby interrupting putative control elements within the first noncoding exon (see Fig. 5.5). Furthermore, the proviral long terminal repeat (LTR) provides a powerful promoter function that augments expression of the endogenous *myc* gene. In a few cases, however, the provirus is positioned downstream from the *myc* gene or in such a way that the promoter is in the opposite orientation to the gene. In these situations an enhancer effect, again due to the viral LTR, appears to augment *myc* transcription. Thus in all studied ALV-induced bursal lymphomas, there is heightened expression of *myc* mRNA.

Insertional activation of an endogenous *myc* proto-oncogene by proviral integration near the gene appears to occur with neoplasia of other species as well.

Table 10.3. Mechanisms of *myc*-oncogene activation.

Mechanism
Viral transduction
Insertional mutagenesis = promoter/enhancer insertion
Gene amplification
Chromosome translocations

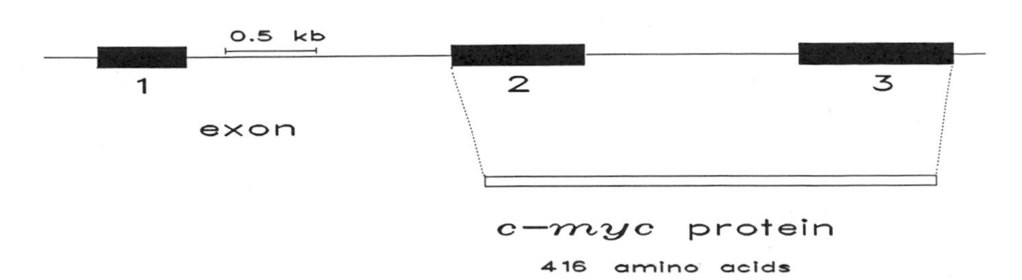

FIGURE 10.1. Structure of the human c-*myc* gene. Exons two and three code for the *myc* protein. This 65-kD nuclear protein may function to regulate its own transcription. Regions within and around the first exon have important transcriptional regulatory functions.

The Moloney murine leukemia virus induces T cell lymphomas/leukemias in mice and is associated with integration of the MLV provirus upstream from, or within, the first *myc* exon of the tumor cells. The feline leukemia virus that causes a common leukemia/lymphoma in cats also inserts within the first exon or first intron of the c-*myc* gene in most tumors. Thus overexpression or inappropriate expression of c-*myc* appears to be important in the development of many tumors (Table 10.3).

myc 5′ Sequences Regulate Gene Function

The frequency with which insertional activation interrupts the *myc* gene suggests that the 5′ portion of the gene has a negative regulatory role. In mammalian species c-*myc* is comprised of three exons, the first noncoding and the other two contributing an open reading frame capable of encoding a protein of about 49 kD (Fig. 10.1; see also Fig. 6.8). The actual in vivo molecular weight is 65 kD, however, reflecting some posttranslational modifications. The conservation of general structure and sequences of all *myc* exons among different species suggests similar functions for the coding and the noncoding portions in all *myc* genes.

As discussed in Chapter 6, there is evidence that the first exon and 5′ sequences provide control signals for the transcriptional and translational expression of c-*myc*. The frequent translocation of this gene in spontaneous lymphoid tumors of mammals (murine and rat plasmacytomas and human Burkitt's lymphoma) is associated with deletions and mutations within the first exon or intron. The result is abnormal *myc* expression. Transgenic experiments, in which different genetic constructions of *myc* are introduced into the germ line of mice, show that the normal *myc* allele with its intact 5′ regulatory sequences does not induce tumors, whereas *myc* genes placed under control of heterologous enhancer/promoters (e.g., immunoglobulin, SV40) do (see Chapters 12 and 6).

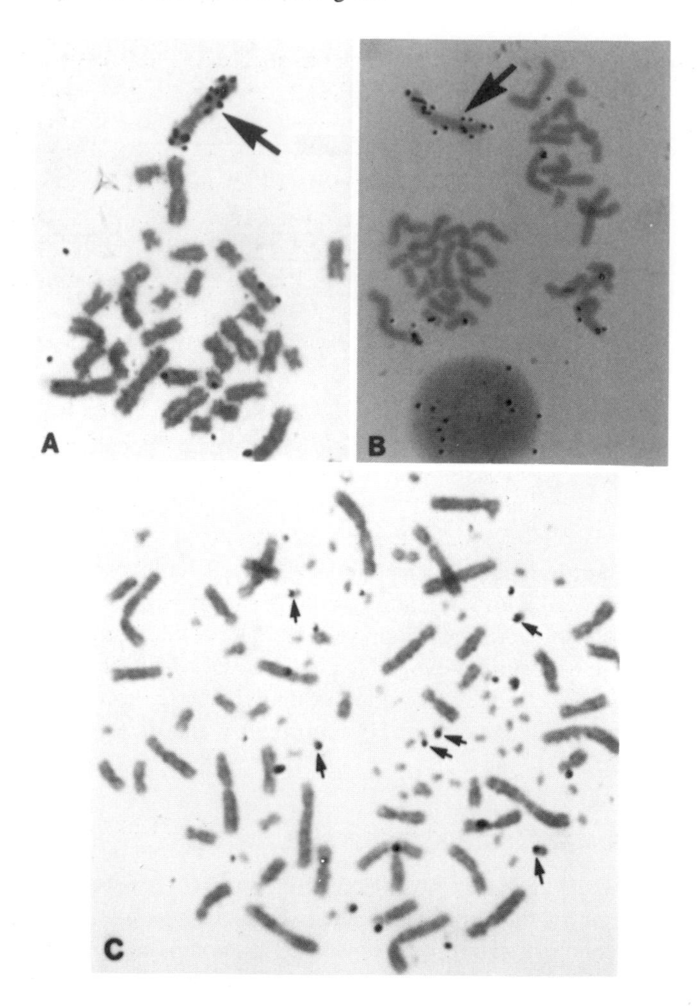

FIGURE 10.2. Chromosome localization of amplified c-*myc* in COLO 320 metaphase spreads. (A) A partial metaphase spread of a COLO 320 HSR cell that was hybridized in situ with an ^{3}H-labeled unique sequence fragment of human c-*myc* DNA. Note that the autoradiographic signal (silver grain) is confined to the homogeneously staining regions in both chromosome arms of the HSR chromosome (arrow). (B) Another representation of a partial metaphase spread of a COLO 320 HSR cell stained with Giemsa stain after the chromosome preparation has been hybridized with the ^{3}H-labeled c-*myc* probe. The HSR chromosome with cluster of silver grain is indicated by the arrow. (C) Metaphase spread of a COLO 320 DM cell that was hybridized in situ with radiolabeled c-*myc* DNA. Some of the DM chromosomes showed the hybridization signal (arrows). (Kindly provided by Dr. C.C. Lin)

Gene amplification of c-*myc* resulting in augmented expression has been reported in human leukemia cell lines (HL60) and human carcinoma cell lines (COLO 320, small-cell lung carcinoma) as well as in primary tumors (Fig. 10.2). Perturbation of *myc* expression thus appears important in tumorigenesis. Perturbation may take several forms including *myc*-oncogene transduction, insertional activation, translocation, amplification, and introduction into transgenic mice (Table 10.3).

Despite the ability of cloned *myc* constructions to transform established fibroblast cell lines in vitro, the exact transforming role of *myc* in vivo remains unclear. Experimental data suggest that events in addition to *myc* activation are required for in vivo tumorigenesis.

The *myc* gene, like other nuclear oncogenes including p*53*, adenovirus E1A, and polyoma large T antigen, can efficiently immortalize primary cells that otherwise would senesce and die during repeated passage. These "immortal" cells are capable of indefinite growth but are not "transformed" in the sense of phenotypical appearance and tumorigenic capacity. The "establishment" or immortalization function requires augmented levels of *myc* gene expression, though probably less than that required for transformation. An extended capacity to divide may predispose primary cells to critical second genetic mutations. In this situation *myc* would act as a "helper" in tumorigenesis rather than as a direct agent. Further experimentation is necessary to better delineate this problem.

Function of the *myc* Protein Is Still a Mystery

Despite considerable interest in the *myc* oncogene, surprisingly little is known about the biochemistry of its protein product. Though the open reading frame of the two coding exons predicts a protein of 49 kD, proteins of 65 kD have been immunoprecipitated from cells derived from human, avian, monkey, hamster, rat, and frog tissues. This result suggests that not only is the nucleotide sequence well preserved among many species but also that the posttranslational processing is similar: These proteins are phosphorylated on serine and threonine.

Site-directed mutagenesis of the cloned human *myc* gene uncovered three regions in the *myc* protein (in exons 2 and 3) that are essential for co-transformation of rat embryo cells. The region most sensitive to mutations appears to be a segment containing a cluster of basic amino acids in the third exon. William Lee, at the University of California, San Francisco, has shown that deletions in this exon abolish not only transforming activity but also nuclear localization (Fig. 10.3; see Plate 2). Thus location of the *myc* protein is critical for the transforming function of *myc*, as is the case for *src* (see Chapter 7).

The nuclear localization of *myc* protein suggests an ability to bind DNA. Though the protein does bind in vitro to single- and double-stranded DNA with moderate affinity, there does not appear to be any sequence specificity. The *myc* protein does not bind to chromatin, and the protein is not released from nuclear preparations after digestion of DNA by DNase I. RNase treatment does release the protein, however, and immunofluorescence techniques show that it co-localizes with small nuclear ribonuclear proteins (snRNPs). These experiments suggest that *myc* may play a role in processing of RNA.

The half-life of normal *myc* protein is short, approximately 30 minutes. Other endogenous nuclear proteins (N-*myc*, p53, *myb, fos*) also have short half-lives, which are well suited to proteins involved in control of the cell cycle and of differentiation functions. Indeed, there is evidence that expression of *myc* and other nuclear oncogenes, e.g., *myb* and *fos*, plays a role in these normal cellular processes.

myc Expression Is Related to the Cell Cycle

During the G_0 phase of the cell cycle, when cells are not proliferating, there is a low level of *myc* expression. However, when cells are stimulated to divide, endogenous *myc* levels increase. This increase is temporally related to the increase of certain other nuclear proto-oncogenes. For example, serum-depleted mouse 3T3 cells express small amounts of *myc* and *fos*, but when stimulated to

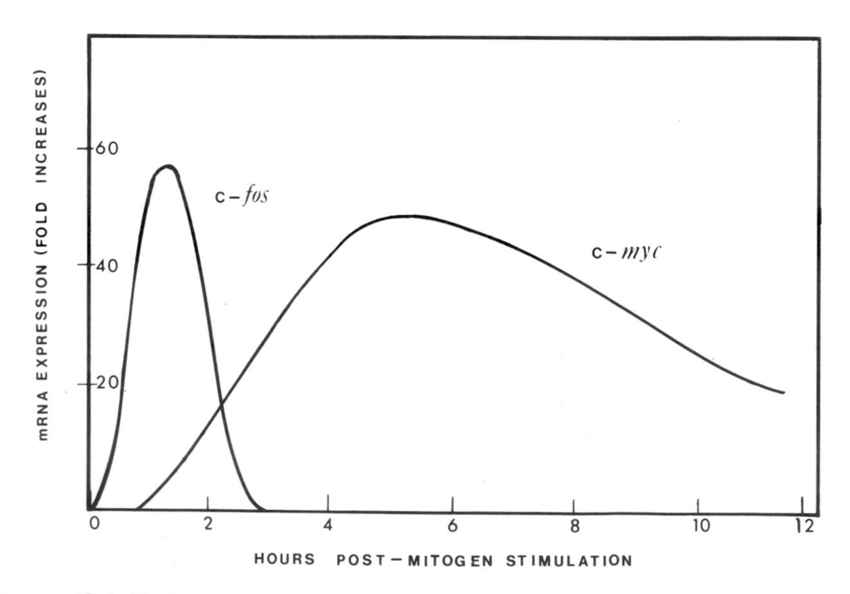

FIGURE 10.4. Nuclear proto-oncogene expression during mitogenesis. There is rapid and transient expression of c-*fos* followed by c-*myc*.

enter G_1 phase by serum, platelet-derived growth factor (PDGF), or 12-*O*-tetradecanoylphorbol-13-acetate (TPA), a transient burst of c-*fos* expression is seen within the first 10 to 30 minutes (Fig. 10.4). Within several hours *myc* levels rise, and cells enter the S phase. Levels of c-*myc* expression are highest during the late G_1 and S phases. This sequence of events is seen also in concanavalin A-stimulated lymphocytes.

The complexity of *myc* regulation is illustrated by the fact that fluctuations in *myc* RNA and protein appear to be due largely to changes in RNA and protein half-lives rather than to increases or decreases in the transcriptional rate. Some of the effects observed during mitogenic stimulation of cultured cells have been reported in vivo as well. For example, partial hepatectomy or exposure to hepatotoxins invariably results in increased *myc* expression within hours of the insult, followed later by hepatic regeneration.

There is a suggestion that the c-*myc* product is critical to allow cells to enter or traverse S phase from G_1 phase. When c-*myc* protein was substantially reduced (by sequence-specific anti-sense oligodeoxyribonucleotides), cells were able to traverse from G_0 to G_1 and into late G_1 following mitogen stimulation but the cells did not enter S phase. Transcription of the c-*myc* gene was specifically inhibited by the oligonucleotide for some 24 to 48 hours following treatment. At later times c-*myc* could again be induced by mitogens.

Differentiation and Oncogene Expression

Proto-oncogene expression also is affected by the state of differentiation of a cell. Experiments with the human promyelocytic cell line HL60 show that treatment with TPA induces the cells to mature to macrophages. This differentiation is associated with augmented c-*fos* and c-*fms* expression but suppression of c-*myc* and c-*myb* expression. HL60 cells form mature granulocytes when stimulated with dimethylsulfoxide (DMSO) or retinoic acid. In this situation, c-*fos* expression is not stimulated, and expression of c-*myb*, c-*myc*, and c-*fms* is reduced. Retinoic acid causes the mouse teratocarcinoma cell line F9 to mature to endoderm, a process associated with increased c-*fos* expression and extinction of c-*myc* expression. Thus the program for proto-oncogene expression in growth and differentiation appears to depend on the type of inducing agent and the terminal phenotype achieved.

The role of *myc* in differentiation was further examined by introducing human *myc* plasmid constructions into a mouse erythroleukemia cell line (MEL). Such introduction resulted in constitutive and high levels of human c-*myc* mRNA. MEL cells normally differentiate into erythroid cells after induction with DMSO, but in clones with high levels of *myc* gene expression differentiation was suppressed. Thus augmented and dysregulated *myc* expression can prevent normal cell differentiation and progression to senescence, thereby "immortalizing" the cells.

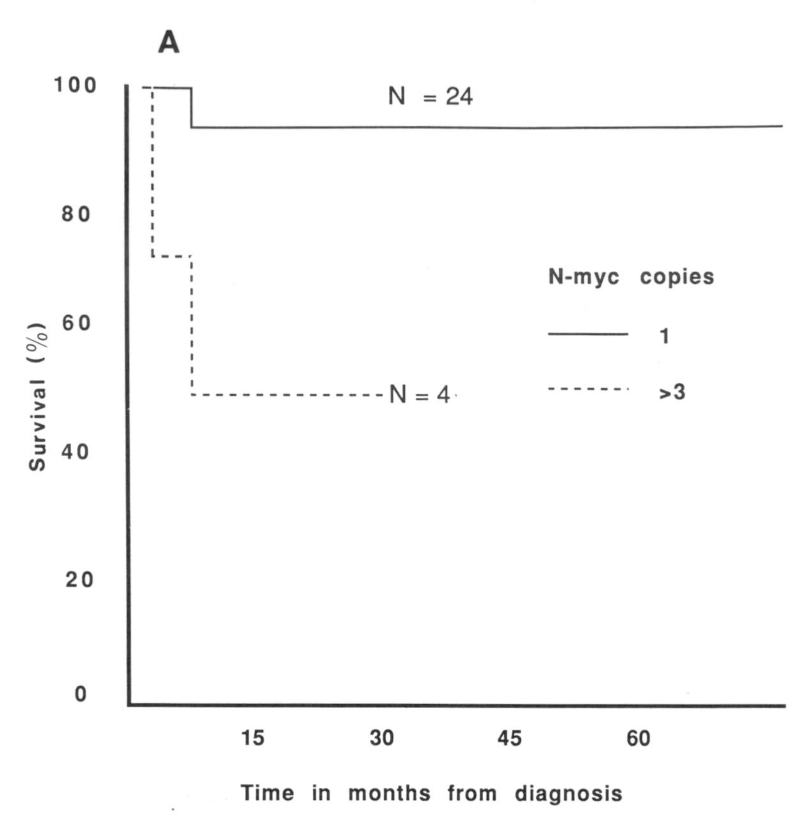

FIGURE 10.5. Amplification of N-*myc* correlates with poor prognosis of neuroblastoma. (A) Survival of infants less than 1 year old at the time of diagnosis. (B) Survival of children older than 1 year at the time of diagnosis. (From Seeger et al. NEJM 313: 1111, 1985)

Other Members of the *myc* Family

Several human genes share considerable sequence homology with the c-*myc* proto-oncogene (Table 10.1). These genes appear to share both structural and functional characteristics with c-*myc*. Each can complement activated *ras* oncogenes to transform primary rat embryo fibroblasts. Best characterized at present is N-*myc*. L-*myc* and R-*myc* comprise the other currently known members of this gene family.

N-*myc*

The N-*myc* gene is most homologous with the second exon of c-*myc*; overall N-*myc* shows as much as 32% identity with c-*myc*. The N-*myc* gene was first identified in human neuroblastomas by its own 100-fold amplification (see

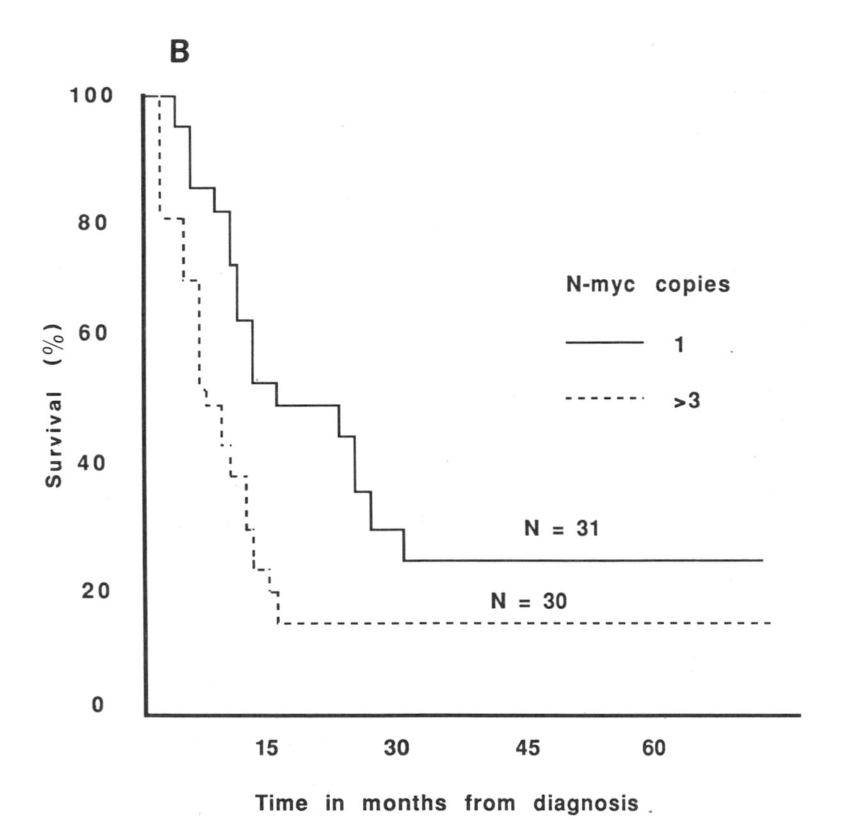

FIGURE 10.5. *Continued.*

Chapter 6). It was found in fresh tumor tissues as well as in their derivative cell lines. The gene locus has been mapped to human chromosome 2p23-p24.

N-*myc* expression has been observed in retinoblastomas (10 of 10 cases) as well as in neuroblastomas. Amplification of N-*myc* occurs in approximately 20% of retinoblastomas. Amplification of N-*myc* in neuroblastomas occurs more frequently in cell lines and in tumors showing more advanced clinical stage than in fresh tissue from less advanced tumors. Thus amplification (and increased N-*myc* expression) may be related to tumor progression (Fig. 10.5).

Research has shown that N-*myc* amplification is associated with decreased expression of class I major histocompatibility complex (MHC) genes. These 45-kD molecules are cell surface glycoproteins that are required for recognition of foreign antigens by cytotoxic T cells. Lower expression of MHC antigens in neuroblastomas with amplified N-*myc* genes may explain the tendency of this tumor to metastasize. Impaired recognition of tumor cells by T cells in this situation could contribute to poor host response against the advancing neoplasm.

A similar decrease in MHC class I gene expression was demonstrated for adenovirus-12-infected cells. This decrease was associated with increased tumorigenicity. Similarly, overexpression of c-*myc* in human melanoma cell lines was inversely correlated with MHC class I expression and more virulent behavior. In addition, a *myc*-like nuclear oncogene, adenovirus E1A, inhibits MHC class I gene expression (see Chapter 3 and below).

N-*myc* expression originally was found to be limited to tumors and cell lines solely of neuroectodermal origin, but other studies suggest a wider range of tissue expression including Wilms' tumors and pre-B cell lines.

The N-*myc* protein has many similarities to the c-*myc* product. N-*myc* protein is a phosphoprotein of 62-64kD, localized to the cell nucleus. This protein is associated with the nuclear matrix and has a relatively short half life.

Antisera to the N-*myc* product may have clinical utility in the differential diagnosis of the small round cell tumors of childhood. This group of tumors, including neuroblastomas, neuroepitheliomas, rhabdomyosarcomas, Ewing sarcoma, and lymphoma, appears very similar histologically, however only neuroblastomas have been found to express appreciable levels of N-*myc*.

L-*myc*

A third member of the *myc* family has been identified in small-cell lung carcinoma cells (SCCL) by its limited homology to both c-*myc* and N-*myc*. This gene maps to human chromosome region 1p32, a region distinct from other *myc* family members. The gene is amplified 10- to 20-fold in various SCCL cell DNAs and in one SCCL patient's tumor at autopsy. A variety of patterns of *myc* family gene amplification have been found in SCCL cell lines.

R-*myc*

The R-*myc* gene has been cloned from a human rhabdomyosarcoma cell line and from normal human cells. This gene shares up to 50% nucleic acid identity with the human c-*myc* gene. A 6.7-kb *Eco* RI fragment containing the R-*myc* gene can immortalize and transform primary mouse macrophages at low frequency. DNA recovered from such transformed cells appears to lack 3′ sequences present in the R-*myc* proto-oncogene. Thus the 3′ truncation of the R-*myc* gene can render it transforming. Transcripts were produced at low levels, suggesting that transforming ability was not due to amplified expression. Regulation of R-*myc* may differ from that of c-*myc* because inhibition of protein synthesis with cycloheximide does not change R-*myc* gene expression.

myc Family: Conclusions

Much work remains to be done to elucidate function and regulation of the *myc*-family of proto-oncogenes. In addition, the *myc* protein remains one of the least well characterized of the major oncogene products. The *myc* protein's association with snRNP suggests that it has a role in RNA processing. All *myc* proteins have a nuclear location; however, it remains to be seen what functions the other members of this family serve. Participation in nucleic acid metabolism remains the most likely role for the products of these ubiquitous genes.

myb ONCOGENE

The *myb* proto-oncogene shares several features with *myc*. Both occupy a location in the nucleus, both cooperate with mutant *ras* genes to transform rat embryo cells, and both share with the adenovirus E1A protein a distant but related structure. *myb* appears to be much more restricted than *myc* in its tissue expression, although both genes are differentially expressed during growth and differentiation.

The *myb* proto-oncogene was first identified in avian myeloblastosis virus (AMV) and avian leukemia virus E26. AMV causes myeloid leukemia, whereas E26 causes erythroblastosis and occasionally mixed myeloid-erythroid leukemia. The differences in the host cell spectrum may be due to the fact that E26 also contains a second oncogene, v-*ets* (see Chapter 11). The v-*myb* protein is 45 kD. Compared to c-*myb*, the viral oncogene is truncated at both 5′ and 3′ ends.

The chicken c-*myb* gene has been the best characterized cellular *myb* gene (Fig. 10.6). It contains at least seven exons. Human c-*myb* cDNA has been cloned and found to contain sequences homologous with the seven known avian exons. The *myb* sequences were conserved among vertebrates and even among invertebrates (see below). Overall, human c-*myb* sequences were 90% identical with murine and 85% identical with avian c-*myb*. Exons 1, 2, and 5 exhibited 97 to 100% identity among the three species, whereas the remaining exons were conserved about 70 to 85%. The human c-*myb* gene has been mapped to chromosome 6q22-24.

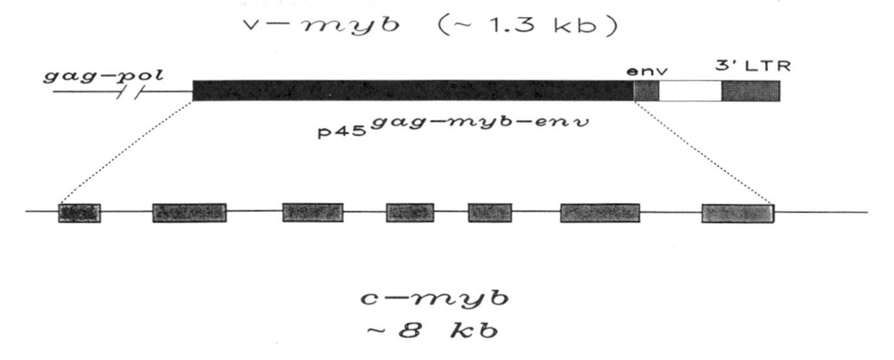

FIGURE 10.6. Structure of the *myb* oncogene. The viral *myb* gene is from the avian myeloblastosis virus. The exon structure is that of chicken c-*myb*.

The *myb* transcripts are detected predominantly in hematopoietic cells. The size of the major transcript has been reported variously as 3.4 kb and 3.8 kb. High levels of these transcripts are found in lymphoid cells and myeloid lineage cells. B cells contain lower amounts of *myb* mRNA, and nonhematopoietic cells generally do not transcribe this gene. An occasional nonhematopoietic cell has been found to contain *myb* transcripts, most notably those derived from colon carcinoma (see below). Cells of the myeloid lineage appear to have two *myb* transcripts, 2.3 and 3.4 (or 3.8) kb. These alternate transcripts are thought to be generated by differential RNA splicing. Alternative splicing of c-*myb* RNA is supported by a study reporting isolation of two separate classes of cDNA clones from AML cell libraries screened with *myb* probes.

The mRNA product of c-*myb* has a short half-life. A study by Thompson et al. demonstrated that c-*myb* expression is associated with cell proliferation. In two cultured cell lines, c-*myb* RNA levels were highest during periods of exponential growth. A tenfold increase in *myb* mRNA was seen in cells traversing the cell cycle compared to nondividing cells. Peak levels of *myb* transcript accumulation were in S phase. No significant variation in the *rate* of c-*myb* transcription was found during the various stages of the cell cycle; instead, the *half-life* of c-*myb* transcripts was significantly increased in late G and S phase cells. Degradation of c-*myb* mRNA appears to depend on a labile protein because the half-life of c-*myb* transcript is markedly extended by cycloheximide. Thus c-*myb*, like c-*myc*, is regulated by posttranscriptional mechanisms.

myb in Invertebrates

A homologue of the vertebrate c-*myb* gene has been identified in *Drosophila*. It shares a conserved domain in which 73% of the amino acids are identical with the chicken c-*myb* gene. This gene is expressed in *Drosophila melanogaster* embryonic tissues and during all major stages of *Drosophila* development. What role, if any, the c-*myb* gene has in the hematopoietic system of *Drosophila* is currently unknown. Attempts to identify c-*myb* and c-*myc* homologues in yeast have not been successful.

myb and Cancer

In chickens *myb* is involved in myeloid leukemogenesis, associated with a phenotypically less mature neoplasm than is *myc*. The v-*myb* transformed cells, however, retain an intact capacity to differentiate.

Transcripts of c-*myb*, 3.8 kb in length, have been detected primarily in immature T lymphoid, myeloid, and erythroid cell lines. No transcripts have been detected in mature T lymphoid, myeloid, or B lymphoid cell lines or in cell lines derived from a wide variety of solid tumors including rhabdomyosarcomas; osteogenic sarcomas; fibrosarcomas; carcinomas of skin, breast, lung, kidney, bladder, and ovary; melanomas; and glioblastomas. In a study of fresh hematopoietic tumor tissues, the 3.8-kb c-*myb* transcript was found in ten of ten acute myelogenous leukemias (AMLs), three of four acute lymphocytic leukemias (ALLs), six of six chronic myelogenous leukemias (CMLs), and none of five chronic lymphocytic leukemias (CLLs). Amplification of the c-*myb* gene was demonstrated in several cell lines derived from a single patient with AML and in two cell lines derived from a single patient with adenocarcinoma of the colon. Although in situ hybridization demonstrated amplified c-*myb* on extra copies of the same marker chromosome, the c-*myb* locus was not rearranged in the colon carcinoma lines. No amplification of the gene was observed in studies of 20 other colon carcinoma cell lines.

It is of interest that many ALL cells exhibit a deletion in the 6q21-qter region, which encompasses the human c-*myb* chromosome locus. In one study, six of six hematopoietic malignancies with the $6q^-$ syndrome had overexpression of the c-*myb* gene. A 6:14 translocation at this location has been described in papillary serous adenocarcinoma of the ovary with the breakpoint at 6q21. Chromosomal translocations and deletions involving the distal portion of chromosome 6q also have been described in various melanomas and lymphomas, but the role, if any, of c-*myb* in these cancers requires further study.

myb Protein Product

The *myb* protein is even less well characterized than the *myc* product. The c-*myb* gene of vertebrates encodes a protein of around 600 amino acids (72–75 kD) localized to the nucleus by immunocytochemical staining. When nuclei from chicken or human cells expressing c-*myb* are fractionated into nucleoplasm, chromatin, and matrix, c-*myb* protein is found to be associated with the nuclear matrix. Because the chromatin fraction contains the DNA, c-*myb* protein is not directly associated with DNA. Nevertheless, nuclear location of this protein is somehow associated with the transformed state; avian myeloblasts induced to differentiate to macrophages show *myb*-product relocated to the cytoplasm rather than in the nucleus.

Abnormal expression of *myb* is most closely associated with leukemia, although admittedly this relation is circumstantial. High levels of c-*myb* expression in leukemic blasts diminish to undetectable levels when the cells are induced to undergo differentiation to mature granulocytes or macrophages. High concentrations of c-*myb* transcripts are found associated with AML in which the patients had more than 190,000 blasts per cubic millimeter. Patients with fewer peripheral blasts showed significantly less c-*myb* expression. The actual role (if any) of *myb* protein on the peripheral blast count as well as its role in the nuclear matrix remain to be elucidated.

fos Oncogene

There is a rapid and transient expression of the c-*fos* gene following stimulation of cells by mitogenic growth factors (Fig. 10.4). The c-*fos* proto-oncogene, along with actin genes, is activated as early as 5 minutes after treatment of 3T3 mouse fibroblasts with PDGF. The level of c-*fos* mRNA increases 20- to 30-fold within

15 to 30 minutes, but proteins and mRNA are undetectable after 2 hours. The subsequent expression of c-*myc* in response to a number of growth stimuli, including fibroblast growth factor, nerve growth factor, insulin, phorbol esters, dibutyl cAMP, and high levels of potassium or calcium, suggests that the ordered expression of these nuclear genes is a general feature of cell growth and differentiation.

The c-*fos* proto-oncogene is the cellular homologue of the v-*fos* oncogene first detected in two murine retroviruses: FBJ murine osteosarcoma virus (FBJ-MuSV) and FBR murine sarcoma virus (FBR-MuSV). Both viruses were isolated from osteosarcomas (spontaneous in the case of FBJ-MuSV and radiation induced for FBR-MuSV). Both viruses induce chondro-osseous sarcomas when injected into newborn mice and focus formation in infected fibroblasts in vitro. Gene structures of FBJ v-*fos* and murine c-*fos* are depicted in Fig. 10.7.

The product of FBJ v-*fos*, $p55^{v\text{-}fos}$, is structurally similar to that of murine c-*fos* aside from a C-terminal frameshift mutation. However the product of FBR v-*fos*, $p75^{gag\text{-}fos\text{-}fox}$, is different including truncations of 24 N-terminal amino acids and 98 C-terminal amino acids with replacement of these sequences with viral gag and cellular fox sequences, respectively. In addition internal deletions and single amino acid substitutions are present in $p75^{gag\text{-}fos\text{-}fox}$. The two v-*fos* products are similarly expressed with similar turnover rates. However the FBR-MuSV is a far more efficient tumorigenic virus than FBJ-MuSV, suggesting that differences in the biological properties of the v-*fos* proteins may be due to the identified structural differences.

Jenuwein and Muller studied the alterations in $p75^{gag\text{-}fos\text{-}fox}$ leading to its increased tumorigenic capacity. They found that 5′ *gag* and 3′ *fox* fusion sequences were not responsible for the enhanced transforming properties nor were truncation of 5′ and 3′ c-*fos* sequences. However, a single amino acid change (Glu→Val) at position 138 activated the immortalizing potential of v-*fos* while internal deletions in the C-terminal half of the protein are responsible for its increased transforming capacity.

Biological studies of the c-*fos* gene, on the other hand, show activation of c-*fos* by deletion of its 3′ noncoding sequences and placement of the gene under transcriptional control of a strong heterologous promoter. The c-*fos* 3′ region appears to have an inhibitory effect on gene expression perhaps by destabilizing the mRNA (as deletion of the 3′ sequences increases mRNA stability); this is another example of activation of a proto-oncogene to a transforming gene by rearrangement outside the coding region.

c-*fos* Expression: A Complex Pattern

The c-*fos* gene product is a 55-kD nuclear protein that is complexed to a cellular protein termed p39. c-*fos* shows a complex pattern of tissue, cell type, and stage-specific expression. During murine development the highest levels of c-*fos* expression occur during midgestation in fetal liver and during late gestation in

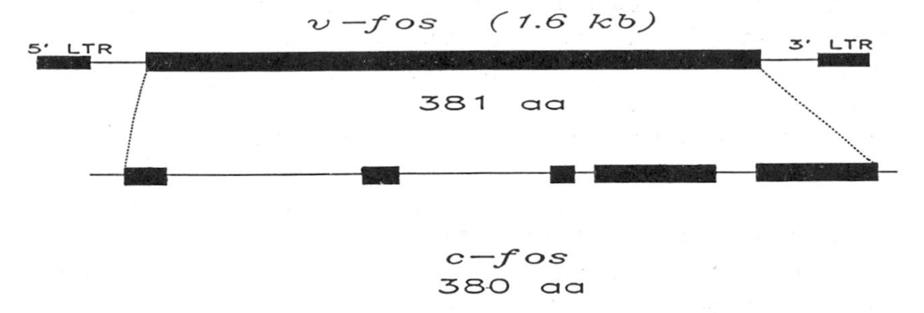

FIGURE 10.7. Structure of the *fos* oncogene. The viral *fos* is from the FBJ murine osteosarcoma virus. The exon structure of murine c-*fos* is shown.

the amnion and yolk sacs. After birth the bone marrow exhibits the highest levels of expression. In the amnion expression of c-*fos* is clearly stage-specific in that the message is low during midgestation and increases as development proceeds. The identity of the particular cells showing high levels of c-*fos* expression in the bone marrow and liver are unknown, although transcripts have been detected in mast cells, differentiated neutrophils, and mature macrophages. The in vivo pattern, then, is a high level of c-*fos* expression during differentiation in a restricted number of cell and tissue types.

Cultured cells have a different pattern: high expression during undifferentiated growth and low expression during terminal differentiation. For example, c-*fos* expression is high in myogenic (muscle) stem cells but decreases sharply during differentiation. The rise in c-*fos* message and c-*myc* message seen during proliferation of erythroid progenitors ends during terminal differentiation of these cells.

Scientists from the National Institutes of Health (NIH) have introduced into NIH 3T3 cells a c-*fos* gene controlled by the steroid-inducible mouse mammary tumor virus promoter. Induction of high-level expression of the c-*fos* gene in the "sense" direction did not alter the growth of these cells. However, when the c-*fos* gene was placed in the opposite orientation (so-called antisense) the growth of the cells was inhibited following steroid induction. Thus although the role of *fos* in differentiation and development is complex, *fos* does appear to have a role in normal cell division.

A small (22 bp) region located 300 bp 5′ to the human *fos* gene is necessary for increased c-*fos* expression in response to serum stimulation. This same sequence appears necessary for stimulation by EGF and phorbol esters as well. Other stimulating factors including PDGF, calcium, and cAMP may act at as many as three distinct regulatory sequences.

The *fos* product has a nuclear location and is known to bind to DNA. It apparently recognizes sequences similar or identical to the consensus sequence recognized by AP-1, a human transactivation factor. This sequence recognition suggests a role for c-*fos* in transcriptional regulation.

ski ONCOGENE

The *ski* oncogene has not yet been well characterized. The Sloan-Kettering viruses (SKVs) are a group of transforming retroviruses that were isolated from chicken embryo cells infected with transformation-defective avian leukosis virus. These viruses encode an oncogene called v-*ski*, which transforms avian cells in vitro. This oncogene is expressed as a p110 or p125 *gag*-fusion protein. It does not share sequence homology with other known retroviral oncogenes. The function of the homologue present in avian and mammalian cells is unknown, although its chromosomal location (1q12→ter) and subcellular location (nucleus) have been determined.

p53: A Cellular Encoded *onc*-PROTEIN

Activation of cellular oncogenes can be mediated by a number of mechanisms, including mutation, translocation, and promoter insertion. This initial step may lead to dysregulation of certain normal cellular proteins, e.g., p53. Augmented synthesis of p53 has been found in a variety of tissue types from several species (human, murine, rat, other rodents) and in tumors induced by DNA and RNA tumor viruses as well as by chemical carcinogens. p53 appears to be a nucleophosphoprotein that results in cell immortalization by itself and cell transformation in conjugation with *ras* (see below).

p53 was described originally as a normal cellular protein that formed a stable complex with the large T antigen of SV40. Subsequently, it was shown to bind the E1B 57-kD protein of adenoviruses as well. This binding increased the half-life of p53 from 30 minutes to 20 hours.

Based on its size and its binding to anti-p53 monoclonal antibodies, p53 of normal cells is considered structurally similar to the p53 found in transformed cells. The major difference observed between normal and tumor cells is quantitative: Tumor cells synthesize large amounts of p53, whereas normal cells make small amounts of the protein. In addition, p53 appears to be located primarily in the nucleus of transformed cells, whereas it is mostly found in the cytoplasm of normal cells.

The role of p53 in tumorigenesis has been studied by Varda Rotter and her associates at the Weizmann Institute. They used a cell line transformed by the Abelson murine leukemia virus (Ab-MuLV), with a p53 gene inactivated by interruption with virus-like sequences. This p53 cell line caused local and regressing tumors when injected into syngeneic mice. However, when an intact p53 gene was introduced into these cells, the resultant clones caused lethal tumors in recipient animals, much like the usual Ab-MuLV transformed $p53^+$ cell lines. Thus expression of p53 appears to be important for manifestation of a fully transformed phenotype.

Studies suggest that normal cell differentiation may be accompanied by a reduction in p53 expression. For this reason it has been hypothesized that p53 is an embryonic protein that is suppressed in adult differentiated cells.

The gene for p53 is located on the long arm of human chromosome 17. The role of changes (if any) in chromosome 17 in the altered expression of p53 in human malignancy is not known.

Antibodies to p53 have been found in the sera of about 10% of breast cancer patients but not in sera obtained from healthy controls. Experiments have shown that in vitro expression of a p53 cDNA causes immortalization of primary cells and renders them sensitive to transformation by an activated *ras* oncogene. Other scientists have shown that p53 can be activated by mutagenesis or rearrangements of its coding sequence. The mutant "activated" p53 was shown to have a longer half-life (24 versus 4 hours), which presumably contributed to cellular immortality.

Although p53 is not a "classical oncogene," its description is included here because it is commonly associated with the transformation process. It is anticipated that as molecular biology becomes more sophisticated other normal proteins associated with the transformed phenotype will be identified. Like p53, some of these proteins may be identified by their association with viral oncogenes. They subsequently may find a place in the oncogene pantheon once it is determined that they can be activated and independently associated with neoplastic transformation.

Nuclear Oncogenes of DNA Tumor Viruses

SV40 Large T/Polyoma Large T Antigens

SV40 and polyoma viruses are papova viruses with transforming potential. Transformation requires the expression of the large T antigen (94 kD) in SV40 and both the large T (100 kD) and middle T (55 kD) antigens in polyoma virus. Polyoma virus large T protein appears to be confined to the nucleus and can induce "immortalization" in primary rodent cells. Polyoma middle T protein is localized in the plasma membrane and is required in conjunction with large T protein for the fully transformed state to occur. In the primary rat embryo fibroblast co-transfection assay (see Chapter 5), neither polyoma large T nor middle T genes alone can transform primary cells, but the co-transfection of the two genes efficiently induces transformed foci. Thus the interaction between polyoma large T and middle T proteins is much like that between *myc* and *ras*. The similarity extends even to the subcellular localization of each gene product, i.e., *myc* and large T protein to the nucleus, *ras* and middle T protein to the cytoplasm.

SV40 large T antigen is required for the viral lytic cycle as well as for transformation. The two functions, however, can be dissociated from one another. SV40 large T is a multifunctional protein found predominantly in the nucleus, but a fraction is found also in the plasma membrane. SV40 large T antigen, unlike polyoma large T antigen, transforms established rodent cell lines and can transform mouse fibroblast cells as well when it is introduced into the cells by a retrovirus vector. Mutants that alter SV40 large T sequences required for transport of the protein from the cytoplasm to the nucleus also transform established fibroblast cell lines. Thus the cytoplasmic protein carries a significant portion of the oncogenic potential. However, in primary rodent cells the requirement of both a nuclear and a cytoplasmic oncogene for the fully transformed phenotype is reiterated with SV40 large T antigen: Co-introduction of another nuclear oncogene is necessary to transform these primary cells infected with an SV40 mutant that is unable to localize to the nucleus.

SV40 large T antigen complexes with the product of p53. The formation of the complex stabilizes p53, dramatically prolonging its normally short half-life and resulting in an increased protein concentration. p53 is thought to be important in cell cycle progression (see above). Furthermore, this protein appears to be important in the tumorigenicity of virally transformed cells. Because no SV40 mutants have been isolated that dissociate the transforming activity of the T antigen from its p53 binding activity, it is possible that the function of SV40 large T antigen is dependent on p53.

Polyoma middle T protein also acts in conjunction with another cellular proto-oncogene, $pp60^{c\text{-}src}$. Polyoma middle T protein and $pp60^{c\text{-}src}$ are tightly bound together, an association that dramatically augments the kinase activity of $pp60^{c\text{-}src}$. The activated *src* protein modifies middle T protein by phosphorylating it on tyrosine. Because increased kinase activity is associated with the transforming activity of v-*src*, it may be surmised that this *src* activation may in part mediate the transforming potential of middle T antigen.

Adenovirus E1A Oncogene

Two early regions of adenovirus, termed E1A and E1B, are clearly involved in neoplastic transformation by the virus (see Chapter 3). E1A is able to transform established cell lines but can only induce "immortalization" of primary fibroblasts. In the primary rat embryo fibroblast system, both E1A and E1B are necessary for full transformation: E1A is localized to the nucleus and performs a *myc*-like function, whereas E1B substitutes for *ras*. Other activities of the E1A product may contribute to its tumorigenic potential. First, E1A inhibits transcription of MHC-I genes in infected or transfected cells. The expression of MHC-I protein appears to be a necessary signal that the cellular immune system needs to recognize "abnormal" cells. MHC antigens are the targets for T-cell-mediated cytotoxicity; their reduction on the surface of tumor cells therefore allows such cells to escape antitumor immune mechanisms. Indeed, research by Tanaka and co-workers shows that tumorigenicity of mouse cells transformed by adenovirus can be abolished by reintroducing a constitutively expressed MHC-I gene. Second, E1A can stimulate as well as suppress expression of many cellular genes, a fact that has prompted speculation that unwarranted gene expression may contribute to the neoplastic phenotype.

The relation between E1A and cellular proto-oncogenes may be structural as well: The *myc, myb*, and p53 proteins appear to have similar protein structures. "Mix and match" experiments between *myc* and E1A show that chimeric genes can act to complement *ras* in the rat embryo fibroblast co-transfection assay.

Papilloma Virus E6 Oncogene

The papilloma virus E6 oncogene is discussed in detail in the papilloma virus section of Chapter 3. In short, this gene is expressed among the papilloma virus early functions and encodes a nuclear and cytoplasmic protein of approximately 15.5 kD. It appears to have an "immortalization" function for primary cells similar to that of *myc* and other nuclear oncogenes.

Bibliography

c-*myc*

Adams JM, Gerondakis S, Webb E, et al. Cellular myc oncogene is altered by chromosome translocation in an immunoglobulin locus in murine plasmacytomas and is rearranged similarly in human Burkitt's lymphomas. Proc Natl Acad Sci USA 80:1982, 1983.

Adams JM, Harris A, Pinkert C, et al. The c-myc oncogene driven by immunoglobulin enhancers induced lymphoid malignancy in transgenic mice. Nature 318:533, 1985.

Alitalo K, Schwab M, Lin CC, et al. Homogeneously staining chromosomal regions contain amplified copies of an abundantly expressed cellular oncogene (c-myc) in malignant neuroendocrine cells from a human colon carcinoma. Proc Natl Acad Sci USA 80:1707, 1983.

Armelin HA, Armelin MCG, Kelly K, et al. Functional role for c-myc in mitogenic response to platelet-derived growth factor. Nature 310:655, 1984.

Ar-Rushdi A, Nishikura K, Erikson J, et al. Differential expression of the translocated and untranslocated c-myc oncogene in Burkitt's lymphoma. Science 222:390, 1983.

Balaban G, Gilbert F. Homogeneously staining regions in direct preparations from human neuroblastomas. Cancer Res 42:1838, 1982.

Battey J, Moulding C, Taub R, et al. The human c-myc oncogene: structural consequences of translocation into the IgH locus in Burkitt's lymphoma. Cell 34:779, 1983.

Bishop JM. Viral oncogenes. Cell 42:23, 1985.

Campisi J, Gray HE, Pardee AB, et al. Cell-cycle control of c-myc but not c-ras expression is lost following chemical transformation. Cell 36:241, 1984.

Colby WW, Chen EY, Smith DH, et al. Identification and nucleotide sequence of a human locus homologous to the v-myc oncogene of avian myelocytomatosis virus MC29. Nature 301:722, 1983.

Collins S, Groudine M. Amplification of endogenous myc-related DNA sequences in a human myeloid leukemia cell line. Nature 298:679, 1982.

Croce CM, Thierfelder W, Erikson J, et al. Transcriptional activation of an unrearranged and untranslocated c-myc oncogene by translocation of a C_λ locus in Burkitt lymphoma cells. Proc Natl Acad Sci USA 80:6922, 1983.

Dalla-Favera R, Bregni M, Erikson J, et al. Human c-myc oncogene is located on the region of chromosome 8 that is translocated in Burkitt's lymphoma cells. Proc Natl Acad Sci USA 79:7824, 1982.

Dalla-Favera R, Martinotti S, Gallo RC, et al. Translocation and rearrangements of the c-myc oncogene locus in human undifferentiated B-cell lymphomas. Science 219:963, 1983.

Dalla-Favera R, Wong-Staal F, Gallo RC. Onc gene amplification in promyelocytic leukemia cell line HL-60 and primary leukemic cells of the same patient. Nature 299:61, 1982.

Evan GI, Hancock DC. Studies on the interaction of the human c-myc protein with cell nuclei: p62 c-myc as a member of a discrete subset of nuclear proteins. Cell 43:253, 1985.

Gonda TJ, Metcalf D. Expression of myb, myc, and fos proto-oncogenes during the differentiation of a murine myeloid leukemia. Nature 310:249, 1984.

Haluska FG, Finver S, Tsujimoto Y, et al. The t(8:14) chromosomal translocation occurring in B-cell malignancies results from mistakes in V-D-J joining. Nature 324:158, 1986.

Hamlyn PH, Rabbits TH. Translocation joins c-myc and immunoglobulin gamma-1 genes in a Burkitt lymphoma revealing a third exon in the c-myc oncogene. Nature 304:135, 1983.

Hann SR, Thompson CB, Eisenman RN. c-myc oncogene protein synthesis is independent of the cell cycle in human and avian cells. Nature 314:366, 1985.

Heikkila R, Schwab G, Wickstrom E. A c-myc antisense oligodeoxynucleotide inhibits entry into S phase but not progress from G_0 to G_1. Nature 328:445, 1987.

Keath EJ, Caimi PG, Cole MD. Fibroblast lines expressing activated c-myc oncogenes are tumorigenic in nude mice and syngeneic animals. Cell 39:339, 1984.

Kelly K, Cochran BH, Stiles CD, et al. Cell-specific regulation of the c-myc gene by lymphocyte mitogens and platelet-derived growth factor. Cell 35:603, 1983.

Klein G. Specific chromosomal translocations and the genesis of B-cell-derived tumors in mice and men. Cell 32:311, 1983.

Land H, Parada LF, Weinberg RA. Tumorigenic conversion of primary embryo fibroblasts requires at least two cooperating oncogenes. Nature 304:596, 1983.

Little CD, Nau MM, Carney DN, et al. Amplification and expression of the c-myc oncogene in human lung cancer cell lines. Nature 306:194, 1983.

Makino R, Kayashi K, Sugimara T. c-myc transcript is induced in rat liver at a very early stage of regeneration or by cycloheximide treatment. Nature 310:697, 1984.

Möröy T, Marchia A, Etiemble J, et al. Rearrangement and enhanced expression of c-myc in hepatocellular carcinoma of hepatitis virus infected woodchucks. Nature 324:276, 1986.

Neil JC, Hughes D, McFarlane R, et al. Transduction and rearrangement of the myc gene by feline leukemia virus in naturally occurring T-cell lymphomas. Nature 308:814, 1984.

Nishikura K, ar-Rushdi A, Erikson J, et al. Differential expression of the normal and of the translocated c-myc oncogenes in B cells. Proc Natl Acad Sci USA 80:4822, 1983.

Payne GS, Bishop JM, Varmus HE. Multiple arrangements of viral DNA and an activated host oncogene (c-myc) in bursal lymphomas. Nature 295:209, 1982.

Persson H, Hennighausen L, Taub R, et al. Antibodies to human c-myc oncogene product: evidence of an evolutionarily conserved protein induced during cell proliferation. Science 225:687, 1984.

Peschle C, Mavilio F, Sposi NM, et al. Translocation and rearrangement of c-myc into immunoglobulin α heavy chain locus in primary cells from acute lymphocytic leukemia. Proc Natl Acad Sci USA 81:5514, 1984.

Pfeifer-Ohlsson S, Goustin AS, Rydnert J, et al. Spatial and temporal pattern of cellular myc oncogene expression in developing human placenta: implications for embryonic cell proliferation. Cell 38:585, 1984.

Rabbits TH, Foster A, Hamlyn P, et al. Effect of somatic mutation within translocated c-myc genes in Burkitt lymphoma. Nature 309:592, 1984.

Ralston R, Bishop MM. The protein products of the oncogenes myc, myb, and adenovirus E1A are structurally related. Nature 306:803, 1982.

Tanaka K, Isselbacher KJ, Khoung G, et al. Reversal of oncogenesis by the expression of a major histocompatibility complex class I gene. Science 228:26, 1985.

Thompson CB, Challoner PB, Neiman PE, et al. Levels of c-myc oncogene mRNA are invariant throughout the cell cycle. Nature 314:363, 1985.

Watt RA, Shatzman AR, Rosenberg M. Expression and characterization of the human c-myc DNA-binding protein. Mol Cell Biol 5:448, 1985.

Zhon RP, Kan N, Papas T, et al. Mutagenesis of avian carcinoma virus MH2: only one of two potential transforming genes transforms fibroblasts. Proc Natl Acad Sci USA 82:6389, 1985.

N-*myc*

Brodeur GG, Seeger RC, Schwab M, et al. Amplification of N-myc in untreated human neuroblastomas correlates with advanced disease stage. Science 224:1121, 1984.

Kohl NE, Legouy E, DePinho RA, et al. Human N-myc is closely related in organization and nucleotide sequence to c-myc. Nature 319:73, 1986.

Lee WH, Murphree AL, Benedict WF. Expression and amplification of the N-myc gene in primary retinoblastoma. Nature 309:458, 1984.

Schwab M, Ellison J, Busch M, et al. Enhanced expression of the human gene N-myc consequent to amplification of DNA may contribute to malignant progression of neuroblastoma. Proc Natl Acad Sci USA 81:4940, 1984.

Schwab M, Varmus HE, Bishop JM. Human N-myc gene contributes to neoplastic transformation of mammalian cells in culture. Nature 316:160, 1985.

Seeger RC, Brodeur GM, Sather H, et al. Association of multiple copies of the N-myc oncogene with rapid progression of neuroblastoma. N Engl J Med 313:1111, 1985.

Slamon D, Boone T, Seeger R, et al. Identification and characterization of the protein encoded by the N-myc oncogene. Science 232:768, 1986.

Wong AJ, Ruppert JM, Eggelston I, et al. Gene amplification of c-myc and N-myc in small cell carcinoma of the lung. Science 233:461, 1986.

L-*myc*

Kirsch O, McBride W, Bertness V, et al. L-myc, a new myc-related gene amplified and expressed in human small cell lung cancer. Nature 318:69, 1985.

myb

Katzen AL, Komberg TB, Bishop JM. Isolation of the proto-oncogene c-myb from D. melanogaster. Cell 41:449, 1985.

Klempnauer KH, Gonda TJ, Bishop JM. Nucleotide sequence of the retrovirus leukemia gene v-myb and its cellular progenitor c-myb: the architecture of a transduced oncogene. Cell 31:453, 1982.

Klempnauer K-H, Symonds G, Evan G, et al. Subcellular location of proteins encoded by viral and cellular myb genes. Cell 37:537, 1984.

Thompson CB, Challoner P, Neiman PE, et al. Expression of the c-myb proto-oncogene during cellular proliferation. Nature 319:374, 1986.

c-*fos*

Curran T, Miller AD, Zokas L, et al. Viral and cellular fos proteins: a comparative analysis. Cell 36:259, 1984.

Franza BR, Rauscher FJ III, Josephs SF, et al. The fos complex and fos-related antigens recognize sequence elements that contain AP-1 binding sites. Science 239:1150, 1988.

Greenberg ME, Ziff EB. Stimulation of 3T3 cells induces transcription of the c-fos proto-oncogene. Nature 311:433, 1984.

Jenuwein T, Muller R. Structure-Function analysis of fos protein: a single amino acid change activates the immortalizing potential of v-fos. Cell 48:647, 1987.

Jenuwein T, Muller D, Curran T, et al. Extended life span and tumorigenicity of nonestablished mouse connective tissue cells transformed by the fos oncogene of FBR-MuSV. Cell 41:629, 1985.

Kruijer W, Cooper JA, Hunter T, et al. Platelet-derived growth factor induces rapid but transient expression of the c-fos gene and protein. Nature 312:711, 1984.

Marx J. The fos gene as "master switch." Science 237:854, 1987.

Miller AD, Curran T, Verma IM. c-fos protein can induce cellular transformation: a novel mechanism of activation of a cellular oncogene. Cell 36:51, 1984.

Mitchell RL, Zokas L, Schreiber RD, et al. Rapid induction of the expression of proto-oncogene fos during human monocytic differentiation. Cell 40:209, 1985.

Muller R, Bravo R, Burckhardt J, et al. Induction of c-fos gene and protein by growth factors precedes activation of c-myc. Nature 312:716, 1984.

Treisman R. Identification of a protein-binding site that mediates transcriptional response of the c-fos gene to serum factors. Cell 46:567, 1986.

Van Beveren C, van Straaten F, Curran T, et al. Nucleotide sequence analysis of gene reveals that viral and cellular fos gene products have different carboxy termini. Cell 32:1241, 1983.

v-*ski*

Li Y, Turck CM, Teumer J, et al. Unique sequence, ski, in Sloan-Kettering avian retroviruses with properties of a new cell derived oncogene. J Virology 57:1065, 1986.

Stavnezar E, Barkas AE, Brennan LA, et al. Transforming Sloan-Kettering viruses generated from the cloned v-ski oncogene by in vitro and in vivo recombination. J Virol 57:1073, 1986.

p53

Eliyahu D, Raz A, Gruss P, et al. Participation of p53 cellular tumour antigen in transformation of normal embryonic cells. Nature 312:646, 1984.

Jenkins JR, Rudge K, Chumakov P, et al. The cellular oncogene p53 can be activated by mutagenesis. Nature 317:816, 1985.

Jenkins JR, Rudge K, Currie GA. Cellular immortalization by a cDNA clone encoding the transformation-associated phosphoprotein p53. Nature 312:651, 1984.

Koeffler HP, Miller C, Nicolson MA, et al. Increased expression of p53 protein in human leukemia cells. Proc Natl Acad Sci USA 83:4035, 1986.

Miller C, Mohandas T, Wolf D, et al. Human p53 localized to short arm of chromosome 17. Nature 319:783, 1986.

Mowat M, Cheng A, Kimura N, et al. Rearrangements of the cellular p53 gene in erythroleukaemic cells transformed by Friend virus. Nature 314:633, 1985.

Parada LF, Land H, Weinberg RA, et al. Cooperation between gene encoding p53 tumour antigen and ras in cellular transformation. Nature 312:649, 1984.

Reich NC, Levine AJ. Growth regulation of a cellular tumor antigen, p53, in nontransformed cells. Nature 308:199, 1984.

Wolf D, Harris N, Rotter V. Reconstitution of p53 expression in a nonproducer Ab-MuLV-transformed cell line by transfection of a functional p53 gene. Cell 38:119, 1984.

SV40

McCormick F, Clark R, Harlow E, et al. SV40 T antigen binds specifically to a cellular 53K protein in vitro. Nature 292:63, 1981.

Rubin H, Figge J, Bladon MT, et al. Role of small T antigen in the acute transforming activity of SV40. Cell 30:469, 1982.

Topp WC, Rifkin DR, Sleigh MJ. SV40 mutants with an altered small-T protein are tumorigenic in newborn hamsters. Virology 119:169, 1981.

Polyoma

Bolen JB, Thiele CJ, Israel MA, et al. Enhancement of cellular src gene product associated tyrosyl kinase activity following polyoma virus infection and transformation. Cell 38:767, 1984.

Courtneidge SA. Activation of the pp60c-src kinase by middle T antigen binding or by dephosphorylation. EMBO J 4:1471, 1985.

Courtneidge SA, Smith AE. Polyoma virus transforming protein associates with the product of the c-src cellular gene. Nature 303:435, 1983.

Garcea RL, Benjamin TL. Host range transforming gene of polyoma virus plays a role in virus assembly. Proc Natl Acad Sci USA 80:3613, 1983.

E1A

Cook JL, Walker TA, Lewis AM Jr, et al. Expression of the adenovirus E1A oncogene during cell transformation is sufficient to induce susceptibility to lysis by host inflammatory cells. Proc Natl Acad Sci USA 83:6965, 1986.

Elsen PVD, De Pater S, Houweling A, et al. The relationship between region E1A and E1B of human adenoviruses in cell transformation. Gene 18:175, 1982.

Ruley HE. Adenovirus early region 1A enables viral and cellular transforming genes to transform primary cells in culture. Nature 304:602, 1983.

11
Additional Oncogenes

Overview

Some transforming genes have been identified in retroviruses or by transfection assays. Others have been associated with chromosome translocations. In previous chapters we described functional groups of these oncogenes. In this chapter we describe several oncogenes that are as yet unclassified or are novel in their apparent mechanisms (Table 11.1). It is anticipated that eventually the function of these genes will be discovered and their role in normal cellular growth control defined.

New Oncogenes Derived from Retroviruses

jun ONCOGENE

A few oncogenes are still being identified in novel retroviruses. One such gene, termed *jun*, is transduced by avian sarcoma virus 17. This virus was isolated from a chicken fibrosarcoma. Unlike many previous oncogenes isolated from sarcoma viruses, the *jun* oncogene sequence bears no relation to sequences encoding tyrosine-specific protein kinases. Proto-oncogene sequences hybridizing to *jun* have been identified in human, chicken, rat, mouse, and quail genomic DNA. The *jun* protein recognizes and binds to the same DNA sequence as the yeast transcriptional activator protein, GCN4, with which it shares 45% amino acid homology to the C terminal 66 amino acids. The human c-*jun* proto-oncogene encodes a DNA binding protein with the structural and functional properties of transcription factor AP-1.

v-*rel* ONCOGENE

The v-*rel* oncogene, which has two subcellular locations, is the oncogene of the reticuloendotheliosis virus, strain T (REV-T). REV-T is a highly oncogenic replication-defective virus that induces an early lymphoid neoplasm in turkeys. The v-*rel* gene is necessary and sufficient for lymphoid cell transformation. Its product is an *env-onc* fusion protein of 503 amino acids. REV-T transforms lymphoid cells but not primary chicken embryo fibroblasts (CEFs) despite the

TABLE 11.1. Additional oncogenes.

New oncogenes derived from retroviruses
jun
rel
Viral oncogenes that modulate the effects of other oncogenes
ets
Oncogenes activated solely by viral insertion
int-1
int-2
pim
Fis-1
fim-1
fim-2
Oncogenes isolated through transfection assays
mas
hst
KS3
lca
ret
thyroid cancer-associated oncogene
Oncogenes associated with chromosome translocation
bcl-1
bcl-2
tcl-1
tcl-2
pvt
Tumor promotion genes
pro

fact that the 59-kilodalton (kD) v-*rel* protein ($p59^{v\text{-}rel}$) is expressed at equal levels in both cell types. In CEF $p59^{v\text{-}rel}$ is localized to the nucleus, but in REV-T transformed lymphoid cells the protein is found only in the cytoplasm. These data suggest that the cytoplasmic presence of $p59^{v\text{-}rel}$ is important in the transformation of cells by REV-T. c-*rel*, the normal cellular homologue of v-*rel*, has been identified but not well characterized.

Viral Oncogenes that Modulate the Effects of Other Oncogenes: *ets*

The *ets* oncogene is associated with erythroid leukemia and possibly with macrophage development. E26 is a replication-defective avian retrovirus that induces erythroblastosis and myeloblastosis in chickens and transforms erythroblasts and myeloid cells in culture. The virus contains two cell-derived sequences, *myb* and *ets*, that are fused with the viral *gag* protein to produce a p135 transforming protein. Because avian myeloblastosis virus (AMV), containing *myb* as its sole oncogene, produces only myeloid leukemias, the *ets* domain of E26 is assumed to contribute to the erythroblast tropism of this virus.

Gene mapping studies have identified two *ets* proto-oncogenes. The *ets*-1 gene maps to chromosome 11q23 in man, 9 in the mouse, and D1 in the cat. The *ets*-2 gene maps to chromosome 21q23 in man, 16 in the mouse, and C2 in the cat. One case of acute myelomonocytic leukemia was found to have a rearranged and 30-fold amplified human *ets*-1 gene. The patient's cells had a homogeneously staining region at 11q23. A case of small-cell lymphocytic lymphoma was found to have an inverted insertion that involved 11q23. Alteration of the Hu-*ets*-1 gene could explain the high frequency of karyotypic abnormalities involving band 11q23 in acute nonlymphocytic leukemia (ANLL) of the myelomonocytic phenotype. Band 11q23 is involved in a number of translocations, deletions, insertions, and homogeneously staining regions in other leukemias, lymphomas, and myeloproliferative syndromes.

The human *ets*-1 locus encodes a single mRNA of 6.8 kilobases (kb) whereas the *ets*-2 locus encodes three mRNAs of 4.7, 3.7, and 2.7 kb. These RNAs encode proteins of 54 kD and 64, 62, and 60 kD, respectively. The latter group of proteins is expressed at high levels in both quiescent and dividing avian macrophages. Their synthesis is specifically and rapidly induced after differentiation of macrophages induced by phorbol esters. Thus c-*ets* genes may play a role in normal macrophage growth and differentiation.

Oncogenes Activated Solely by Viral Insertion

Table 5.3 listed cellular proto-oncogenes activated by viral insertion, and several of these genes were discussed in previous chapters. Five unclassified oncogenes activated in this fashion are described below.

int GENES

The *int* genes are associated with mammary tumors. During the 1930s John Bittner, at The Jackson Memorial Laboratory in Bar Harbor, Maine, showed that the development of mouse mammary tumors could be transmitted via milk. Moreover, development of mammary carcinoma was hormonally dependent because male mice of the high tumor strain rarely developed the tumors unless treated with estrogens. During the early 1950s B-type particles (105 nm in diameter with eccentric nuclei) were identified in the milk. Thus mammary tumors in these mice were shown to be transmitted by retroviruses. Subsequent research showed that there are multiple murine mammary tumor viruses (MMTVs). They often exist in the mouse germ line as integrated proviruses, similar in gene structure to the more prevalent type C retroviruses.

A number of laboratories have investigated the mechanism of oncogenesis by MMTV. The virus does not carry a transduced oncogene but, instead, acts as an insertional mutagen. With most MMTV-induced tumors, a provirus can be detected in either or both of two distinct loci, *int*-1 and *int*-2, situated on mouse chromosomes 15 and 7, respectively. The *int*-1 and *int*-2 genes are unrelated and *int*-1 has no appreciable homology to other known oncogenes. The insertion site of MMTV is always outside (i.e., 5′ or 3′ to) the coding region of the *int*-1 gene.

Infection of mammary epithelial cells by retroviral vectors carrying the *int*-1 gene produces striking morphological changes. No such changes are noted when fibroblast are so infected. The mammary cells grow to a higher saturation density and produce foci of piled cells that do not exhibit contact inhibition.

The *int*-1 gene is comprised of four exons coding for a glycoprotein of 370 amino acids. This protein has a hydrophobic leader sequence and a high content of cysteine residues suggesting that it may be an extracellular membrane protein. Alternatively, it may be secreted. In adult mice, expression of *int*-1 is restricted to the testis of sexually mature mice. During murine development *int*-1 transcripts are detectable in the embryo from 8.5 to 12.5 days from conception. Using a more sensitive in situ hybridization assay, Wilkinson and co-workers detected *int*-1 transcripts in a subset of neural plate cells from 9.0 to 14.5 days post conception. A specific pattern of expression was detected indicating regional differentiation of the neuroepithelium prior to neural tube closure. The precise role of *int*-1 in normal development as well as its role in mammary tumorigenesis remains to be elucidated.

The *int*-2 gene is even less well characterized. Its expression appears limited to early embryogenesis in normal murine development. Its predicted amino acid sequence bears a resemblance to the basic form of bovine fibroblast growth factor.

pim Oncogene

The *pim* proto-oncogene first was identified as an insertionally activated sequence in murine T cell lymphomas induced by retroviruses. It has an open reading frame encoding a protein of 313 amino acids. Proviruses always integrate outside the coding domain, with most insertions in the 3′ terminal exon. Although the function of the gene is unknown, it shares homology with serine-protein kinases.

The human homologue, h*pim*, maps to human chromosome 6p21. The h*pim* gene is expressed as a 3.2-kb message in various human cell lines of hematopoietic lineage, especially the K562 erythroleukemia cell line, which contains a cytogenetically demonstrable rearrangement in the 6p21 region. The reciprocal translocation [t(6;9)(p21;q33)] has been described in myeloid leukemia and may involve the h*pim* gene.

Friend Murine Leukemia Virus Integration Sites: *fis*-1, *fim*-1, *fim*-2

The Friend murine leukemia virus (F-MuLV) induces a wide variety of leukemias, T and B cell lymphomas, and myeloblastic leukemias. In vitro this virus transforms myelomonocytic cells in a multi-step process that mimics in vivo leukemogenesis.

By cloning integration sites of F-MuLV, two groups have discovered unique potential oncogenes. A 1.5 kb region common to 4 of 35 F-MuLV induced tumors (two myelogenous leukemias and two lymphomas) was mapped to murine chromosome 7. This region, designated *Fis*-1, was distinct from three other murine proto-oncogenes previously mapped to chromosome 7. Two different integration regions were detected with probes to host-virus junction sequences. One probe detected rearrangements in 2 of 42 screened myelogenous leukemia cell lines (*fim*-1) while the second detected rearrangements in 6 of the 42 cell lines (*fim*-2). Each rearrangement resulted from integration of an F-MuLV provirus. Each region was less than 3 kb in length and distinct from previously described proto-oncogenes or viral integration sites including *int*-1, *int*-2, *Fis*-1, or *pvt* (see below).

Oncogenes Isolated Through Transfection Assays

mas ONCOGENE

The *mas* oncogene, encoding a novel transmembrane protein, was identified by a variation of the 3T3 transfection technique. Genomic DNA derived from a human epidermoid tumor was transfected into NIH 3T3 cells, which then were injected into a nude mouse. From the resulting tumor, scientists at Cold Spring Harbor Laboratory identified a novel gene that apparently had been activated by rearrangement of its 5′ untranslated region (thought to control gene expression) during the original transfection process.

The *mas* cDNA encodes a protein of 325 amino acids. It is notable for having seven transmembrane hydrophobic regions. In this regard the gene is unique among oncogenes and is similar to membrane proteins, including the acetylcholine receptor, the beta-adrenergic receptor, and visual rhodopsins. The gene, in fact, shares sequences with red and blue opsin and rhodopsin. The tissue and tumor expression of this novel oncogene have not yet been determined.

Only one transforming gene has previously been described that encodes a protein with multiple transmembrane domains. This gene, called LMP, is found in Epstein-Barr virus. It encodes a plasma membrane protein that can transform rat-1 cells, but it differs structurally from the *mas* protein and is unable to transform NIH 3T3 cells.

It is tantalizing to speculate that, like rhodopsin, the *mas* gene product may interact with GTP binding proteins as part of a growth regulatory pathway or as an ion channel.

hst ONCOGENES

Carcinoma of the stomach is among the most common cancers found in Japan. A group of Japanese researchers has identified a novel oncogene associated with stomach cancer, termed *hst*, by transfection of NIH 3T3 cells. From a total of 21 human stomach cancers, 16 stomach cancer lymph node metastases, and 21 apparently normal noncancerous specimens of stomach mucosa, these scientists found three samples of DNA with transforming activity. All samples contained the same DNA fragment. The novel gene identified is apparently unrelated to the 24 previously identified oncogenes, including members of the *ras* family. It is of interest that one of the positive samples was from normal stomach mucosa derived from a patient with moderately well differentiated adenocarcinoma. Cloning of cDNA from a 3.0 kb mRNA transcript predicted an amino acid sequence for the protein very similar to that of fibroblast growth factor.

KS3: An Oncogene Associated With Kaposi's Sarcoma

Kaposi's sarcoma (KS) is generally considered to be an angiosarcoma of endothelial/mesenchymal origin. A transforming gene was isolated from a KS specimen by transfection of 3T3 cells. Subsequently the gene was cloned and found to express two transcripts of 3.5 kb and 1.2 kb. The 1.2 kb transcript encodes a protein of 206 amino acids with considerable sequence homology to both basic and acidic fibroblast growth factor. The KS3 protein was sufficient to transform cells in culture and represents a third example of a fibroblast growth factor related *onc*-protein, the others being *hst* and *int*-2 (see above).

lca Oncogene

The *lca* (liver cancer) oncogene was first identified by transformation of NIH 3T3 cells with DNA from a primary hepatocellular carcinoma. The transformed cells demonstrated anchorage-independent growth, had a unique morphology, and were tumorigenic in nude mice. Subsequently an identical *lca* gene was isolated from a second patient with hepatocellular carcinoma.

The *lca* gene maps to human chromosome 2, which also carries the N-*myc* and *rel* genes. However, the cloned gene isolated from a lambda phage library showed no cross-hybridization with any known oncogene.

The mechanism of activation of this gene is unclear. No gross rearrangements are found when restriction patterns of tumor and placental (germ line) DNA are compared. The role of the *lca* gene in liver cancer is also unclear. The gene is unable to transform primary cultures of human or rat hepatocytes. Its relation, if any, to the hepatitis B virus is not known.

ret Oncogene

The *ret* oncogene is a human transforming gene activated during transfection. Many oncogenes have been identified by the 3T3 focus formation assay (see Chapter 6). Genes activated during the transfection procedure have been described frequently, the activation possibly occurring via artificial rearrangement during transfection (see above). Included in this category is *ret*, an oncogene isolated after transfection of 3T3 cells with human lymphoma DNA. The *ret* proto-oncogene spans 37 kb of DNA and consists of two segments that are unlinked in both the lymphoma DNA used to isolate the gene and in normal human DNA. The structure and function of this artificially recombined gene is still undefined.

Thyroid Cancer-Associated Oncogene

An un-named oncogene was isolated by transfection of NIH-3T3 cells with high molecular weight DNA isolated from five thyroid papillary carcinomas and two of their respective lymph node metastases. No such gene was recovered from normal thyroid DNA. It is of interest that this gene lacks homology with any previously described oncogene yet the same sequence was activated in five different papillary tumors; this result suggests tissue-specific activation. Little more is known about the gene at this time.

Cancer-Associated Chromosome Translocation

bcl-1 and *bcl*-2 in B Cell Lymphomas

Certain malignancies are associated with specific chromosome translocations. Studies of sequences adjacent to such translocation sites can be used to demonstrate potential new oncogenes. For example, mapping of such breakpoint regions has directed scientists to novel genes that may play a role in lymphoid neoplasia. With non-Hodgkin's lymphomas and chronic lymphocytic leukemias, translocations between chromosomes 11 and 14 [t(11;14)(q13–q32)] and between chromosomes 14 and 18 [t(14;18)(q32–q21)] occur frequently. Tsujimoto et al. and Cleary used the immunoglobulin heavy chain locus at chromosome 14q32 as a probe to clone the breakpoints on chromosomes 11 and 18 involved in these translocations. The breakpoints were determined to cluster in particular regions on chromosomes 11 and 18, and a novel transcriptional unit expressed from these regions was demonstrated. The new genes were named *bcl* for "B cell lymphoma"; *bcl*-1 is on chromosome 11q13, and *bcl*-2 is on chromosome 18q21. Their exact role is presently unclear, but their importance is underscored by the frequency of their involvement in the translocations seen in B cell tumors.

The *bcl*-2 gene was further characterized by cloning and sequencing *bcl*-2 cDNA. At least 2 exons and three different transcripts were found. All three transcripts (8.5 kb, 5.5 kb, 3.5 kb) overlap within the first exon but only the largest two contain sequences of the second exon. The 8.5 kb and 5.5 kb transcripts carry two overlapping reading frames. One codes for a protein of 239 amino acids (*bcl*-2α:p26) and the other a protein of 205 amino acids (*bcl*-2β:p22). The transcripts appear to use different polyadenylation sites. The predicted amino acid sequences show neither leader peptide nor transmembrane domain suggesting they are not secreted nor localized to the plasma membrane. It is of interest that the *bcl*-2 products in follicular lymphomas appear identical to the normal *bcl*-2 products.

tcl-1 and *tcl*-2 in T Cell Lymphomas

A set of T cell lymphoma genes, *tcl*-1 and *tcl*-2, also has been identified. A translocation of the beta locus of the T cell receptor from chromosome 7(q32–q34) to chromosome 14 appears to have "activated" *tcl*-1. A translocation between chromosome 11(p13) and the T cell receptor alpha locus at chromosome 14q11 appears to have "activated" *tcl*-2.

It is emphasized that although these novel genes are associated with B and T cell lymphomas their role in oncogenesis is not known, and their unaided oncogenic potential has not been tested. Because oncogenesis is a multistep process, and because these and similar changes are acquired by lymphoma cells, it is assumed that they contribute to the overall process.

The technique of cloning genes activated by translocations will undoubtedly uncover many new genes associated with the process of oncogenesis.

pvt-1 Locus

The *pvt*-1 locus is associated with plasmacytoma varient translocations. Most translocations associated with Burkitt's lymphoma, non-Burkitt's lymphoma, and murine plasmacytomas involve the c-*myc* gene (chromosome 8 in humans, chromosome 15 in mice) and the immunoglobulin heavy chain locus (chromosome 14 in humans, chromosome 12 in mice). Varient translocations occur some 10–20% of the time and involve chromosome 8 (humans) or 15 (mice) and light chain loci (chromosomes 2 (κ) or 22 (λ) in humans, chromosome 6 (κ) in mice). These latter events have been less well characterized. In murine plasmacytomas, analysis of a t(6;15) translocation revealed that 5 of 9 cases did not involve the c-*myc* gene directly but instead translocated a 4.5 kb region designated *pvt*-1 on murine chromosome 15 to the joining (J) region of the κ constant region gene on murine chromosome 6. Despite its significant distance from the c-*myc* gene, translocation of *pvt*-1 appeared to activate c-*myc* suggesting a role for *pvt*-1 in regulation of the c-*myc* gene.

Tumor Promotion Genes

pro Oncogenes

Two nonhomologous genes, denoted *pro*-1 and *pro*-2, were cloned from a mouse cell line, JB6, which is 100-fold more sensitive to the promotion of neoplastic transformation (i.e., JB6 is a P^+ cell line) by phorbol esters than similar "P^-" cell lines. These genes confer sensitivity to phorbol esters when introduced into P^-

cells. Although transcripts of these genes have been identified in both mouse and human carcinoma cell lines, their structure and function are unknown. The meaning of the provocative finding of specific genes related to tumor progression is not known at present.

Summary

We have discussed several novel oncogenes in this chapter. Although all of these genes are associated with the process of oncogenesis, many are not, strictly speaking, "oncogenes" (i.e., they have not been shown to transform cells). However, cancer is a multistep process, and genes involved in this process are important. Some of these genes eventually may prove to be important cancer-associated genes.

Bibliography

jun

Bohman D, Bos TJ, Admon A, et al. Human proto-oncogene c-jun encodes a DNA binding protein with structural and functional properties of transcription factor AP-1. Science 238:1386, 1987.

Maki Y, Bos TJ, Davis C, et al. Avian sarcoma virus 17 carries a new oncogene, jun. Proc Natl Acad Sci USA 84:2848, 1987.

Maki Y, Davis CT, Starbuck M, et al. A biologically active clone of avian sarcoma virus 17 and the sequence of its one gene, jun. Presented at the Second Annual Meeting on Oncogenes, Hood College, Frederick, MD.

Struhl K. The DNA binding domains of the jun oncoprotein and the yeast GCN4 transcriptional activator protein are functionally homologous. Cell 50:841, 1987.

rel

Gilmore TD, Temim HM. Different localization of the product of the v-rel oncogene in chicken fibroblasts and spleen cells correlates with transformation by REV-T. Cell 44:791, 1986.

Wilhelmsen K, Eggleton K, Temim H. Nucleic acid sequences of the oncogene v-rel in reticuloendotheliosis virus strain T and its cellular homologue, the proto-oncogene c-rel. J Virol 52:172, 1984.

ets

Chen JH. The proto-oncogene c-ets is preferentially expressed in lymphoid cells. Mol Cell Biol 5:2993, 1985.

Rovigatti U, Watson DK, Yunis JJ. Amplification and rearrangement of Hu-ets-1 in leukemia and lymphoma with involvement of 11q23. Science 232:398, 1986.

pim

Cuypers H, Selten G, Bernes A, et al. Assignment of the human homologue of pim-1, a mouse gene implicated in leukmogenesis, to the pter-q12 region of chromosome 6. Hum Genet 72:262, 1986.

Nagarcijan L, Louis E, Tsujimoto Y, et al. Localization of the human pim oncogene (pim) to a region of chromosome 6 involved in translocations in acute leukemias. Proc Natl Acad Sci USA 83:2556, 1986.

Selten G, Cuypers HT, Loelens W, et al. The primary structure of the putative oncogene pim-1 shows extensive homology with protein kinases. Cell 46:603, 1986.

int-1 AND *int*-2

Casey G, Smith R, McGillivray D, et al. Characterization and chromosome assignment of human homologue of int-2, a potential proto-oncogene. Mol Cell Biol 6:502, 1986.

Dickson C, Peters G. Potential oncogene product related to growth factors. Nature 326:833, 1987.

Dickson C, Smith R, Brookes S, et al. Tumorigenesis by mouse mammary tumor virus: proviral activation of a cellular gene in the common integration region int-2. Cell 37:529, 1984.

Fung Y, Shackleford G, Brown A, et al. Nucleotide sequence and expression in vitro of cDNA derived from mRNA of int-1, a provirally activated mouse mammary oncogene. Mol Cell Biol 5:3337, 1985.

Moore R, Casey G, Brooks S, et al. Sequence, topology, and protein coding potential of mouse int-2: a putative oncogene activated by MMTV. EMBO J 5:919, 1986.

Nuuse R, Varmus HE. Many tumors induced by the mouse mammary tumor virus contain a provirus integrated in the same region of the host genome. Cell 31:99, 1982.

Van Ooyen A, Nuuse R. Structure and nucleotide sequence of the putative mammary oncogene int-1; proviral insertions leave the protein-encoding domain intact. Cell 39:233, 1984.

Varmus HE. Recent evidence for oncogenesis by insertion mutagenesis and gene activation. Cancer Surv 2:309, 1982.

Wilkinson DG, Bailes J, McMahon A. Expression of the proto-oncogene int-1 is restricted to specific neural cells in the developing mouse embryo. Cell 50:79, 1987.

fim-1 AND *fim*-2

Sola B, Fichelson S, Bordereaux D, et al. fim-1 and fim-2: two new integration regions of Friend murine leukemia virus in myeloblastic leukemias. J Virol 60:781, 1986.

Fis

Silver J, Kozak C. Common proviral integration region on mouse chromosomes 7 in lymphomas and myelogenous leukemias induced by Friend murine leukemia virus. J Virol 57:526, 1986.

mas

Young D, Waitches G, Birchmeier C, et al. Isolation and characterization of a new cellular oncogene encoding a protein with multiple potential transmembrane domains. Cell 45:711, 1986.

hst

Sakamoto H, Mori M, Taira M, et al. Transforming gene from human stomach cancers and a noncancerous portion of stomach mucosa. Proc Natl Acad Sci USA 83:3997, 1986.

Taira M, Yoshida T, Miyagawa K, et al. cDNA sequence of human transforming gene hst and identification of the coding sequence required for transforming activity. Proc Natl Acad Sci USA 84:2980, 1987.

KS3

Delli Bovi P, Basilico C. Isolation of a rearranged human transforming gene following transfection of Kaposi's sarcoma DNA. Proc Natl Acad Sci USA 84:5660, 1987.

Delli Bovi P, Curatola A, Kern F, et al. An oncogene isolated by transfection of Kaposi's sarcoma DNA encodes a growth factor that is a member of the FGF family. Cell 50:729, 1987.

lca

Ochiya T, Fujiyama A, Fukushige S, et al. Molecular cloning of an oncogene from a human hepatocellular carcinoma. Proc Natl Acad Sci USA 83:4993, 1986.

ret

Takahasi M, Ritz J, Cooper GM. Activation of a novel human transforming gene, ret, by DNA rearrangement. Cell 42:581, 1985.

Thyroid Cancer-Associated Oncogene

Fusco A, Grieco M, Santoro M, et al. A new oncogene in human thyroid papillary carcinomas and their lymph nodal metastases. Nature 328:170, 1987.

bcl-1 and *bcl*-2

Bakhshi A, Jensen JP, Goldman P, et al. Cloning the chromosome breakpoint of the t(14;18) human lymphomas: clustering around Jh on chromosome 14 and near a transcriptional unit on 18. Cell 41:899, 1985.

Cleary ML, Sklar J. Nucleotide sequence of a t(14;18) chromosomal breakpoint in follicular lymphoma and demonstration of a breakpoint cluster region near a transcriptionally active locus on chromosome 18. Proc Natl Acad Sci USA 82:7439, 1985.

Cleary ML, Galili N, Sklar J. Detection of a second t(14;18) breakpoint cluster region in human follicular lymphoma. J Exp Med 164:315, 1986.

Tsujimoto Y, Cossman J, Jaffe E, et al. Involvement of the bcl-2 gene in human follicular lymphoma. Science 228:1440, 1985.

Tsujimoto Y, Jaffe E, Cossman J, et al. Clustering of breakpoints on chromosome 11 in human B-cell neoplasms with the t(11;14) chromosome translocation. Nature 315:340, 1985.

Tsujimoto Y, Croce C. Analysis of the structure, transcripts, and protein products of bcl-2, the gene involved in human follicular lymphoma. Proc Natl Acad Sci USA 83:5214, 1986.

tcl-1 AND *tcl*-2

Erickson J, Williams DL, Finan J, et al. Locus of the alpha-chain of the T-cell receptor is split by chromosome translocation in T-cell leukemias. Science 229:784, 1985.

Isobe M, Erikson J, Emanuel BS, et al. Location of gene for beta subunit of human T-cell receptor at band 7q35, a region prone to rearrangements in T cells. Science 228:580, 1985.

Morton CC, Duby AD, Eddy RL, et al. Genes for beta chain of human T-cell antigen receptor map to regions of chromosomal rearrangement in T cells. Science 228:582, 1985.

pvt

Cory S, Graham M, Webb E, et al. Varient (6:15) translocation in murine plasmacytoma involve a chromosome 15 locus at least 72 kb from the c-myc oncogene. EMBO J 14:675, 1985.

Webb E, Adams J, Cory S. Varient (6:15) translocation in a murine plasmacytoma occurs near an immunoglobulin K gene but far from the myc oncogene. Nature 312:777, 1984.

pro

Garrity R, Hegamyer G, Colburn N. Progene expression in promotion of neoplastic transformation by TPA. Presented at the Second Annual Meeting on Oncogenes, Hood College, Frederick, MD.

12
Transgenic Mice: Direct In Vivo Assay for Oncogenes

Overview

The evidence supporting the crucial role of a small number of proto-oncogenes in the development of cancer is substantial. Most of the studies using human tissue and human cell lines give circumstantial support to the hypothesis. However, until recently it was not possible to test these genes directly in vivo independently of their association with acute retroviruses. The development of methods to introduce genes into the germ lines of mice and other species, thereby producing so-called transgenic animals, provides the means to test directly the function of cellular oncogenes in vivo. Oncogenes introduced into the mouse germ line contribute to oncogenesis in a tissue-specific manner depending on the promoter with which they are linked.

Oncogenes Introduced into Mouse Germ Line

In experiments in which oncogenes are introduced into the mouse germ line (Fig. 12.1), a cloned gene is microinjected into the male pronucleus of a fertilized ovum. The injected eggs then are implanted into the fallopian tube of a pseudopregnant foster mother. A normal gestation ensues (approximately 3 weeks), and typically one-fourth of the injected eggs develop into viable pups. DNA is extracted from the tails of the pups, and Southern blots are performed to confirm the presence of the injected gene. In most cases the gene integrates randomly into the host genome and is acquired by all tissues, including the germ cells of the transgenic mice. Thus the introduced DNA becomes a mendelian gene that can be transmitted to the recipient's offspring. Lines of mice carrying the gene of interest can be bred to study the effects on development, aging, normal physiology, and the genesis of various diseases such as cancer.

Introduced Genes Are Under Tissue-Specific Regulation

Specific regulatory elements surrounding the coding sequence of the introduced gene usually direct its expression in a tissue-specific manner. These regulatory sequences include promoters and enhancers. At times other sequences flanking

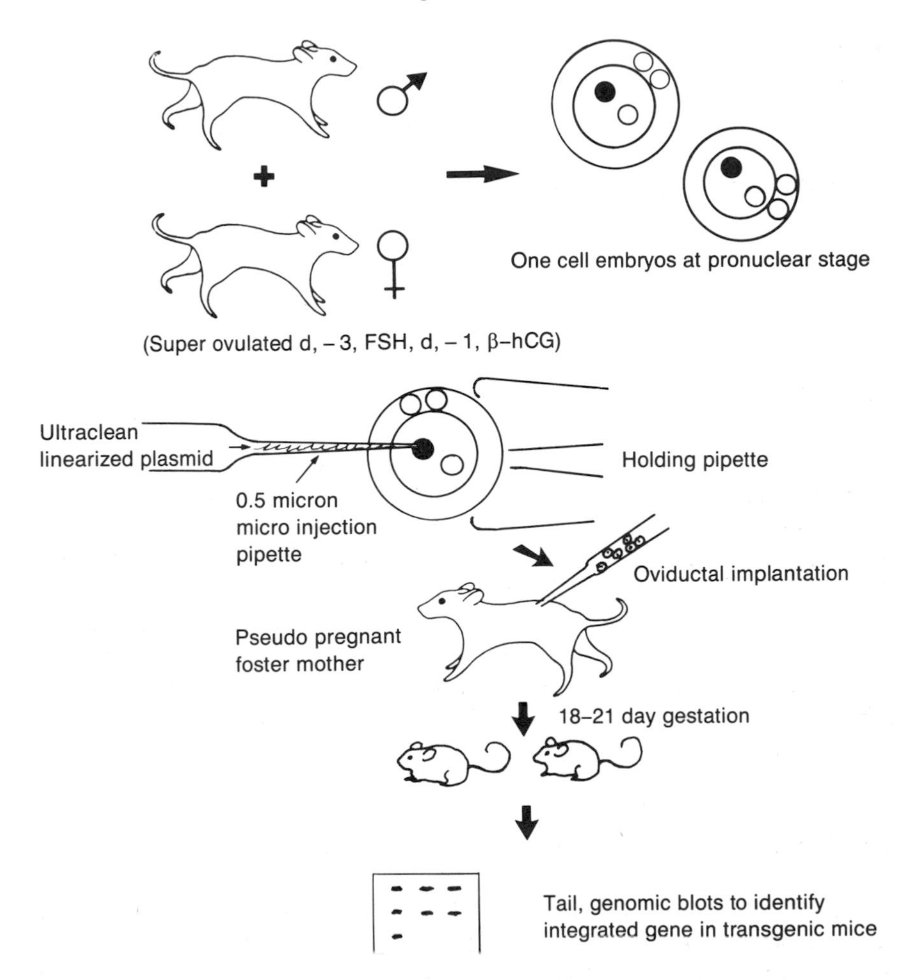

FIGURE 12.1. Transgenic mice: introduction of genes into the germ line. Female mice are superovulated with follicle-stimulating hormone (FSH) and beta human chorionic gonadotropin (β-hCG) given 3 days and 1 day, respectively, prior to egg harvest. Mice are mated overnight, and fertilized eggs are harvested. Plasmid DNA containing the gene of interest is microinjected into the male pronucleus. The eggs are transferred into the oviducts of pseudopregnant (mated with vasectomized male mice) foster mothers. Mouse pups are born after a normal gestation period. The presence of the introduced gene is established by Southern blotting of DNA extracted from their tails. These mice are observed for phenotypic effects of the gene of interest (e.g., development of tumors), and tissue is used to study its expression. Mature mice can be bred to establish lineages of mice carrying a particular gene because the injected genes are present in germ cells.

TABLE 12.1. Tissue-specific gene expression in transgenic mice.

Introduced gene	Cell/tissue type	Expression (%)[a]
Human β-globin	Erythroid progenitors	100
Rat myosin light chain	Skeletal muscle	500
Rat elastase 1	Pancreas	1200
Mouse α-fetoprotein	Liver/yolk sac	25
Mouse κ light chain	B lymphocytes	50
Mouse μ heavy chain	T and B lymphocytes	85[b]

[a] mRNA produced relative to the mRNA of the endogenous gene.
[b] Rearranged immunoglobulin heavy chain transcripts also reported in T cells.

and within a given introduced gene may influence the tissue in which the gene is expressed. For example, introduced immunoglobulin genes are expressed only in the lymphoid cells of transgenic mice (Table 12.1).

Introduced Oncogenes Cause Specific Types of Tumors in Transgenic Mice

The introduction of both viral and cellular oncogenes have been shown to promote characteristic tumors in transgenic mice influenced by tissue-specific control elements (Table 12.2). One well studied example is the tumor type induced by the SV40 large T antigen (see Chapter 3). When the native SV40 72-base pair enhancer is used, tumors of the choroid plexus develop. When the regulatory region of the insulin gene is combined with the large T antigen, pancreatic tumors develop. When the metallothionein-I control region is used, the mice develop demyelinating peripheral neuropathies, hepatocellular carcinomas, and islet cell adenomas. The tissue tropism shown in the latter experiments is less clear than in the first two examples (Table 12.2). Interestingly, an SV40-related polyoma virus (BK virus) causes the human central nervous system demyelinating disease called progressive multifocal leukoencephalopathy (PML). Transgenic mice made with BK virus DNA alone (no known oncogenes) exhibit many of the features of PML.

In addition to SV40 large T antigen, oncogenes derived from other DNA tumor viruses have been introduced into the mouse germ line. For example, transmission of bovine papilloma virus-1 (BPV-1) (see Chapter 3) through the mouse germ line results in the heritable formation of fibropapillomas of the skin. This phenotype is analogous to that observed in cattle as a result of natural BPV-1 infection. Such tumors in the transgenic mice arise at 8 to 9 months of age, particularly in areas prone to wounding. All of the tumors have extrachromosomal (i.e., episomal) DNA, whereas normal tissue has only integrated DNA.

Several studies of transgenic *"myc"* mice have greatly increased our understanding of the role this oncogene plays in the genesis of tumors (see Chapter 10). Neither the introduction of the normal c-*myc* gene nor the introduction of the truncated gene lacking the first exon leads to tumors. However, linking this gene

TABLE 12.2. Oncogenes in transgenic mice.

Oncogenes	Regulation	Tumors/ transgenic mice	Comments
c-*myc*	Normal gene	0/5	–
	Truncated c-*myc* (i.e., exons 2 and 3)	0/11	–
	Ig μ enhancer	13/15	Aggressive B lymphomas.
	Ig κ enhancer and SV40 promoter	6/15	Died at median age of 11 weeks.
	SV40 promoter	3/21	Lymphosarcoma, renal carcinoma, fibrosarcoma.
	LTR (Soule MLV)	1/13	Thymic lymphoma.
	MMTV-LTR	1/13	Novel transcripts in salivary glands.
		1/13	One animal with novel transcripts in all organs. Mice derived from this mouse developed testicular, breast, and T, B, and mast cell tumors (mice developed normally).
		2/13	Spontaneous mammary adenocarcinoma after pregnancies.
SV40 Large T antigen	SV40 enhancer (72 base pairs repeat)	7/8	Choroid plexus papillomas.
		3/8	Thymic hyperplasia.
		1/8	Nephritis.
		1/8	Polycystic kidneys.
		1/8	Hepatocellular carcinomas.
	Metallothionein/ human growth hormone	13/16	Developed weakness or paralysis.
		11/16	Developed hepatic carcinomas.
		8/16	Islet cell adenomas.
		1/16	Choroid plexus papillomas.
c-H-*ras*	Normal (codon 12-glycine) gene with elastase 1 promoter	0/5	Normal pancrease at 5–9 months.
		0/5	Morphologically abnormal pancrease in 4/5 after 11 months.
	Activated (codon 12-valine) gene with elastase 1 promoter	20/20	Pancreatic acinar tumors.
tat	HTLV-1 promoter	3/8	Soft tissue tumors.
		5/8	Thymic involution, early death.

to enhancers from the immunoglobulin heavy (μ) chain or the light (κ) chain resulted in the development of aggressive B cell lymphomas in almost two-thirds of the mice. Using the SV40 promoter alone resulted in a low frequency of sporadic tumors, whereas regulation by the long terminal repeat (LTR) of a murine T cell leukemia virus produced thymic lymphomas in 1 of 13 transgenic mice. The c-*myc* oncogene linked to the regulatory region of the mouse mammary tumor virus (MMTV) LTR produced hormone-dependent mammary carcinomas. Female mice developed mammary cancer only after sustaining the hormonal stimulus associated with several pregnancies. A subline of "*myc*" mice expressing the introduced c-*myc* gene in almost all tissues tested was particularly prone to developing tumors of multiple types. Mice derived from this strain developed testicular and hematopoietic tumors (T cell, B cell, and mast cell), as well as breast carcinomas. Despite the high oncogenic tendency, these mice develop into normal-appearing adults.

Introduction of activated human H-*ras*-1 genes (valine substituted for glycine at position 12) coupled to promoter and enhancer elements of the rat elastase-1 gene resulted in pancreatic acinar tumors in 20/20 transgenic mice. Fifteen of the 20 died or were moribund as newborns. The other five survived to adulthood; four of these were proven mosaics. None of 10 transgenic animals carrying the normal H-*ras*-1 gene developed pancreatic tumors; however four of the 10 sacrificed after 11 months of age showed histologic abnormalities in pancreatic acinar cells. Earlier experiments had shown the elastase-1 control elements to specifically regulate gene expression in pancreatic acinar cells and that timing of expression was developmentally regulated similarly to the endogenous elastase-1 gene.

On the other hand, introduction of a murine c-*fos* gene under control of the human metallothionein promoter resulted only in interference with normal bone development without inducing overt neoplasia. This result is of interest in light of the induction of osteosarcomas by v-*fos* genes (see Chapter 10).

Interesting experiments performed by Nerenberg and coworkers at the NIH establish that HTLV-1 transactivator (*tat*) gene (see Chapter 4) may be oncogenic in transgenic mice. The *tat* gene under control of the HTLV-LTR was introduced; two types of animals resulted. One type (three of eight founder animals) appeared to develop normally and reached adulthood, however all developed multiple soft tissue tumors between 13 and 17 weeks of age. Histologically the tumors were composed of spindle cells with granulocytic infiltration. Cells expressed elevated levels of *tat* protein in the tumors compared to other tissue. The other type (five of eight animals) appeared normal at birth but by 10 to 14 days were noted to be slower growing than their normal littermates. The five all died by 6 weeks of age; necropsy showed extensive thymic involution. Further studies on this genotype could not be performed as all of these mice were too young to breed.

Transgenic Oncogenesis: A Multistep Process

Most of the activated oncogene-bearing mice develop tumors at a young age and inherit the tendency just as they would any other autosomal dominant gene. A critical issue addressed by these studies is whether a single oncogene can cause malignant transformation. Acute retroviruses bearing the v-*myc* oncogene (e.g., avian retrovirus MC-29) directly promote transformation of cells in vivo, yet studies in vitro suggest that transformation of normal cells requires combinations of activated oncogenes, i.e., *myc* plus *ras* (see Chapter 5). The *myc*-bearing transgenic mice develop clonal tumors after a latency of 3 to 5 months. Thus dysregulated *myc* expression is necessary but not sufficient for tumorigenesis.

It is assumed that elevated constitutive *myc* expression results in a greater proportion of cells in mitosis. Mitotic B lineage cells would be expected to accrue further chromosomal changes that could produce a truly malignant and potentially metastatic clone. At present the nature of the co-oncogenic changes are not known. In vitro studies suggest that other oncogenes (e.g., *ras*) are probably activated in addition to those injected into the germ line.

Direct evidence that *myc* plus *ras* act synergistically in vivo was obtained by Sinn et al. They developed two transgenic lines: (1) c-*myc* under control of the mouse mammary tumor virus (MMTV) promoter and (2) v-H-*ras* again under control of the MMTV promoter. Transgenic *myc* animals developed mammary tumors and occasional lymphomas consistent with a necessary but not sufficient role for c-*myc* in tumorigenesis. The *ras* animals similarly developed mammary adenocarcinomas and occasional lymphomas and in addition salivary gland tumors and Harderian gland hyperplasia. Animals carrying both *myc* and *ras* transgenes were obtained by mating *myc* and *ras* parental strains. Dual carriers did not differ significantly in the *types* of tumors they developed but the *kinetics* of tumor appearance differed dramatically. Fifty percent of parental *myc* animals developed tumors by 325 days of age while 50% of parental *ras* animals developed tumors at 168 days of age. In contrast 50% of *myc/ras* dual carriers develop tumors by 46 days of age and 100% by 163 days. (Only female animals were assayed as male *myc* animals do not develop tumors in this system.) These results illustrate that *myc* and *ras* indeed work synergistically during in vivo oncogenesis.

The use of transgenic mice is a powerful technique to test subsets of oncogenes for their ability to promote cancer in vivo. Already several activated oncogenes have been shown to promote tumors in a tissue-specific pattern. Clearly, this area is an exciting one to further elucidate the mechanisms of tissue-specific oncogenesis.

Bibliography

Adams JM, Harris AW, Pinkert CA, et al. The c-myc oncogene driven by immunoglobulin enhancers induces lymphoid malignancy in transgenic mice. Nature 318:533, 1985.

Grosschedl R, Weaver D, Baltimore D, et al. Introduction of an immunoglobulin gene into the mouse germ line: specific expression in lymphoid cells and synthesis of functional antibody. Cell 38:647, 1984.

Lacey M, Alpert S, Hanahan D. Bovine papilloma-virus genome elicits skin tumors in transgenic mice. Nature 322:609, 1986.

Leder A, Pattengale P, Kuo A, et al. Consequences of widespread deregulation of the c-myc gene in transgenic mice: multiple neoplasms and normal development. Cell 45:485, 1986.

Messing A, Chen HY, Malmiter RD, et al. Peripheral neuropathies, hepatocellular carcinomas and islet cell adenomas in transgenic mice. Nature 316:461, 1985.

Nerenberg M, Hinrichs S, Reynolds R, et al. The tat gene of Human T-Lymphotrophic Virus type 1 induces mesenchymal tumors intransgenic mice. Science 237:1324, 1987.
Palmiter RD, Brinster RL. Transgenic mice. Cell 41:343, 1985.
Palmiter RD, Chen HY, Messing A, et al. SV40 enhancer and large-T antigen are instrumental in development of choroid plexus tumors in transgenic mice. Nature 316:457, 1985.
Quaife CJ, Pinkert C, Ornitz D. Pancreatic neoplasia induced by ras expression in acinar cells of transgenic mice. Cell 48:1023, 1987.
Rüther U, Garber C, Komitowski D, et al. Deregulated c-fos expression interferes with normal bone development in transgenic mice. Nature 325:412, 1987.
Sinn E, Muller W, Pattengale P, et al. Coexpression of MMTV/v-H-ras and MMTV/c-myc genes in transgenic mice: synergistic action of oncogenes in vivo. Cell 49:465, 1987.
Stewart T, Pattengale P, Leder P. Spontaneous mammary adenocarcinomas in transgenic mice that carry and express MTV/myc fusion genes. Cell 38:627, 1984.
Storb U, O'Brien RL, McMullen MD, et al. High expression of cloned immunoglobulin κ genes in transgenic mice is restricted to B lymphocytes. Nature 310:238, 1984.

13
Potential Diagnostic Uses of Oncogenes

Overview

Qualitative changes in the structure of proto-oncogenes or their products and quantitative changes in their expression have been documented for several cancers. With chronic myelogenous leukemia, for example, the *abl* oncogene is translocated to chromosome 22 in the vicinity of the *bcr* gene. A cancer-specific fusion protein qualitatively different from parent cell proteins is thus produced and is an ideal cancer marker. Mutant ras genes have now been implicated in the earliest stages of human leukemias and colon cancers. The detection of these mutations in defined premalignant states therefore could provide valuable prognostic information for clinicians.

Using family studies of restriction fragment length polymorphisms genetically linked to proto-oncogenes, it may be possible to identify cancer-prone individuals. With this scheme oncogenes and their products will become more precisely associated with stage and pathogenesis of cancers of all types. The development of specific antibodies and probes will facilitate this work, and the use of novel DNA amplification techniques and the cell sorter to study populations of cells will permit cancer to be diagnosed earlier and followed more closely.

In this chapter we present current theory and practice regarding the use of proto-oncogene DNA and protein products as specific markers for early detection and monitoring of malignancy. Experiments are in progress in many laboratories to investigate further uses of both DNA and antibody probes for detection of differences between normal and neoplastic tissue. Cancer diagnosis may well be as revolutionized by our advancing knowledge of oncogenes as was the diagnosis of infectious disease by technical advances in microbiology.

Early Detection of Cancer

It is believed that cancer detected early is more amenable to therapy than cancer detected late. This idea is logical as an early cancer has fewer malignant cells, implying a smaller tumor size—one that may be more easily resected by surgery. Fewer malignant cells also means lower probability of metastasis for most types of cancer. Metastatic disease is a major factor contributing to morbidity and mortality for many cancer patients.

Early detection of cancer is important in two general situations: (1) detection of primary disease in asymptomatic, previously undiagnosed individuals; and (2) detection of cancer recurrence in previously diagnosed and treated patients. In clinical series comparing cancer stage and survival, patients with less advanced disease uniformly have longer survival times. Cure rates are highest in patients with small, localized tumors. For patients with recurrence of previously diagnosed disease, detection during clinically asymptomatic phases may allow less aggressive and therefore better tolerated treatment regimens.

Many oncologists and other scientists have attempted to design tests for the early detection of cancer. Such tests have employed biochemical, cytological, and genetic techniques. So far, however, these methods have had only limited success.

Current Cancer Detection Tests

Current cancer tests are nonspecific and of limited clinical application. Perhaps the most widely known cancer detection test in current use is the papanicolaou smear, a cytological test named for George Nicolas Papanicolaou. Cells obtained from the uterine cervix are streaked onto a microscope slide by the physician. The cells are then fixed and stained, and a trained cytopathologist scours the slide for atypical cells. Although the test is not perfect, it has resulted in a significant increase in the early diagnosis of cervical cancer with a resultant decrease in deaths caused by this type of cancer.

A biochemical test, widely used for both diagnosis and monitoring of cancer, measures levels of carcinoembryonic antigen (CEA) (Fig. 13.1; see Plate 1). CEA is one of the so-called oncofetal antigens detectable in large amounts in embryonal tissue but in small amounts in normal adult tissues. It was found that serum of patients with certain gastrointestinal cancers contains elevated CEA levels that can be measured by immunological methods. The amount of CEA in serum correlates with the remission or relapse of these tumors, with the levels decreasing abruptly after surgical removal of the tumor. The return of elevated CEA levels signifies a return of malignant cells, and examination of the patient frequently reveals new evidence of malignancy. CEA, however, is also a normal glycoprotein found at low levels in nearly all adults. Moreover, this protein can be elevated with several nonmalignant conditions and is not elevated in the presence of many cancers. Therefore although it is better than nothing, it is far from an ideal marker.

A similar oncofetal tumor marker is α-fetoprotein, an embryonic form of albumin. Again, the antigen is detectable in high amounts in embryonal tissue and in low amounts in normal adults. It is elevated in a number of gastrointestinal malignancies including hepatoma. Like CEA, a decrease correlates with remission of the cancer and a reelevation with relapse. Again, however, there is insufficient sensitivity and specificity to make this marker useful for screening for malignancy or for monitoring previously diagnosed cancer in any but a few selected cases.

The ideal tumor marker for screening, diagnosis, or monitoring should have at least the following characteristics.

1. *Specific* for malignant cells only
2. Detectable *early* in the malignant process
3. Detectable in an *easily obtainable* body fluid (e.g., blood, urine, cerebrospinal fluid) or detectable on cells obtained by simple invasive procedures, e.g., needle biopsy
4. Quantity of marker detectable *proportional to the extent of the malignancy.*

One currently available marker that meets many of these criteria is β-human chorionic gonadotropin (β-hCG). This protein is present in the serum of pregnant women because it is released by the trophoblastic cells of the developing placenta. In the normal situation it is a useful serum or urine marker for pregnancy. However, it is also produced in high quantities by choriocarcinoma, a malignancy derived from cells of the placenta. Sensitive serum tests measuring β-hCG have guided the therapy of this cancer. Although highly malignant, choriocarcinoma has proved to be one of the more curable cancers, in part because of the ability to monitor the extent of disease by β-hCG levels and to apply therapy accordingly.

Use of Oncogenes in Cancer Diagnosis

Cancer diagnosis may be revolutionized by creative uses of oncogene reagents. Oncogenes and their products represent a potential leap forward in our diagnostic capabilities with respect to cancer. The c-*onc* protein counterparts of normal proto-oncogene products represent potential ideal tumor markers. In addition, qualitative and quantitative differences between oncogene and normal DNA and RNA structure represent other potential diagnostic parameters. Theoretically, differences between normal and oncogene DNA, RNA, and protein should be amenable to molecular, biochemical, or immunological detection techniques.

Screening for Malignancy: DNA Restriction Fragment Length Polymorphisms

Differences in chromosome and gene structure have been described for malignancies associated with oncogenes and for those where oncogene associations have not yet been elucidated. Perhaps the best known are the chromosome deletions associated with Wilms' tumor and retinoblastoma by cytogenetic analysis, a technique that unfortunately has limited sensitivity. Use of restriction fragment length polymorphism (RFLP) analysis, however, provides a much more sensitive technique for diagnosing susceptible individuals.

The RFLP analysis (Fig. 13.2) is based on the capacity of restriction enzymes to cut DNA at specific sequences leading to a characteristic pattern of bands on a Southern blot after hybridization with DNA probes (see Chapter 2). The human genome is highly redundant, which would imply that nucleotide changes may occur within individuals in the population without affecting the function of a particular gene. Because such changes do not affect the viability of offspring, they

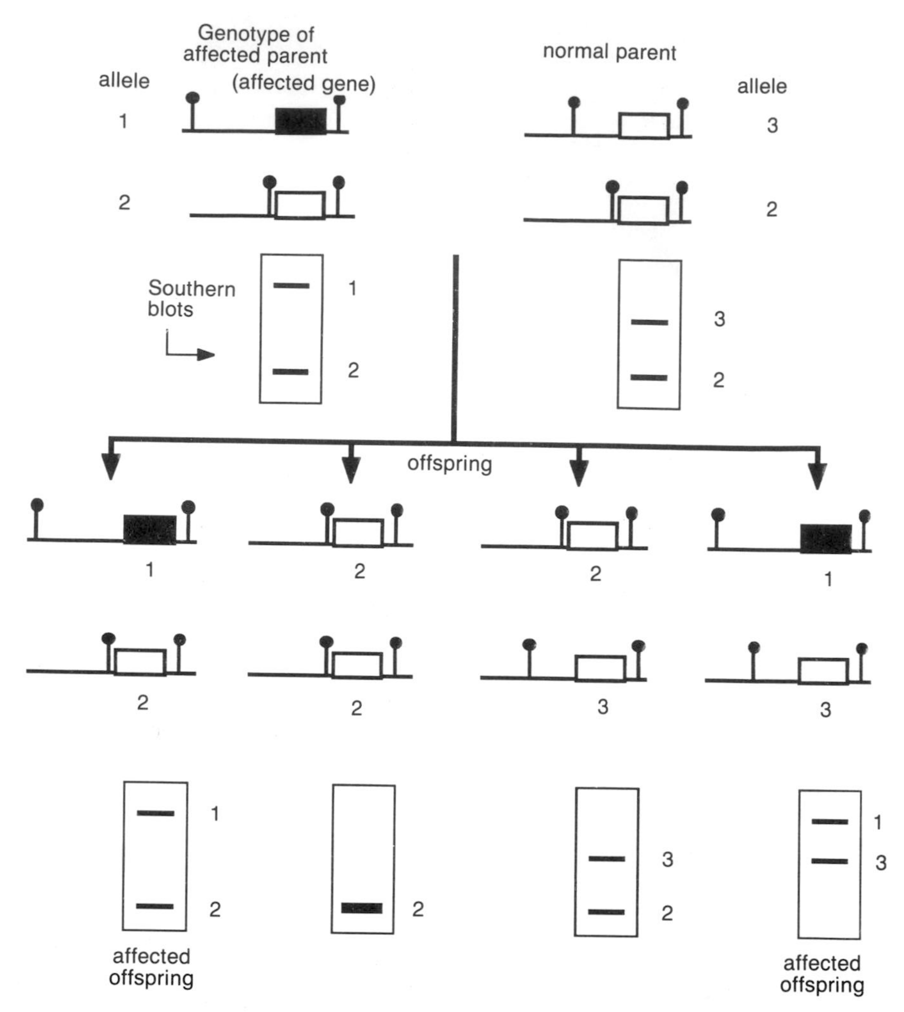

FIGURE 13.2. Principle of restriction fragment length polymorphism. The location and restriction enzyme sites for a hypothetical gene are shown. The normal parent (right) has two alleles designated 2 and 3; they are identified by their slightly different sizes on a Southern blot. The parent bearing the abnormal gene has one normal allele (band 2) and the larger abnormal allele designated by the black square and band 1. Offspring of this couple can inherit any of four combinations of these alleles. These possibilities are detected by banding patterns on Southern analysis. The presence of band 1 segregates with the disease.

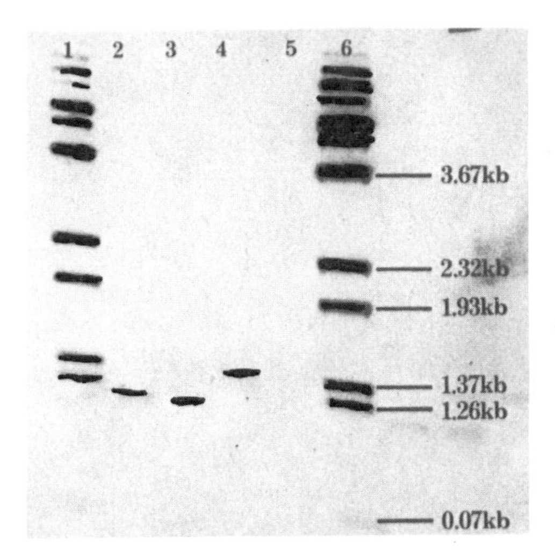

FIGURE 13.3. Use of RFLP for detection of sickle cell trait. In this experiment, 2 μg of DNA from each genotype was digested by restriction endonuclease CvnI and fractionated by agarose minigel electrophoresis, transferred to a nylon membrane, and hybridized to a specific sickle cell probe. Lanes 1 and 6 contain DNA size markers. Lane 2 contains DNA for the normal homozygous globin allele ($\beta^A\beta^A$). Lane 3 contains DNA heterozygous for the sickle cell allele ($\beta^A\beta^S$) (sickel cell trait). Lane 4 contains DNA that is homozygous for the sickle cell allele ($\beta^S\beta^S$) (sickle cell disease). Lane 5 contains DNA that is homozygous for the δβ-globin deletion. (Kindly provided by Dr. J. Sninsky)

can be carried genetically in a random fashion. If such a base change happens to occur within a restriction enzyme recognition site, Southern analysis reveals a different banding pattern for that particular gene or segment of DNA. Scientists have used this phenomenon to "tag" an affected gene or chromosome, thereby enabling it to be identified.

A prime use of this technology is the prenatal diagnosis of sickle cell anemia (Fig. 13.3). It was found in many (but not all) cases that the abnormal hemoglobin gene is associated with a particular banding pattern on Southern analysis. If the affected parent carries such a genetic signature, the presence of this pattern in the offspring suggests that the individual is a carrier of the disease.

The identification of abnormal oncogene alleles by this method is still in its infancy; however, as discussed below, RFLPs of the c-H-*ras* locus may provide a clue to individuals susceptible to some cancers. Another important application of RFLP analysis is to determine not the presence of an abnormal allele but the loss of critical "suppressor" genes (see Chapter 6). Tables 13.1 and 13.2 list a number of lymphomas, leukemias, and solid neoplasms with acquired genetic deletions. The disorders for which "anti-oncogenes" are best characterized are

TABLE 13.1. Loss of constitutive heterozygosity with lymphomas and leukemias.[a]

Tumor histology	Chromosome marker	Smallest common deletion
B cell prolymphocytic lymphoma	del 3p13	3p13
Chronic myelomonocytic leukemia	del 5/12	5q31,12p11
Acute nonlymphocytic leukemia (French-American-British classification M1, M2, M4, M5, M6)	del 5	5q12-q32
Multiple myeloproliferative disorders	del 9 (8:21)	9q
Multiple cutaneous T cell lymphoma	del 11	11p
Acute nonlymphocytic leukemia/ dysmyelopoietic syndrome	del 12	12p
Hairy cell leukemia	del 14	14q
Acute myelomonocytic leukemia	del 16	16q22 or inv(16)(13q22)
Acute lymphoblastic lymphoma	Hypodiploid	–

[a] This list of genetic deletions does not indicate whether the deletion is a primary or an ancillary event, nor does it specify whether the deletion occurs during tumorigenesis, tumor evolution, or metastasis. Inclusion in the list does not mean that the deletion is causative but merely acknowledges the association of the deletion with the tumor.

Reprinted with permission from Fuscaldo KE, Krueger LJ. Loss of heterozygosity and the anti-oncogene hypothesis: deletion of genetic material as an early event in tumorigenesis. Intern Med Specialist 10:119, 1986.

retinoblastoma and Wilms' tumor. With these diseases, deletions in critical chromosome regions (13q14 and 11p13, respectively) define susceptible individuals.

The RFLP analysis may be outlined as follows: If the affected parent has banding pattern "1,2" (representing the two alleles of the DNA fragment used as probe) and the normal parent has a banding pattern "2,3," the offspring may have the banding combinations (1,2), (2,2), (2,3), and (1,3). If it has been determined by pedigree analysis that allele 1 represents the abnormal chromosome, the offspring with patterns containing pattern 1–i.e., (1,2), (1,3)–are at higher risk for developing cancer. The parents then can be advised if their fetus is at high risk for developing neoplasia. If they choose to have the baby, the infant can be closely monitored clinically for signs of malignancy.

Cancer Susceptibility: Use of RFLPs with Retinoblastoma

Several lines of evidence suggest that a locus associated with retinoblastoma is found on human chromosome 13 at band q14: (1) Retinoblastomas are commonly found in patients with the 13q-deletion syndrome. These patients have chromosome rearrangements resulting in the loss of DNA including band 13q14 and a clinical presentation consisting in mental retardation and congenital malformations. (2) Deletions in chromosome 13 found in karyotypes of many

TABLE 13.2. Loss of constitutive heterozygosity with solid neoplasms.[a]

Tumor histology	Chromosome marker	Smallest common deletion
Melanoma	del 1 Nonspecific (del)	1p
Neuroblastoma	del 1	1p31pter
Hereditary colorectal adenomatosis	RFLP 2q21.3	–
Neuroblastoma	del 2	2p21
Non-small-cell lung cancer	del 3	3p
Small-cell lung cancer	del 3	3p14-p23
Ovarian adenocarcinoma	del 3	3p
Renal cell carcinoma	del 3	3p
Rhabdomyosarcoma	del 3	3p21
Familial adenomatous polyposis; colon cancer	del 5	5q12-q22
Pancreatic carcinoma secondary to refractory macrocytic anemia	del 5	5q
Metastatic melanoma	del 6	6q
Ovarian carcinoma	del 6	6q
Prostatic carcinoma	del 7	7q22
Aniridia, Wilms' tumor	del 11	11p13
Beckwith-Wiedemann syndrome	del 11p	–
Ewing's tumor	del 11; hypodiploid	11p –
Superficial transitional cell bladder cancer	del 11	
Colorectal carcinoma	del 12	12p
Osteosarcoma	del 13	13q22
Retinoblastoma	del 13	13q14.11
Colonic carcinoma	rear 17q11, del 13	−18
Multiple cutaneous leiomyomas	del 18	18pter
Multiple endocrine neoplasia	del 20	20p12.2
Meningioma	del 22	−22

[a]This list of genetic deletions does not indicate whether the deletion is a primary or an ancillary event, nor does it specify whether the deletion occurs during tumorigenesis, tumor evolution, or metastasis. Inclusion in the list does not mean that the deletion is causative but merely acknowledges the association of the deletion with the tumor.

Reprinted with permission from Fuscaldo KE, Krueger LJ. Loss of heterozygosity and the anti-oncogene hypothesis: deletion of genetic material as an early event in tumorigenesis. Intern Med Specialist 10:119, 1986.

retinoblastoma patients were found by retrospective analysis to have a common deletion at band q14. (3) The familial form of retinoblastoma segregates with an enzyme marker called esterase D, which maps to chromosome 13q14. Esterase D has several forms called isoenzymes. Each isoenzyme is coded for by a different esterase D allele. Persons may be homozygous or heterozygous for particular alleles of esterase D, and a particular pattern of esterase D expression can be linked to the risk of developing retinoblastoma (see also Chapter 6).

Studies of retinoblastoma kindreds by Cavenee and co-workers demonstrated the potential use of restriction fragment length polymorphisms to increase our understanding of cancer genetics. Chromosome 13-associated RFLPs from retinoblastoma tumor cell DNA show a homozygous pattern, and corresponding DNA from the patient's white blood cells shows a heterozygous pattern. It can be concluded from these data that homozygosity for a mutant allele, presumably at 13q14, is associated with and might be a prerequisite for development of retinoblastoma. In other words, the "retinoblastoma-associated" alleles on *both* chromosomes must be adversely affected in order for malignancy to occur. Such genes, in which a homozygous loss results in malignancy, have been called "anti-oncogenes" or "recessive oncogenes" in the sense that their products appear to act functionally as repressors of tumor formation.

By using a combination of esterase D isoenzyme analysis and specific RFLPs, Cavanee et al. have been able to follow individual parental chromosomes (so-called haplotypes) predisposing to retinoblastoma. Applying this analysis to fetal cells obtained by amniocentesis, they were able to successfully predict in four of five cases if a fetus at risk would later develop the malignancy. This feat represents the first demonstration of the feasibility of predicting if a person at risk for a dominantly inherited cancer actually has the disease.

Oncogenes as Tools in Genetic Counseling

If proto-oncogene variations can be shown to be associated with particular malignancies, screening of persons at risk may be done pre- or postnatally. Heritable cancers appear to fall into several major categories.

1. Each affected individual in a kindred has the same type of cancer; examples include retinoblastoma and familial polyposis. Such cancers appear to have autosomal dominant inheritance.
2. Affected individuals may have cancers of one or more types often as a part of a multisystem syndrome; examples include neurofibromatosis and the multiple endocrine neoplasia (MEN) syndromes. As a group, these cancers probably have a more complex inheritance than simple autosomal dominant or recessive involvement of one particular gene. However, in a particular kindred they may have the appearance of either dominant or recessive inheritance; for example, MEN II appears to have an autosomal dominant pattern of inheritance in several families that have been studied.
3. Affected individuals have cancers of multiple types clearly associated with autosomal dominant inheritance. For example, neurofibromatosis is inherited as an autosomal dominant with variable penetrance. The "cancer families" described by Li and Fraumeni fall into this category.
4. Affected individuals have rare autosomal recessive syndromes with increased risk for developing a variety of cancer types; examples include Fanconi's anemia, Bloom's syndrome, Wiskott-Aldrich syndrome, and ataxia telangiectasia. Chromosome abnormalities including instability with multiple exchange events have been described with some of these disorders, most notably Bloom's syndrome.

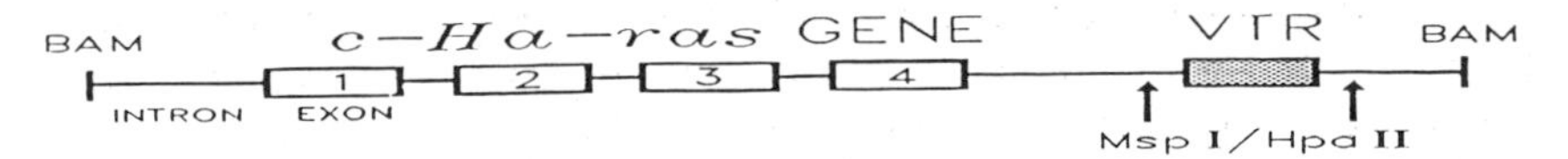

FIGURE 13.4. Location and organization of variable tandem repeat (VTR) relative to the H-*ras* gene.

Differences in DNA primary sequence are well established for *ras* proto-oncogenes (see Chapter 9). Not unexpectedly, restriction enzymes have been found that differentiate among various *ras* alleles. For example, c-Ha-*ras* genes can be differentiated by the size of Bam H1 restriction fragments. This restriction fragment length polymorphism results from variation in the size of a DNA sequence located 1.5 kilobases (kb) downstream from the 3′ terminus of c-Ha-*ras* coding sequences (Fig. 13.4). This downstream sequence is called a variable tandem repetition (VTR) because it consists of a 28 base pair consensus sequence repeated a variable number of times and clustered in tandem to form the longer VTR. The VTR is flanked on either side by restriction sites for Msp or Hpa II endonucleases.

Using a probe from the VTR region, Krontiris and co-workers analyzed leukocyte DNA from 115 healthy people and 82 individuals with either myelodysplasia or malignancy. Some 20 alleles of the variable tandem repetition were defined. Four of the alleles comprised 93% of the total, and the 16 "rare" alleles had individual frequencies of 0.2 to 1.3%. These researchers showed the rare alleles to be significantly more frequent in the normal cells of cancer patients than in the normal cells from unaffected individuals. Because these alleles are found in normal, nonmalignant cells, the implication is that they must have been genetically inherited rather than acquired. How might the VTRs influence the development of cancer? The answer is not known explicitly; however, transforming activity of a mutant c-H-*ras* gene lacking the VTR region is 5- to 10-fold lower than when the gene contains this region. The lower expression of p21 without the VTR in these experiments suggests that this region might act as an enhancer for c-H-*ras*. Thus it may prove possible to differentiate persons at higher risk for development of malignancies associated with abnormal *ras* gene structures by examination for RFLPs in the VTR of this gene.

A similar approach was utilized by Lidereau and colleagues. They showed that 6 of 75 breast tumor DNAs had an extra 5-kb *Eco* RI restriction fragment hybridizing with a human c-*mos* probe. In addition, this *Eco* RI fragment was found in three of those six patients' lymphocytes. No such band was found in 69 control DNAs from an unaffected population. On the other hand, 1 of 73 leukemia cell DNAs exhibited the 5-kb band. Thus particular DNA polymorphisms can be found in patients with a variety of cancers. It should prove possible to extend the RFLP technique to screen persons at risk for other malignancies.

At this time a limited number of clearly defined chromosome loci are identifiable by RFLP analysis. Most of these loci are not correlated with known oncogenes. Thus the technology is limited to the heritable tumors associated with specific chromosome sites rather than to defined oncogenes. This group includes, in addition to retinoblastoma (discussed above), neuroblastoma (chromosome 1), certain renal cell carcinomas associated with rearrangements of chromosome 3, Wilms' tumor on chromosome 11, multiple endocrine neoplasia type 2 (chromosome 20), and meningioma (chromosome 22). As loci for other familial cancers are identified, it will be possible to provide genetic counseling for families predisposed to these cancers in a manner similar to that provided currently for families with better characterized genetic illnesses, such as cystic fibrosis, phenylketonuria (PKU), or the hemoglobinopathies.

The logical aim of this type of analysis is identification of the genes associated with the RFLPs. Molecular cloning and biochemical characterization of these genes may give insights into general mechanisms of carcinogenesis. How such genes are related to known oncogenes is a question for future research.

Another use of RFLP analysis is to identify loss of allelic heterozygosity as a marker of genetic instability. It has been noted that tumor DNA has a propensity to lose one of the two alleles of particular genes when compared to fibroblast DNA from the same patient. The loss of an allele may be seen on occasion as a gross deletion within a chromosome or, using RFLP analysis, as the loss of a band on a Southern blot (Fig. 13.5). Loss of allelic heterozygosity by RFLP analysis has been seen with the H-*ras* gene in some human breast cancers; such loss is associated with a more aggressive tumor. Analysis of this sort may be helpful for localizing potential "suppressor" genes or "anti-oncogenes" specific for certain tumors. Using highly polymorphic probes localized to specific chromosomes, researchers have seen the consistent loss of one allele in many of the tumors discussed above, including chromosome 1 in neuroblastomas, chromosome 10 in glioblastoma multiforme (a brain tumor), chromosome 13 in retinoblastoma, chromosome 22 in meningiomas, and chromosome 19 in Ewing's sarcoma. Thus RFLP analysis may provide diagnostic and prognostic information for a variety of human malignancies.

Detection of Tumor DNA Rearrangement

A variation on the theme of polymorphism is the oncogene DNA rearrangement associated with a number of tumors. Restriction pattern differences are detectable with Burkitt's lymphoma showing rearranged c-*myc* genes (see Chapter 10). Hu and co-workers have shown that clonal immunoglobulin gene rearrangements can be detected in DNA extracted from peripheral blood lymphocytes of patients with low-grade B cell lymphomas. Clonal proliferation of B cells exceeding the 1% threshold of detection of the Southern blot hybridization procedure is seldom if ever seen *except* with neoplasia. Follicular lymphomas exhibit a chromosomal abnormality t(14;8) which involves the translocation of the immunoglobulin heavy chain locus on chromosome 14 to a position on chromosome 18 near a gene termed *bcl*-2. Rearrangements of this putative proto-oncogene, *bcl*-2, can be detected by Southern blot analysis using the appropriate probes. Weiss and

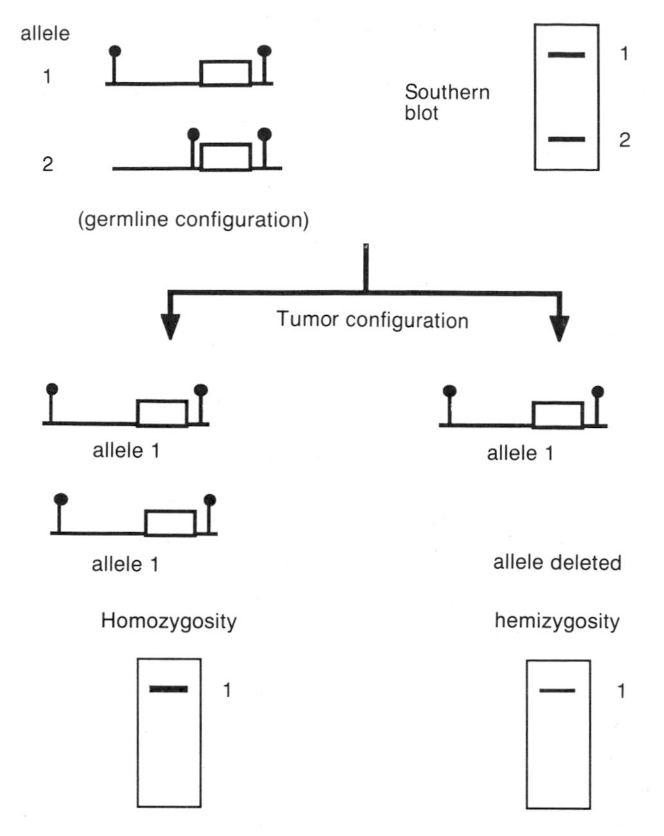

FIGURE 13.5. RFLP analysis: loss of heterozygosity in tumor cells. Two alleles are detected by two different-size bands on Southern analysis. The loss of one allele can occur by two mechanisms. (Left) Chromosome loss and reduplication can lead to homozygosity. (Right) Chromosome loss or gene deletion leads to hemizygosity.

co-workers found rearrangements of *bcl*-2 in virtually all cases of follicular lymphomas, and in 28% of diffuse large cell lymphomas. This last finding is of interest since diffuse large cell lymphomas and follicular lymphomas differ in prognosis and in treatment. Thus, perturbations of *bcl*-2 may represent a molecular marker to better classify patients with lymphomas.

Chronic myelogenous leukemia (CML) represents another example of use of oncogene DNA rearrangements detectable by Southern analysis to diagnose human disease. Details of the molecular pathogenesis of CML are discussed in Chapter 6. Probes specific for the breakpoint cluster region (bcr) of chromosome 22 can detect the characteristic translocation of the *abl* gene from chromosome 9 to chromosome 22. In some cases cytogenetic techniques are unable to detect such a translocation, although Southern blotting can. The latter technique is important for determining prognosis in acute leukemias (particularly acute lymphocytic leukemia), where up to 10% of patients may have the (9:22) translocation. These patients do not respond to therapy as well as other leukemia patients and have short durations of remission. The presence of the (9:22) translocation argues that more aggressive forms of therapy such as bone marrow transplantation should be tried early in the course of the disease.

Gene Amplification as a Useful Tumor Marker

Yet another difference between normal and neoplastic tissue is the amplification of oncogene DNA in some of the latter. When present, oncogene amplification serves as a specific cancer marker of the somatic cell genetic type. The amplified DNA can be detected by chromosome banding techniques and by differences in restriction pattern using oncogene DNA probes. For example, with childhood neuroblastoma, increasing levels of N-*myc* gene amplification correlate with increasing severity of the disease (see Chapter 6). Detection and quantitation of N-*myc* gene amplification may serve as a useful tumor marker for staging and treatment of neuroblastoma. Amplification of the c-*myc* gene has been found in cell lines associated with high-grade, aggressive small-cell lung carcinomas, and amplification of *erb* B-2 (*neu*) has been found in advanced breast carcinomas (see Chapter 6).

The genetic markers so far mentioned can be used to provide information helpful to clinicians. Certain cancers have an unidentified morphology such that diagnosing the type of cancer is difficult. This represents a therapeutic dilemma since different cancers respond best to different therapies, and some cancers are even curable if properly treated. Thus if such a tumor shows a clonal rearrangement of the immunoglobulin gene, and a rearrangement of *bcl*-2, then it is a lymphoproliferative disorder and most likely, a non-Hodgkin's lymphoma. If however, the tumor exhibits no immunoglobulin rearrangement, but has amplification of the N-*myc* gene, then the diagnosis of neuroblastoma, or small cell carcinoma should be considered.

Polymerase Chain Reaction

Kary Mullis and co-workers at Cetus Corporation have developed a novel method for the amplification of DNA sequences (Fig. 13.6). The technique has the potential to revolutionize detection of genetic aberrations present in a small subpopulation of cells. The basis of the method is hybridization to particular chromosome DNA sequences by specific primers of 20 to 25 base pairs. A pair of primers are used to hybridize to opposite strands of the DNA duplex, a distance of 300 to 1000 bases apart. The primed DNA is used as a template for DNA polymerase, which copies the corresponding strand. After a few minutes the mixture is heated to separate the strands. The separated strands are cooled and then bind the appropriate primer. The polymerization reaction subsequently proceeds in the

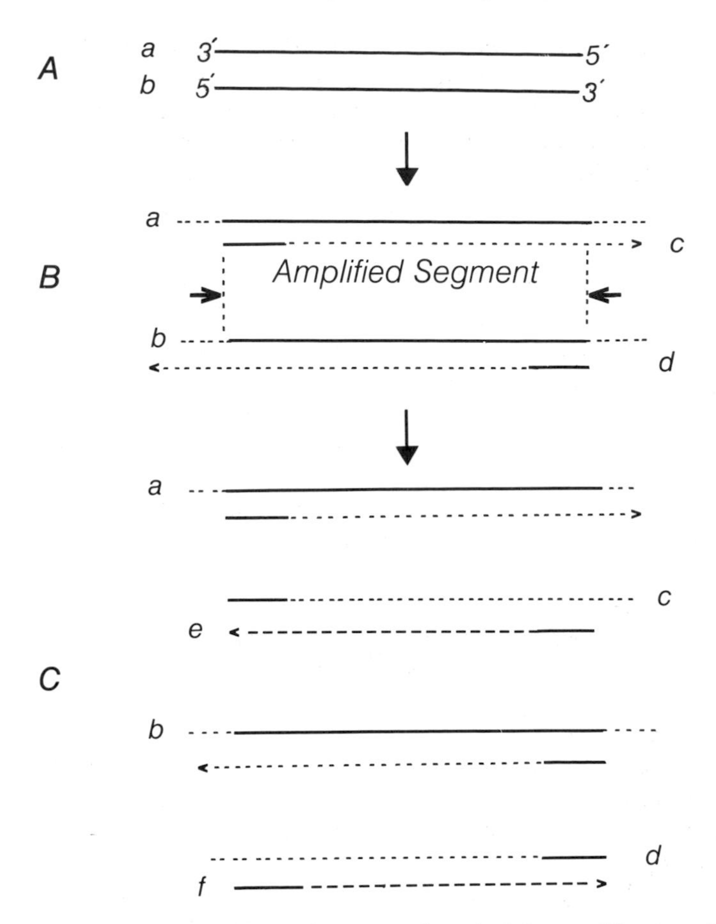

FIGURE 13.6. Polymerase chain reaction: a novel method for amplification of selected segments of nucleic acids. (A) Representation of two antiparallel strands of DNA. (B) The strands of DNA are separated, and two primers, 20 bases in length binding to the ends of the region to be amplified, are annealed to their respective strand of DNA. Using these primers, DNA polymerase makes a copy of each strand (c,d). (C) The process is repeated. When fragments c or d are primed, the DNA polymerase copies only to the end of these fragments. Thus the region between the fragments is amplified each time the DNA is melted, primed, and copied.

opposite direction. Because the polymerase can read only to the end of the first strand, amplification of the region between the primers occurs, and the amount of this DNA grows exponentially. Theoretically, 20 cycles of the polymerase chain reaction should yield approximately one million copies of the target sequence: $2^{20} \approx 10^6$.

Using specific probes for the hemoglobulin locus, it has been possible to detect the presence of the sickle cell trait using small numbers of cells. Recently, several groups have used the polymerase chain reaction to determine the presence of single base substitutions of the *ras* genes in human tumors and have shown this technique to be a reasonable screening method. As discussed in Chapters 6 and 9, point mutations in codons 12, 13, and 61 convert the *ras* proto-oncogenes into transforming genes. The standard transfection methods used to examine human tumors for these mutations require large amounts of tumor DNA and is labor intensive. With the polymerase chain reaction, comparatively small amounts (1-10 μg), of genomic DNA are needed and large numbers of tumor samples can be screened simultaneously. The importance of this technology can be seen in human myelodysplasia (see Chapter 6). Several groups have shown that the presence of a mutant *ras* allele in patients with this disorder defines a subgroup who will subsequently develop acute leukemia. Thus the detection of *ras* mutations in myelodysplastic patients will provide important prognostic information to the clinician.

Oncogene RNA in Diagnosis

There has been much documentation of altered expression of c-*onc* mRNA in tumor cells compared with their nonmalignant counterparts. Theoretically one might be able to detect an increased level of mRNA (indicating enhanced gene expression) or altered mRNA (i.e., fusion or deletion) transcripts in tumor tissue. However, the assays currently available to measure RNA are not easily amenable for screening or monitoring purposes. The techniques are cumbersome and entail isolation of RNA from tumor cells or tissue. Thus RNA studies are, for the foreseeable future, confined to the research laboratory where enhanced levels or altered copies of particular mRNAs provide unequivocal evidence of oncogene expression.

DNA Probes

RNA and DNA abnormalities (RFLPs, abnormal messages, rearranged genes, etc.) can now be detected by nonisotopic means. Figure 13.7 illustrates this principle.

Tumor Markers: Oncogene Products

In many cases the c-*onc* protein counterpart of a normal proto-oncogene product represents a qualitatively different protein. When *onc*-proteins are structurally distinct from their normal counterparts (e.g., the products of *ras, abl*, and *neu* genes) antibody probes specific for the tumor proteins may be generated. The *onc* proteins can then be detected by immunological techniques, e.g., enzyme-linked immunosorbent assay (ELISA), radioimmunoassay (RIA), or radioreceptor assay (see Chapter 2), if they are present in fluid accessible for sampling. The most easily detected proteins are those secreted into blood or urine by the

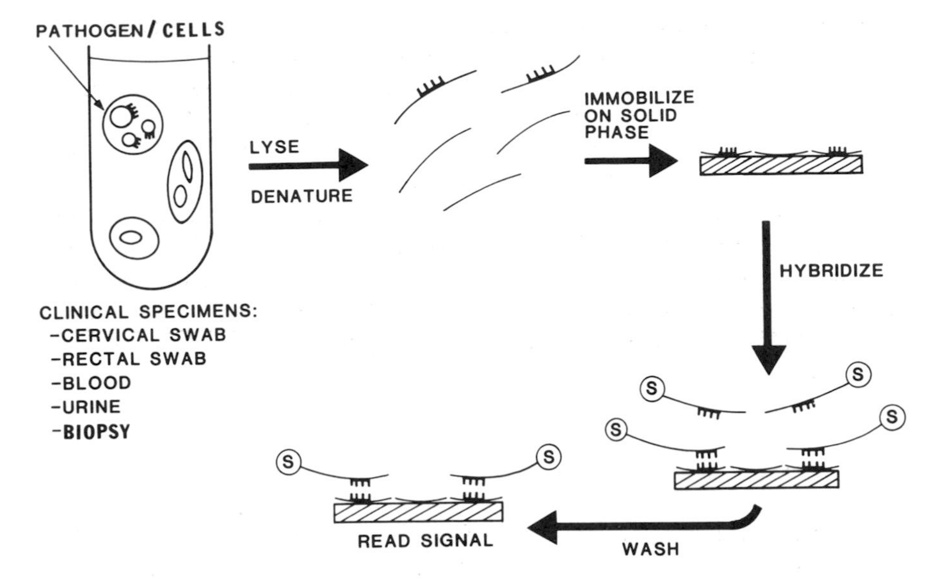

FIGURE 13.7. Diagnosis by DNA probe. RFLPs or other genes can be detected by non-isotopic methods. DNA is extracted and immobilized on a solid support nylon membrane. Probes used for hybridization are labeled with biotinylated bases. The blot is developed using strepavidin enzyme conjugates. When substrate is added, it precipitates on the nylon membrane where the specific probe has hybridized. S = signal generation system on the probe.

tumor. One example of a secreted protein is the *sis*-oncogene product resembling the platelet-derived growth factor (see Chapter 8). However, to date the expression of the *sis* oncogene product in human tumors has been limited, and the usefulness of this marker remains to be determined. The *onc* proteins might also be accessible to immunological reagents if they are embedded on the cell surface. Altered receptor molecules, e.g., the c-*erb* B product, might conceivably be detected on tumor cells shed into accessible body fluids. Some c-*onc* products, e.g., mutated *ras* and *bcr-abl*, fit at least one criterion for an ideal tumor marker—that of specificity for the tumor cells.

Detection of Altered Oncogene Products by Specific Antibodies

The point mutants of some *ras* oncogenes or the *bcr-abl* fusion protein characteristic of CML are potential targets for antibody-based diagnostics. In both of these cases the oncogene products appear to be specific markers of human cancer, as these oncogenic mutations have not been found in normal tissue. These markers therefore may be useful for the detection and monitoring of cancer in clinical laboratories.

With CML it is possible that detection of P210, the unique fusion protein translated from Philadelphia chromosome mRNA, will prove to be a more convenient means of detecting and monitoring the disease than the chromosome analysis currently employed. Detection of P210 may be achieved by several means, including use of antibody reagents. For example, an antibody that reacts with the c-*abl* protein would detect the normal 150,000-dalton protein in nontransformed cells as well as the normal and 210,000-dalton forms in CML tissue. Normal c-*abl* protein expressed in CML cells presumably comes from the unrearranged chromosome. The two products of the rearranged and normal c-*abl* genes are different sizes and thus separable by sodium dodecyl sulfate (SDS)-polyacrylamide gel electrophoresis. Visualization of the proteins is possible using Western immunoblot technology.

The *ras* oncogene products differ from their normal counterparts by single amino acid changes, as discussed in Chapters 6 and 9. These proteins show slight differences in mobility on SDS gels, but the differences are too small to be used conveniently to detect mutant forms. Antibodies that distinguish some *ras* oncogene proteins from the normal *ras* gene product p21 have been developed (Fig. 13.8). These antibodies may prove to be specific for tumor cells. The ultimate usefulness of detecting mutant *ras* gene products in clinical situations remains to be determined.

A synthetic peptide derived from one *ras* gene activated in T24 bladder carcinoma has been used to generate monoclonal antibodies that react with colon and mammary carcinomas. The level of reactivity was shown to be minimal in normal colon or mammary epithelia, benign colon tumors, and inflammatory or dysplastic colon tissue. Malignant colon tumors and infiltrating ductal breast carcinomas reacted at a higher level. Furthermore, the more invasive the carcinoma, the stronger the reactivity of the monoclonal antibodies in the histological sections. Thus it is possible that antibodies to the products of oncogenes will prove useful for grading tumors in terms of degree of cellular malignancy.

Identifying Cancer-Prone Individuals: Chromosome Fragile Sites

Fragile sites are particular locations on chromosomes that are less stable than other chromosome sites and thus theoretically more susceptible to breaks and recombination than other areas. Fragile sites have been identified on all chromosomes. There are two general types of site: (1) those occurring commonly in many individuals without regard to inheritance and (2) those (more rare) sites that appear to be inherited in mendelian fashion.

As described in preceding chapters, a number of human tumors, particularly leukemias and lymphomas, are characterized by nonrandom chromosome abnormalities. Several proto-oncogenes have been mapped to the chromosome breakpoints involved in these specific rearrangements. There is good experimental evidence that altered proto-oncogenes play an integral role in tumorigenesis; the rearrangements often “activate” or otherwise modify expression of the adjacent

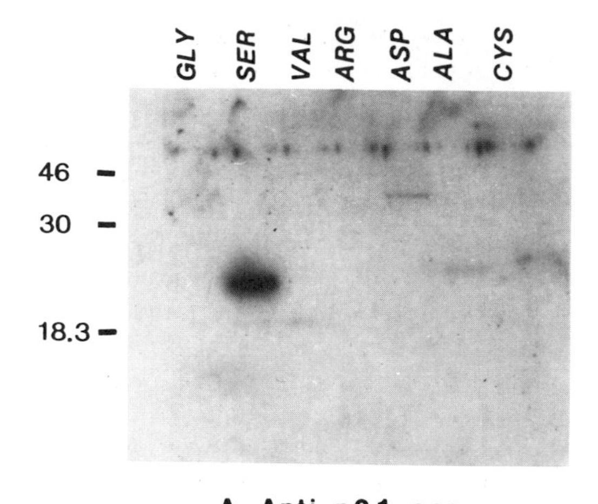

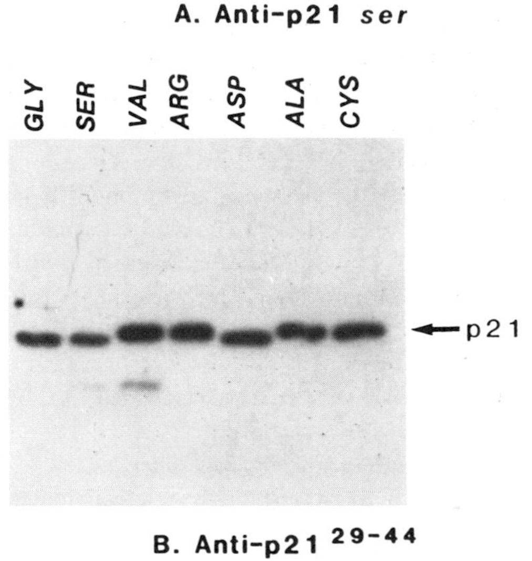

FIGURE 13.8. Antibodies can be used to distinguish mutant *ras onc*-proteins. (A) Specific antiserum recognizing only the serine mutant p21 protein in an immunoblot. (B) Antiserum recognizing all p21 mutants.

genes. In addition, there appears to be remarkable concordance between breakpoints found in chromosome rearrangements and the location of both heritable and common fragile sites. Thus there is a provocative potential role for fragile sites in the genetic predisposition to development of neoplasia. This idea is strengthened by the finding that certain individuals with particular malignant diseases characterized by specific chromosome abnormalities are carriers of one or another fragile site where chromosome rearrangement has occurred.

As correlations between specific fragile sites and predispositions to cancer are strengthened, it should prove possible to predict individuals at higher risk for developing particular tumors. This information, in conjunction with information about environmental mutagens, would permit physicians to counsel patients regarding exposure to potentially cancer-causing substances.

DNA Repair and Cancer

Repair of DNA following chemical or physical damage limits the rate of genetic mutation. The *inability* to properly repair DNA is associated with dramatically increased cancer rates. The prototype of such syndromes is xeroderma pigmentosum (XP), a rare autosomal recessive disease characterized by severe sensitivity to ultraviolet (UV) irradiation from sunlight leading to a high incidence of skin tumors. Cultured XP fibroblasts exposed to UV irradiation are unable to repair UV-induced pyrimidine dimers in their DNA, and it is thought that the resultant mutations contribute to tumor formation in these patients. Although the UV repair gene(s) responsible for this disease have not been isolated (genes presumably similar are well characterized for bacteria) and although no oncogenes associated with these tumors have yet been described, one can imagine that an inability to repair mutations in proto-oncogenes might contribute to the neoplastic process in this disease. DNA repair can be quantitated in isolated amniotic cells and thus permit prenatal diagnosis of XP. Fetal test cells, isolated by amniocentesis, are UV-irradiated and incubated with radiolabeled thymidine. Normal cells repair thymidine dimers by excision of the dimers and replacement with the labeled, nondimerized thymidine. XP cells cannot repair such dimers and thus do not incorporate the labeled thymidine. Fetal cells are compared with known standards for the uptake of radiolabeled thymidine in the nuclei. Xeroderma pigmentosa can be diagnosed with this biochemical test and the parents advised should the fetus turn out to have the disease.

Detection of Human Cancer Viruses

Several human viruses have been linked to the genesis of human cancers (see Chapters 3 and 4). The strongest candidates are the following.

1. HTLV-I and HTLV-II, with leukemias/lymphomas
2. Human papilloma viruses (HPV), with cervical cancer
3. Epstein-Barr virus (EBV), with nasopharyngeal carcinoma and African Burkitt's lymphoma
4. Hepatitis B virus (HBV), with hepatomas
5. Cytomegalovirus (CMV), with Kaposi's sarcoma

It soon should be possible to identify patients with atypical clinical findings using virus-specific antibody and nucleic acid probes. For example, a number of patients with unusual histories of lymphoma have been found to be infected with one or another of the human T lymphotropic retroviruses.

Study of cervical cells has revealed a link between some strains of human papilloma virus and dysplasia and early preneoplastic changes. Measurement of HPV infection and integration eventually may become as routine as the Papanicolaou smear, and patients infected with known oncogenic strains could be followed more closely. Detection and monitoring of other viruses or their associated proteins also may be important. Ultimately, development of vaccines for

TABLE 13.3. Uses of flow cytometry in oncology.

Flow cytometry parameter[a]	Comment[b]
Cancer diagnosis	
DNA index	Abnormal in about 80% of solid tumors, 80% of myeloma, 50% of lymphoma, and 20% of leukemia
Nucleolar antigen	To be examined as a cancer marker for DNA–diploid disease
Double-stranded RNA	Same as for nucleolar antigen; seems to identify residual leukemia in remission
Differential diagnosis	
DNA index	Near-diploid in hematological cancers
	Higher values commonly in solid tumors
	Indolent vs. aggressive lymphoma
	Recurrence of same vs. second neoplasm
RNA index	Distinguishes 70–80% of ALL vs. AML
	Identifies nearly 100% of myeloma
S phase	Lower with active leukemia than leukemia in remission
	With DNA index distinguishes indolent and aggressive lymphomas
cIg	Monoclonality identified myeloma and pre-B cell disease (IgM)
smIg	Monoclonality identified B lymphoma
Monoclonal light chain excess	Suspicious of lymphoma
T antigens	Identify T cell lymphomas and leukemias
TdT, ER, CEA, etc.	Under investigation
Prognosis	
DNA index	High values; unfavorable for survival; exception (?) ALL
	Hypodiploid leukemia, myeloma with poor prognosis
S phase	High values; favorable for response in ALL and AML patients < 50 years of age, unfavorable for response duration and survival
RNA index	High values; favorable for response in myeloma, under investigation in leukemia
Monoclonal light chain excess	Unfavorable during remission of aggressive lymphomas
Treatment	
DNA index	Persistence in leukemia remission marrow: more treatment?
	Persistence in bladder irrigates or effusions; more treatment
	Selection of patients without abnormalities for autologous marrow support
S phase	May become major criterion for treatment choices for lymphoma and other solid tumors
RNA index	Low values with myeloma; investigational therapy
Surface markers	Selection of monoclonal antibodies for treatment

[a] cIg = cytoplasmic immunoglobulin. smIg = surface membrane immunoglobulin. TdT = terminal deoxynucleotidyl transferase. CEA = carcinoembryonic antigen.
[b] ALL = acute lymphocytic leukemia. AML = acute myelogenous leukemia. IgM = immunoglobulin M.
Adapted from Barlogie B, Raber MN, Schumann J, et al. Flow cytometry in clinical cancer research. Cancer Res 43:3982, 1983.

tumor-associated viruses will be the most useful result of this research (see Chapter 14). In the meantime, testing for the presence of these tumor-associated viruses is being investigated for monitoring and evaluating the prognosis of selected malignancies.

Flow Cytometric Analysis of Tumor Cells

The fluorescence-activated cell sorter is a powerful tool for analyzing numerous parameters on vast numbers of individual cells in a short time. Quantitative data can be accumulated regarding cell size, DNA and RNA content, cell surface receptors, and protein antigens using a large number of fluorescent probes and antibodies. Table 13.3 lists several parameters that can be evaluated by flow cytometry to aid cancer diagnosis and treatment.

Leukemia cells can be stained directly; however, solid tumors require enzymatic dispersal prior to study. Using flow cytometry, several laboratories have been able to follow the reappearance of leukemia cells during disease relapse and to monitor cell cycle arrest by chemotherapeutic drugs. It has also become possible to document tumor aneuploidy and to study gene expression directly at the single-cell level.

An abnormal nuclear DNA content is closely correlated with malignancy. In a study of almost 5000 cancer patients, more than two-thirds had aneuploid cells detected by cell sorter analysis. Furthermore, the proportion of cells in S phase increases with increasing DNA content and permits discrimination of low and high grade malignancy, i.e., those containing few and many (rapidly) dividing cells. The use of fluorescein-labeled anticytokeratin antibodies to differentiate malignant from nonmalignant epithelial cells and the determination of relative cell DNA content are already routine procedures in many cancer centers. In the future, flow cytometry will replace many of the tasks currently performed by cytopathologists. Of particular interest is the study of oncogenes or their products in single cells. For example, the expression of the *bcr-abl* p210 fusion protein in individual CML cells might correlate with the development of blast crisis. Using flow cytometry one can simultaneously study the expression of several oncogenes in a single cell. Unforeseen correlations between cell cycle, oncogene expression, and malignant behavior may result from these studies.

Bibliography

Aurias A, Rimbaut C, Buffe D, et al. Chromosomal translocations in Ewing's sarcoma. N Engl J Med 309:496, 1983.

Barlogie B, Raber, MN, Schumann J, et al. Flow cytometry in clinical cancer research. Cancer Res 43:3982, 1983.

Berger R, Bloomfield CD, Sutherland GR. Report of the committee on chromosome rearrangement in neoplasia and on fragile sites. Cytogenet Cell Genet 40:490, 1985.

Bos J, Fearon E, Hamilton S, et al. Prevalence of *ras* gene mutations in human colorectal cancer. Nature 327:293, 1987.

Cavenee WK, Murphree AL, Shull MM, et al. Prediction of familial predisposition to retinoblastoma. N Engl J Med 314:1202, 1986.

Cline M, Slamon DJ, Lipsick JS. Oncogenes: implications for the diagnosis and treatment of cancer. Ann Intern Med 10:223, 1984.

Fearon ER, Feinberg AP, Hamilton SH, et al. Loss of genes on the short arm of chromosome 11 in bladder cancer. Nature 318:377, 1985.

Friend, SH, Bernards R, Rogelj S, et al. A human DNA segment with properties of the gene that predisposes to retinoblastoma and osteosarcoma. Nature 323:640, 1986.

Fuscaldo KE, Krueger LJ. Loss of heterozygosity and the anti-oncogene hypothesis: deletion of genetic material as an early event in tumorigenesis. Intern Med Specialist 7:119, 1986.

Hanawalt PC, Sarasin A. Cancer-prone hereditary diseases with DNA processing abnormalities. Trends Genet 1:124, 1986.

Hirai H, Kobayashi Y, Mano H, et al. A point mutation at codon 13 of the N-ras oncogene in myelodysplastic syndrome. Nature 327:430, 1987.

Horan Hand, P, Thor A, Wunderlich D, et al. Monoclonal antibodies of predefined specificity detect activated ras gene expression in human mammary and colon carcinomas. Proc Natl Acad Sci USA 81:5227, 1984.

Hu E, Trela M, Thompson J, et al. Detection of B-cell lymphoma in peripheral blood by DNA hybridization. Lancet 2:1092, 1985.

Knudson AG Jr. Mutation and cancer: statistical study of retinoblastoma. Proc Natl Acad Sci USA 68:820, 1971.

Krontiris TG, DiMartino NA, Colb M, et al. Unique allelic restriction fragments of the human H-ras locus in leukocyte and tumor DNAs of cancer patients. Nature 313:369, 1985.

Le Beau MM. Chromosomal fragile sites and cancer-specific rearrangements. Blood 67:849, 1986.

Li FP, Fraumeni JF Jr. Soft tissue sarcomas, breast cancer, and other neoplasms: A familial syndrome? Ann Int Med 71:747, 1969.

Lidereau R, Mathieu-Mahul D, Theillet C, et al. Presence of an allelic EcoRI restriction fragment of the c-mos locus in leukocyte and tumor cell DNAs of breast cancer patients. Proc Natl Acad Sci USA 82:7068, 1985.

Liu E, Hjelle B, Morgan R, et al. Mutations of the Kirsten-ras proto-oncogene in human preleukemia. Nature 330:186, 1987.

Müller R, Trembay JM, Adamson ED, et al. Tissue and cell type-specific expression of two human c-onc genes. Nature 310:249, 1983.

Saiki RK, Scharf S, Faloona F, et al. Enzymatic amplification of β-globin genomic sequences and restriction site analysis for diagnosis of sickle cell anemia. Science 230:1350, 1985.

Seizinger BR, Martuza RL, Gusella JF. Loss of genes on chromosome 22 in tumorigenesis of human acoustic neuroma. Nature 322:644, 1986.

Sparkes RS, Murphree AL, Lingua RW, et al. Gene for hereditary retinoblastoma assigned to human chromosome 13 by linkage to esterase D. Science 219:971, 1983.

Thor A, Hand PH, Wunderlich D, et al. Monoclonal antibodies define different ras gene expression in malignant and benign colonic diseases. Nature 311:562, 1984.

Weiss LM, Warnke RA, Sklar J, Cleary ML. Molecular analysis of the t(14;18) chromosomal translocation in malignant lymphomas. N Engl J Med 19:1185, 1987.

14
Potential Therapeutic Applications of Oncogenes

Overview

Research on oncogenes and their products is motivated in part by the belief that more fundamental understanding of the mechanisms of cancer causation and maintenance will lead to more rational means of treating malignancies. Therapeutics potentially may be directed against the genes themselves, RNA transcripts, or protein products. One obvious approach is to directly replace the mutated activated oncogene with its normal counterpart. However, present experimental capabilities are not advanced enough to accomplish this goal. Current technology is best suited to antiprotein therapeutics. Potential methods include antibodies against immunoglobulin idiotypes, growth factors, and other factors associated with the transformed phenotype, e.g., enzymes or proto-oncogene products such as *ras* p21. In addition, antibody-directed humoral toxins and cytotoxic cells might act against neoplasia. Oncogenic nucleic acid sequences might be targeted by antisense oligonucleotides of DNA or RNA and by nucleotide antimetabolites such as dideoxy- or methylphosphonate-modified nucleotides.

Another approach is to vaccinate against tumors or their known etiological agents. In particular, vaccines against tumor viruses in animals (feline leukemia, Merek's disease of chickens) and potentially in humans (hepatitis B virus) may significantly reduce the risk of virus-associated neoplasia. Tumor cell surface antigens and mutant products of activated oncogenes are additional potential targets of vaccination.

Although antiviral vaccines are in use and several antitumor or antioncogene product antibodies are being tested, much of the work on oncogene therapeutics is in a preliminary or even speculative theoretical stage. Enthusiasm for therapeutics of these types is tempered by the realization that oncogene products probably are of fundamental importance to normal cell growth and differentiation, and that mutant oncogenes and their products often differ from their normal counterparts only in minor details. Thus it is possible that antioncogene therapeutics might prove as, or more, toxic than currently available therapeutic modalities. In addition, with few exceptions, the technical means to achieve these therapeutics range from the primitive and investigational to the currently impossible, making practical realization of many of these treatments a good number of years in the future.

Oncogene Research and Cancer Treatment

The intense research activity on oncogenes has been stimulated in no small part because of an underlying belief that understanding the fundamental mechanisms of oncogenesis will lead to novel and more rational strategies for treatment of cancer. Despite the limitations of our current knowledge, several groups are actively engaged in antioncogene therapeutic approaches (Table 14.1). In this chapter we describe some of this ongoing work and some more speculative proposals as well.

Therapeutics against cancer, by analogy with those against infectious diseases, can take two general forms: *treatment* of an affected individual or *prevention* of the disease in the first place, e.g., by vaccination or avoidance of the environmental conditions leading to the disease. We focus primarily on potential approaches to treatment of cancer but mention some approaches to its prevention later in the chapter.

Oncogene therapeutics, as well as diagnostics, potentially can be directed against DNA, RNA, or protein. With our current technology, interference with oncogene protein products will be the most easily realized modality. Antibodies can already be made that specifically recognize *onc*-proteins. Therapeutically, such antibodies could be used to interfere with the actions of *onc*-proteins or to direct cytotoxins against cells expressing *onc*-proteins on their surfaces. Given nucleotide differences between oncogenic and nononcogenic genes, antibodies to the abnormal gene products might prevent cancer cell replication or metastasis. Theoretically, replacement of a defective oncogenic gene with its normal counterpart is the most specific anticancer treatment. If done properly, gene replacement also would have the least potential perturbation of normal cellular functions. For example, introduction of the normal Rb-1 allele into retinoblastoma cells might be expected to suppress tumor progression. At our current state of knowledge, it is difficult to imagine specific gene replacement therapy; however, normal genes may even now be introduced into malignant cells in a crude way.

In addition to merely removing the malignant cells themselves, one would like to maximize the beneficial effects of therapy while minimizing potentially harmful effects, i.e., to achieve the best possible therapeutic index. Current cancer therapies suffer from a low therapeutic index. Both irradiation and chemotherapy produce unwanted symptoms, e.g., diarrhea by affecting rapidly dividing gastrointestinal tract cells, alopecia from effects on hair follicles, and bone marrow suppression with resultant anemia, granulocytopenia (suscept-

TABLE 14.1. Clinical applications of oncogenes.

Diagnostic classification of tumors
Improved cancer detection
Development of tests to predict susceptibility to cancer
More effective therapy

TABLE 14.2. Associations: oncogenes and human cancer.

Oncogene	Cancer
Unequivocal association	
abl-bcl	Chronic myelogenous leukemia
c-*myc*	Burkitt's lymphoma
N-*myc*	Neuroblastoma
erb B-2	Breast carcinoma
Moderate association	
ras	Epithelial tumors, sarcomas, hematopoietic malignancies
EGF-R/*erb* B	Gliomas, squamous cell carcinomas
erb B-2	Breast carcinomas, salivary gland adenocarcinoma
myb	Acute leukemias
bcl 1, *bcl* 2	Non-Hodgkin's lymphomas
Speculative association	
src	Increased protein kinase activity in some colon carcinomas
fms	$5q^-$ Syndrome (hematopoietic), malignant histocytosis

ibility to infections), and thrombocytopenia (tendency toward bleeding). In addition, both irradiation and chemotherapeutic drugs are mutagenic to normal cells, and induction of a second, more refractory malignancy can occur as a consequence of these traditional therapeutic modalities.

Table 14.2 presents several of the best associations of oncogenes and human cancer. Unfortunately, many common cancers have not yet made the list.

Originally it was hoped that therapy directed against oncogenes or their products would be more specific to cancer and less harmful to the unrelated normal tissues of the body. However, in most cases the ubiquitous nature of an oncogene product precludes specific inactivation of the particular cancer-causing entity. For example, *src*-related protein kinases were among the earliest recognized oncogene products, and at first specific inhibitors of tyrosine kinases (sparing other cellular kinases) seemed like a logical approach to anticancer therapy. Now it is known that tyrosine kinases carry out numerous essential cellular functions. It is anticipated, therefore, that blanket prohibition of tyrosine phosphorylation would have as low a therapeutic index as traditional anticancer-cell drugs. In addition, as described in Chapters 7 and 8 it is the *lack* of a tyrosine substrate at the end of the cytoplasmic tails of several tyrosine kinase receptors that accounts for their failure to autoregulate. The tyrosine kinase portions of these oncogenes seem to differ only slightly from their normal proto-oncogene counterparts. Thus truly specific anti-*onc* protein inhibitors may be difficult to realize in practice.

Anti-oncogene Therapeutics

Despite limitations of both current technology and what may be theoretically possible, work is in progress in many laboratories to develop antioncogene therapeutics.

Anti-*onc*-Protein Therapeutics

The *onc*-proteins that are expressed on the cell surface make ideal targets for specific therapeutics. Such proteins may be products of oncogenes per se, or they may be products of normal genes that are overexpressed or differentially expressed on malignant cells. Particular "receptors" used for cellular interaction with other cells and with the environment are among the more easily defined cell surface proteins.

Antiidiotype Monoclonal Antibodies

Probably the most specific cell surface receptors are those of T and B cells. Each T or B lymphocyte is clonally derived, and each B cell surface immunoglobulin receptor or T cell surface receptor has unique antigenic determinants called idiotypes. Scientists at Stanford University have shown that so-called antiidiotype antibodies can identify a malignant clone of T or B cells (i.e., cells in a lymphoma or leukemia derived from these lymphoid cells) (Fig. 14.1). In a limited number of cases, treatment of B cell lymphoma patients with antiidiotype antibodies has resulted in partial to complete remission after failure of more conventional cytotoxic drug therapy. Evidence suggests, however, that the tumor cells can "escape" the antiidiotype antibody by accumulating mutations within the receptor variable region genes. This action generates new clones of cells resistant to the original antiidiotype monoclonal antibodies. In addition, bi- or multiclonal tumors are more difficult to control by this method. Escape of the aberrant clones has been a consistent problem for this novel approach to cancer therapy. Tumor cell heterogeneity and mutation/selection of variant receptors on the cell surface may be a generic problem with therapies directed to these targets.

Immunotoxins

Immunotoxins are a promising approach to cancer therapy (Figs. 14.2 and 14.3). These novel molecules are composed of an antibody and an attached toxin, drug, or radionuclide. In most cases a monoclonal antibody identifies a target structure specific for tumor cells, e.g., a tumor antigen on the cell surface. Then, together with an attached toxin, the antibody is internalized and released into the cytoplasm. In the case of an attached ricin A chain toxin, the toxin is released within the endosome and undergoes a conformational change that facilitates its release into the cytoplasm. Ricin A chain kills cells by ribosylating and thus inactivating elongation factor II (EF II), which is necessary for cellular protein synthesis. Because ricin is an enzyme, few ricin molecules are required to kill a target cell. Those oncogenes that are expressed on the cell surface are possible targets for toxin-associated monoclonal antibodies.

Antibodies binding the *neu* oncogene product inhibit the growth of cells that express this protein in vitro and in vivo. Anti-EGF receptor antibodies also appear to make good immunotoxins. Transferrin receptor expression is generally increased on rapidly growing cells and on some malignant cells in particular. Several research groups have shown that antitransferrin receptor monoclonal

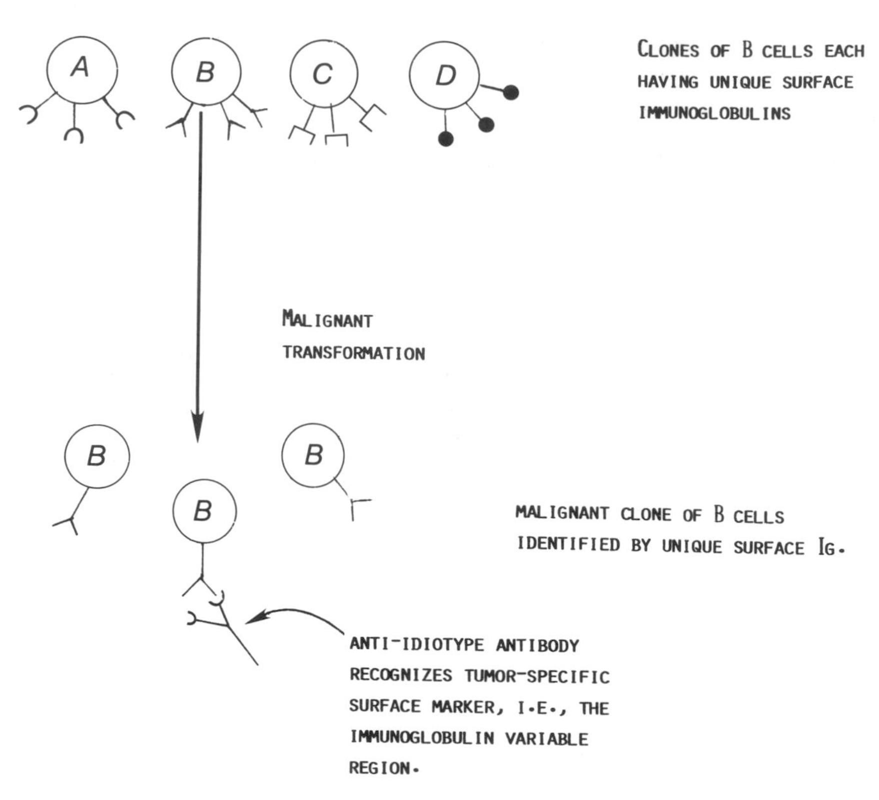

FIGURE 14.1. Antiidiotype monoclonal antibody therapy. B and T lymphocytes have surface receptors that are unique markers for individual clones. They are surface immunoglobulin (B cells) and the antigen receptor of T cells. Monoclonal antibodies directed to the variable portions of these molecules, idiotypes, are tumor-specific.

antibodies with attached toxins (e.g., ricin or diphtheria toxin) are capable of killing cells overexpressing the transferrin receptor in vitro. Such immunotoxins also slow the growth of human tumors implanted in nude mice.

Phase I clinical trials of immunotoxins directed against melanoma, lymphoma, breast cancer, ovarian cancer, and other solid tumors are in progress. In general, these novel therapeutics have been well tolerated, and there have been a few documented responses by melanoma patients. In addition, antibodies carrying radionuclides or conventional chemotherapeutic agents such as adriamycin appear promising. In the latter case there is a much improved therapeutic index of the antibody-directed drug.

Monoclonal Anti-Growth Factor Antibodies

Antibodies may also be exploited for their ability to bind autocrine growth factors. For example, antibodies have been shown to block the growth in vitro and

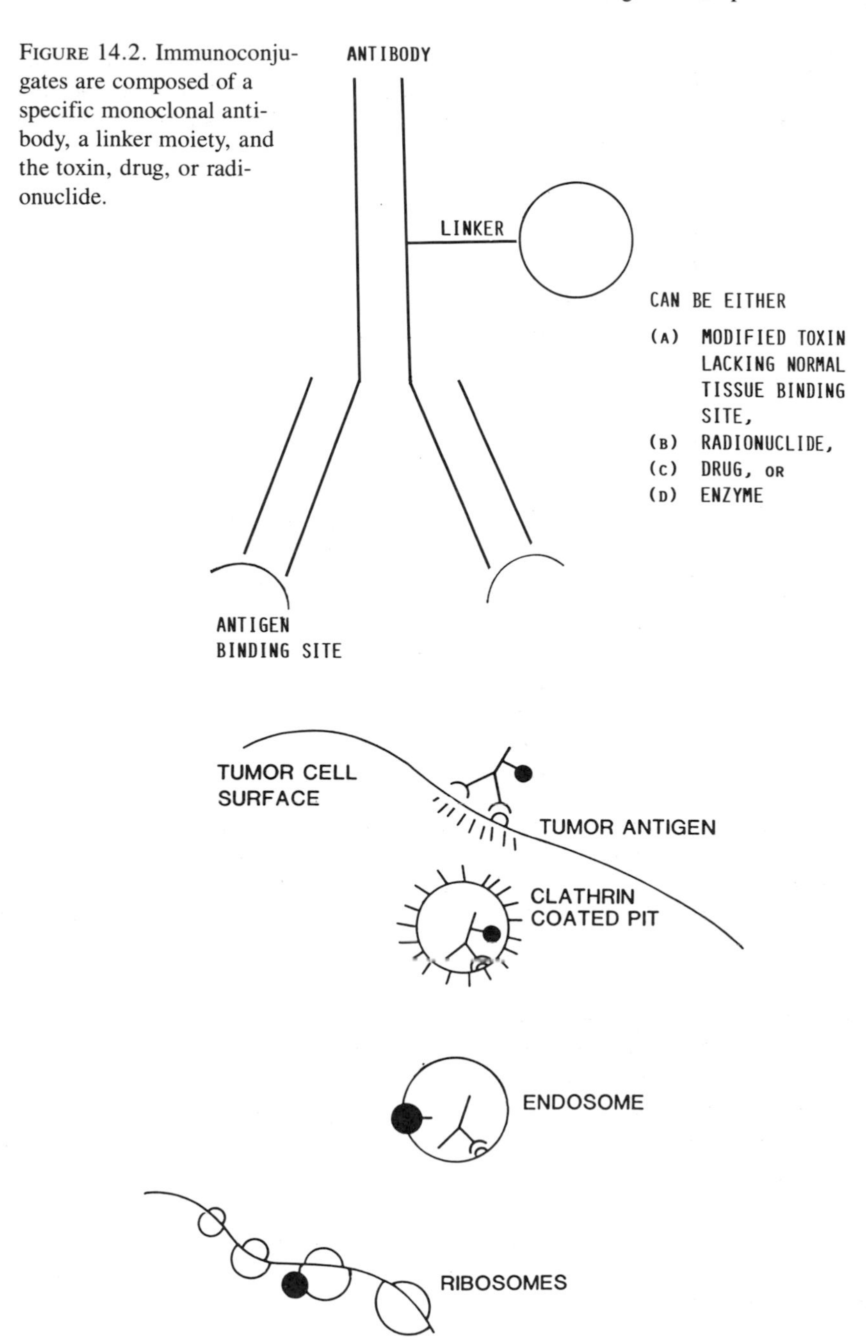

FIGURE 14.2. Immunoconjugates are composed of a specific monoclonal antibody, a linker moiety, and the toxin, drug, or radionuclide.

FIGURE 14.3. Pathways involved in the internalization and subsequent killing of cells by ricin immunoconjugates. The conjugate binds to a cell surface tumor antigen and is endocytosed. The ricin (●) is released from the antibody; it then ADP-ribosylates elongation factor II, a necessary component of the ribosomal translation apparatus.

in vivo of small-cell lung carcinoma cells by binding bombesin, a known autocrine growth factor released by these cells. Because bombesin possesses various normal metabolic functions, it remains to be seen what the side effects of these monoclonal antibodies are. Autocrine growth factors also have been identified for breast carcinoma cells and lymphoid cells. It is anticipated that the growth of cells from many lineages is directed by autocrine growth factors and that inhibition of such growth factors will block cell proliferation.

Obviously, if an attack could be directed against the products of oncogenes that are qualitatively different from normal host growth factors, the therapeutic index would be increased. Antibodies to the *sis*-gene product are in this category. However, in some cases antiautocrine growth factor antibodies have not blocked continued cell growth. This lack of effect may be the result of *intra*cellular stimulation of growth by autocrine growth factors. For example, anti-PDGF antibodies inhibit only a portion of v-*sis*-containing cell lines.

Neutralization of Other Factors Associated with the Transformed Phenotype

Numerous studies have demonstrated increased levels of proteolytic enzymes associated with malignant tumors in vivo and malignant transformation in vitro. Activities of collagenases, various cathepsin enzymes, and serine proteases have been shown to be elevated in malignant cells.

An important proteolytic activity associated with neoplastic cells results from the cascade initiated by generation of plasmin from serum plasminogen by the serine protease, plasminogen activator (PA). Plasmin is an important protease can activate tumor-associated PA and procollagenase. PA production and the generation of plasmin correlate with growth in agar, tumorigenicity of viral transformants, and tumorigenicity of malignant melanoma cells; however, increased PA is not an absolute requirement for malignancy. Sullivan and Quigley from Downstate Medical Center (Brooklyn) showed that low concentrations of a monoclonal antibody selected for its ability to inhibit PA inhibited the overgrowth and morphological changes associated with Rous sarcoma virus transformation of chick embryo fibroblasts. The monoclonal antibody inhibited the degradation of extracellular matrix, an effect mediated by PA released from the tumor cells. This research demonstrated that "pathogenic enzymes" released by tumor cells can be neutralized by monoclonal antibodies. One can surmise that other substances released by tumor cells that promote their growth or damage the host's cells could be neutralized in similar fashion by specific antibodies.

Monoclonal Targeting of Cytotoxic Cells

Although many cancer patients have defective cell-mediated immunity, their lymphoid cells often can be shown to have antitumor cell activity when tested in vitro. Research by David Segal at the National Institutes of Health (NIH) suggested that heteroconjugates of two monoclonal antibodies, one directed to a cytotoxic cell surface antigen and the other to the tumor-selective antigen, may greatly improve the specificity and lethality of these cells (see Perez et. al.). The

NIH group has linked anti-CD3 monoclonal antibodies (recognizing a pan T lymphocyte antigen marker that fortuitously activates the cytotoxic response of the T cell that binds it) together with monoclonal antibodies recognizing ovarian, breast, and lung carcinomas and melanomas. These "franked" (for "Frankenstein!") T cells are lethal to tumor cells in vitro. These and other scientists have conjugated monoclonal antibodies recognizing monocyte and natural killer Fc receptors to antitumor antibodies and have directed these activated cells to their tumor targets with excellent results.

Another group at the NIH led by Steven Rosenberg has generated so-called lymphokine-activated killer (LAK) cells by co-culture of peripheral blood lymphocytes with interleukin-2. This procedure generated active tumor-lethal cells. In a phase I/phase II trial of this adoptive immunotherapy, the in vitro activated LAK cells infused back into cancer patients produced regression of several types of cancer including breast cancer, melanoma, and renal cell carcinomas. In Segal's research, culture of the antibody-directed cells with interleukin-2 markedly increased their lethality. Thus a combination of an appropriate antibody or conjugate with lymphokine-activated killer cells may provide a promising therapeutic approach for the future. Furthermore, cells derived directly from the tumors (tumor-infiltrating lymphocytes) of affected patients have proved to be the best targets for interleukin-2 stimulation. The important issue is to define oncogene-related surface targets or other cancer-specific targets to which monoclonal antibodies can be generated and to define the best way to activate the host's own defense mechanisms.

Other Cancer Cell Surface Targets

A number of monoclonal antibodies have been identified by their increased selectivity for tumor cells compared to nonmalignant cells. The targets recognized by antibodies selected in this manner are not well defined, and none is known to be the product of oncogenes. Although expression of these target molecules appears to be altered in cancer cells, in several analyzed cases the proteins were not qualitatively different from the ones expressed on normal cells, a feature that explains the cross-reactivity of the "tumor-specific" monoclonal antibodies with some normal cells. Nevertheless, it is hoped that further study will identify oncogene-encoded cell surface products with qualitative differences in expression between malignant and normal cells. These proteins would be ideal targets for immunoconjugates.

A promising cell surface target for antibody-directed therapy is a membrane protein found to be present in cells that are resistant to multiple chemotherapeutic drugs. This 170 kD protein is expressed primarily in drug-resistant transformed cell lines and primary tumors. Antibodies directed to this target might help destroy drug-resistant tumor cells, which would significantly improve the efficacy of current chemotherapies.

Several monoclonal antibodies selectively recognizing cancer cells have been tested in animal models and are being evaluated in clinical trials. The dramatic

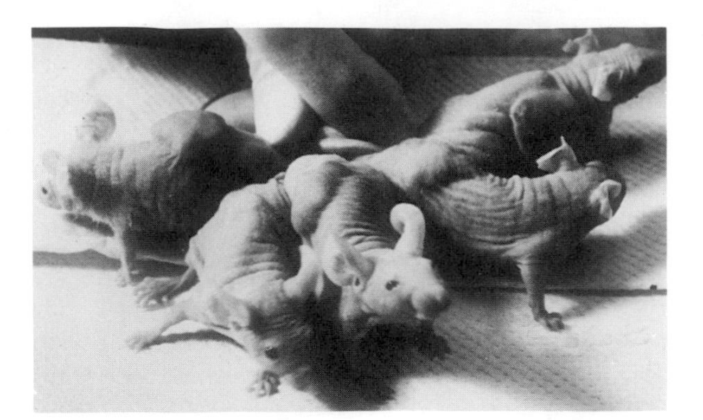

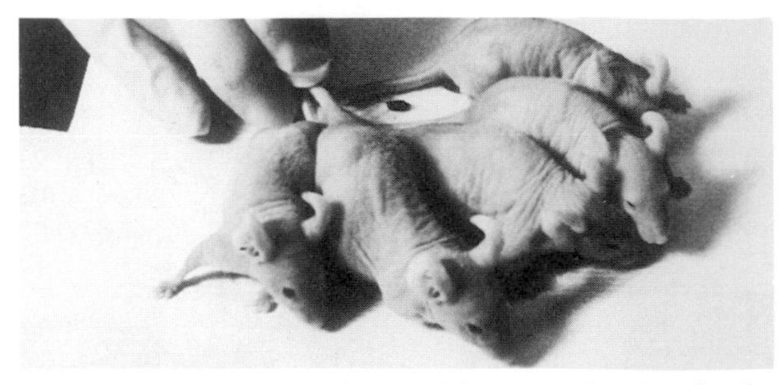

FIGURE 14.4. Inhibition of human tumor cell growth by monoclonal antibodies. (A) Nude mice were inoculated subcutaneously with 5×10^6 human colon carcinoma cells SW 948 and injected intraperitoneally with 200 μg antiinfluenza virus control antibody once daily for 5 days started immediately after tumor cell inoculation. Colon carcinoma growth at 8 weeks after tumor inoculation is shown. (B) Nude mice were inoculated with tumor cells and treated with anticolon carcinoma monoclonal antibody CO 17-1A as described above. The antibody completely inhibited tumor growth in three of five mice, as shown at 8 weeks after tumor cell inoculation. In two mice (second and fourth mice from the left) small tumor nodules approximately 50 mm^3 in size started to grow. (Kindly provided by Dr. Dorothee Herlyn)

ability of one of these monoclonal antibodies to inhibit the growth of human tumor cells growing in immunodeficient nude mice is shown in Figure 14.4.

Specific Reversal of the Transformed Phenotype by Antibodies

The normal *ras* protein, p21, differs in defined molecular ways from the *ras* proteins coded by *ras* oncogenes (see Chapter 9). Feramisco and co-workers exploited that difference by microinjecting an antibody capable of binding *onc*-p21 into living cells transformed by *ras* oncogenes. The result (Fig. 9.8) was reversion of the transformed cells to a normal phenotype. Because the antibody did not bind the normal form of p21, the revertant cells continued to grow normally. As the microinjected antibody was degraded, *onc* p21 was reexpressed and the cells returned to their transformed state. This experiment showed that an agent capable of specifically inactivating an oncogene product is able to reverse a transformed morphology. However, microinjection of such agents into malignant cells in vivo offers little hope as a practical anticancer vehicle with current biotechnology.

ANTI-*onc* NUCLEIC ACID THERAPEUTICS

Oncogene DNA and RNA differ in nucleotide sequence from normal proto-oncogene DNA and RNA. It is theoretically conceivable that specific antisequence molecules could be designed to block replication or expression of oncogene DNA or translation of oncogene mRNA (Fig. 14.5). For example, by analogy with the microinjection of anti-*ras onc* p21 antibodies cited above,

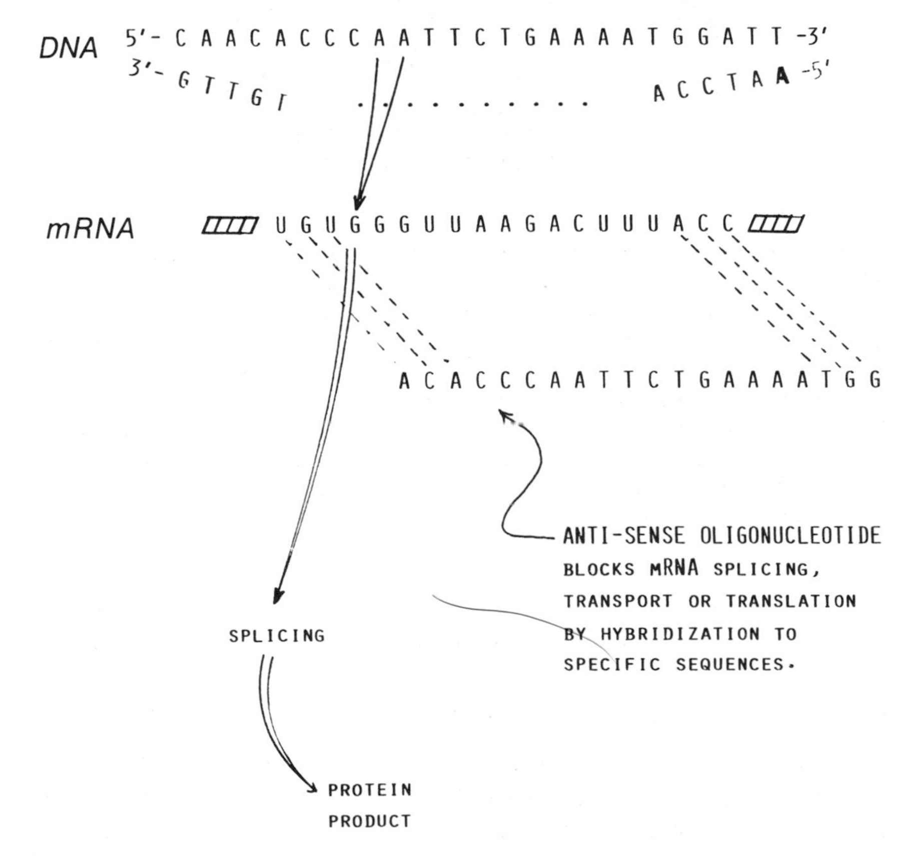

FIGURE 14.5. Principle of antisense therapeutics. Oligonucleotides or retroviral encoded antisense mRNAs bind to critical control regions (e.g., splice acceptor sites) and prevent the processing, extranuclear transport, and translation of selected transcripts.

microinjection of antisense DNA sequences might bind to and block expression of *onc*-DNA. Such "blocking DNA" could perhaps incorporate an anti-DNA toxin toxin (e.g., alkylating agent) to specifically destroy that portion of the chromosome encoding the oncogene. Similarly, an antibody against the oncogenic RNA sequence might bind and prevent that mRNA molecule from being used by ribosomes, or it might inhibit elongation by blocking the tRNA from binding to the *onc*-sequence anticodon. At our current level of technology, these suggestions are admittedly speculative.

A more elegant and permanent, though at present not practical, way of reversing the transformed phenotype would be to replace the altered, mutant oncogene with its normal counterpart or to introduce a dominantly acting antioncogene. Several groups have tried to reverse the transformed phenotype by expressing large amounts of mRNA from the DNA strand opposite the one coding an aberrant oncogene protein using "antisense RNA" techniques. In the cells' nuclei the two complementary strands hybridize to form a double-stranded structure that effectively prevents transcription of the mRNA. Results to date have been equivocal. When our ability to "target" genes and control their expression in a tissue-specific manner improves, this approach may be a practical method for "turning off" cancer genes.

Presently, several laboratories are testing retroviruses encoding antisense RNA expressed in a tissue-specific manner by use of appropriate enhancer (and other control) regions attached to the anticancer genes. A better understanding of retroviral tissue tropism mediated by cell surface receptors is important to make this approach practical. Before the idea of infecting a cancer patient with a retroviral vector carrying an antisense gene is acceptable, its efficacy and most importantly its safety must be proved.

Oligonucleotide Therapeutics

A novel therapeutic avenue being explored is the use of oligonucleotides. Some naturally occurring (e.g., viral) or synthetic [e.g., poly (I)-poly (C)] polynucleotides have been successfully used as antiviral and antitumor agents, perhaps owing to their interferon-inducing properties.

A number of successful chemotherapeutic agents used in cancer chemotherapy, immunological suppression, and viral inhibition are analogues of nucleosides. This approach has previously been limited by cell toxicity owing to in vivo conversion of the nucleoside analogues to nucleotides that become incorporated into cellular DNA, RNA, and nucleotide enzymes. However, the use of dideoxynucleosides for treatment of HIV infections has been promising and so far has been associated with low cell toxicity.

Smith et. al. and Zamecnik and collaborators have produced exciting results using synthetic methylphosphonate-modified oligodeoxynucleotides to inhibit herpes and HIV virus replication. These investigators synthesized oligo-deoxynucleotides that inhibited infected cells in culture by "hybridization competition." These molecules were termed "hybridons." The selected bases of the synthetic oligonucleotides were complementary to important segments of target mRNA. Sequences of 12 to 20 bases were used. Particularly high (88%) inhibition was noted using sequences close to mRNA splice acceptor sites. A specific sequence of 20 bases is expected to occur at random on the average of one time in 10^{12}. Thus the specificity of a given fragment of this length is high given the approximately 3×10^9 to 4×10^9 bases in the human genome. Clearly, it is a promising therapeutic approach to thwart cancer cell growth. Hybridons specifically binding to oncogenes might prove efficacious for blocking *onc*-expression providing these genes were expressed at higher levels in transformed cells. The potential problem of oligonucleotide degradation by nucleases could be overcome by using liposome or biodegradable polymer delivery systems, daily parenteral administration, or synthesis of nonhydrolyzable linkages (e.g., the methylphosphonates cited above).

Monitoring Cancer Therapy

It is important to monitor the effects of chemotherapeutic or immunotherapeutic agents on tumor cell growth. The effects of these agents on the expression of specific oncogenes in vitro might be useful in predicting a tumor's response in vivo. Few studies of this nature have been carried out, although interferon is known to suppress transcription of several proto-oncogenes including c-Ha-*ras*, c-*src*, and c-*myc* in transformed or tumor cell lines. In some cases all growth was inhibited; in others it was not. Just as expression of *onc*-proteins may permit novel staging of tumors, it may also guide therapy.

Cancer Vaccines

The most desirable and cost-effective way to prevent a disease is to eliminate its causes or to protect the population at risk by measures such as vaccination. For those cancers caused by defined agents such as viruses this approach is already within our technical grasp (Table 14.3).

The most common cancer worldwide is hepatocellular carcinoma. This disease has been linked epidemiologically to chronic infection of hepatocytes by the

TABLE 14.3. Potential cancer vaccines.

Cancer associated agent	Malignancy
Hepatitis B virus	Hepatomas
Cytomegalovirus	Kaposi's sarcoma
HTLV-I, HTLV-II viruses	Adult T cell leukemia and related malignancies
Epstein-Barr virus	Burkitt's lymphoma, nasopharyngeal carcinomas
Various papilloma viruses	Cervical and penile cancers
Surface *onc*-proteins	Unknown

hepatitis B virus (HBV) (see Chapter 3). The development of effective hepatitis B vaccines should not only reduce the prevalence of acute hepatitis but the prevalence of hepatocellular carcinomas following chronic infection as well. Although the current vaccine (prepared from the blood plasma of chronic hepatitis carriers) is effective, it is still expensive. Hopefully, the production of hepatitis B vaccines using recombinant DNA technology will reduce the cost and improve the worldwide distribution of this potential cancer vaccine. Because of the long latency period between HBV infection and development of hepatoma, observation over several years is necessary before the efficacy of these vaccines can be established.

In the meantime, development of veterinary vaccines may demonstrate the potential of human cancer vaccines. For example, recombinant feline leukemia virus vaccines have been developed. Antiwart vaccines against papilloma viruses have been available to veterinarians for a number of years. Effective vaccines have been developed against two herpes virus-induced diseases: Marek's disease of chickens and a primate lymphoma. Conceptually, it should be possible to vaccinate against any animal virus that has been associated with cancer induction (Table 14.3). Examples of these viruses (as discussed in more detail in Chapters 3 and 4) include Epstein-Barr virus (B cell malignancies, nasopharyngeal carcinoma), cytomegalovirus (Kaposi's sarcoma), papilloma viruses (warts; cervical, vulvar, penile, and laryngeal papillomas, condylomas, and carcinomas), human T cell lymphotropic viruses (adult T cell leukemia and other miscellaneous leukemias and lymphomas), and herpes viruses (possibly cervical carcinoma). Unfortunately, the major hurdles for development of such vaccines are legal and financial rather than technical. A reduction of the current litiginous atmosphere is required before pharmaceutical companies will be willing to invest the time and effort required to develop these therapeutics. Hopefully, the World Health Organization and the NIH will continue to research these promising avenues of investigation.

Can the protein products of oncogenes be used for immunogens? When the behavior of oncogenes in cancer is better understood, it should be possible to predict which translocations and mutations occur most commonly. In principle, it should be possible to vaccinate high risk individuals, particularly if the *onc*-proteins are expressed on the cancer cell surface. In at least one model case, surface expression of the *onc*-protein was not necessary (see below) for successful vaccination. On the other hand, the discovery of idiotype heterogeneity in B cell lymphomas suggests that surface antigen variation might limit development of useful cancer vaccines.

Scientists at the NIH have shown that in utero administration of the chemical carcinogen ethylnitrourea (ENU) followed by the tumor promoter butylated hydroxy toluene (BHT) in mice leads to development of lung adenocarcinomas and T cell lymphomas. These tumors are associated with the expression of *raf*-1, an oncogene coding for an intracellular protein kinase. When animals were immunized with the v-*raf* protein, the latency period for tumor development after birth was prolonged, but the final number of deaths in the vaccinated versus the control (unvaccinated) group was equal. Research continues using a number of model systems to investigate the feasibility of oncogene-based cancer vaccines.

Numerous cancer-associated proteins appear to be expressed on tumor cell surfaces. Although few of these are truly "cancer-specific," their predominant distribution on transformed cells makes them useful not only for monoclonal antibody-based therapeutics (see above) but also for cancer vaccines. Cloned surface antigens expressed in greater numbers on tumor cells would make ideal vaccines if problems of immunogenicity could be overcome. The development of novel adjuvants and carriers for subunit recombinant vaccines is important in this regard. A better understanding of the function of proto-oncogenes should elucidate further the possibility of vaccines based on these molecules.

Antiidiotype Vaccines

Another approach to the cancer vaccine problem is development of human monoclonal antibody, "antiidiotype" vaccines. With this scheme antibodies recognizing tumor-associated antigens are used as the immunogens. An "antiidiotype" response, i.e., an antibody against the antibody used as immunogen, or an "anti-antiidiotype" response might have a direct antitumor effect (Fig. 14.6). It has been shown that antiidiotype antibodies can produce effective cellular immunity. Thus it may be possible to augment or promote both the humoral and cellular immune responses to cancer using this method.

In clinical trials using murine monoclonal antibodies recognizing cancer-restricted cell surface-bound antigens, scientists at the Wister Institute showed that those patients developing an antiidiotype response to the treatment monoclonal antibody had the best overall clinical response. Studies by other scientists are in progress to exploit the antiidiotype concept.

Investigators at Bionetics Research have tried using autologous cancer cell vaccines. Colon cancer cells removed from patients at the time of surgery were dissociated and reinjected with bacille calmette Guérin (BCG) adjuvant. In phase I/II trials, some colon cancer patients vaccinated in this manner have shown detectable clinical responses. B lymphocytes from these patients have also been used for generation of cancer-specific human monoclonal antibodies. Such antibodies are now being used for tumor imaging (diagnostically) and antitumor therapy.

Richard Cote and his group at Memorial Sloan-Kettering Institute have shown that human hybridomas derived from B cells isolated originally from cancer patients produce human monoclonal antibodies that recognize cytoplasmic antigens predominantly. This result suggests that the human humoral immune response to *surface* antigens is limited, probably because of immunological tolerance and active immune suppression. This group also showed that human monoclonal antibodies generated from persons without known disease or those with autoimmune diseases also recognized cancer cell antigens. In fact, some of the monoclonal antibodies demonstrated broad cancer-specific binding, whereas others showed more restricted binding. Clearly, we know little about the

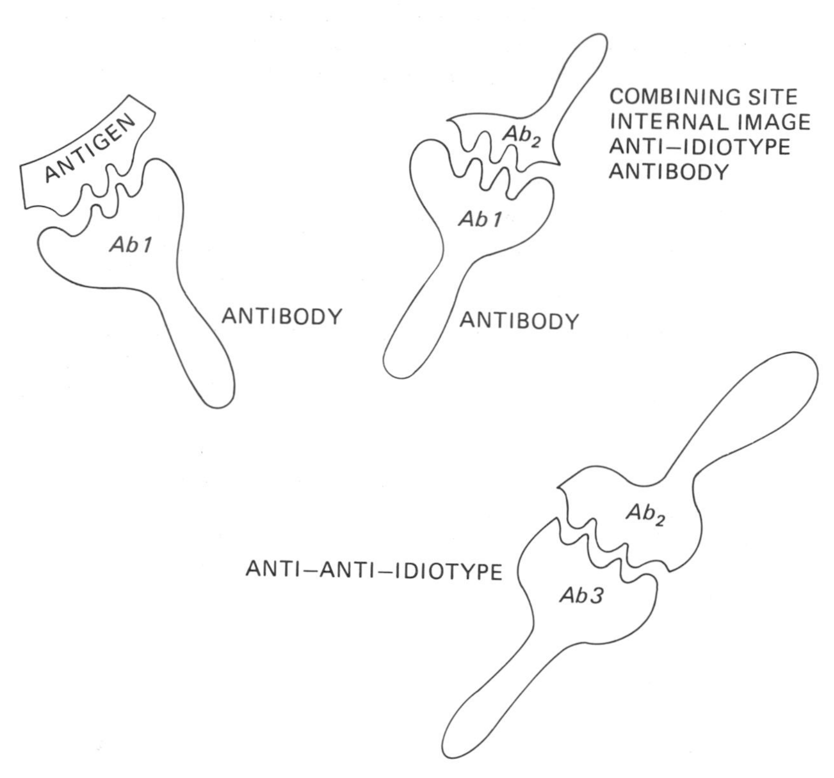

FIGURE 14.6. Antiidiotype antibodies: potential vaccines. Antibody 1 recognizes an antigenic determinant on the cell surface. When this antibody is used as an immunogen, antibodies are generated that recognize multiple epitopes on antibody 1. A subset of these antibodies, so-called antiidiotype antibodies, bind to the variable region of antibody 1. A subset of these antivariable region antibodies, so-called internal image antibodies, recognize the combining site of antibody 1. Antibodies made to these AB2 antibodies can recognize the original antigen if they are also antiidiotype internal image antibodies.

immune response to cancer and less about how to manipulate it. Further elucidation of the fundamental principles of the immune system might make practical prevention or treatment of cancer by vaccination possible.

Prospective Outlook

We have discussed several examples of potential anticancer therapeutics that may be developed by exploiting differences between normal proto-oncogenes and their protein products compared to cancer-causing genes and *onc*-proteins. The inspiration for this work comes historically from successful antimicrobial therapy developed by utilizing differences between prokaryotic microorganisms and their eukaryotic target cells. In the case of cancer therapeutics there are few obvious differences between normal and malignant eukaryotic cells. We have only begun to recognize the complexity of the problem and the subtlety of the answer.

BIBLIOGRAPHY

Bjorn MJ, Groetsema G, Scalapino L. Antibody-Pseudomonas exotoxin A conjugates cytotoxic to human breast cancer cells in vitro. Cancer Res 46:3262, 1986.

Cline M, Slamon DJ, Lipsick JS. Oncogenes: implications for the diagnosis and treatment of cancer. Ann Intern Med 101:223, 1984.

Cote RJ, Houghton AN. The generation of human monoclonal antibodies and their use in the analysis of the humoral immune response to cancer. In: Engleman EG, Foung SKH, Larrick JW, Raubitschek A (eds), Human Hybridomas and Monoclonal Antibodies. New York: Plenum Press, 1985, p. 189.

Douillard JY, Lehur PA, Aillet G, et al. Immunohistochemical antigenic expression and in vivo tumor uptake of monoclonal antibodies with specificity for tumors of the gastrointestinal tract. Cancer Res 46:4221, 1986.

Drebin JA, Stern DF, Link VC, et al. Monoclonal antibodies identify a cell-surface antigen associated with an activated cellular oncogene. Nature 312:545, 1984.

Epenetos AA, Snook D, Durbin H, et al. Limitations of radiolabeled monoclonalantibodies for localization of human neoplasms. Cancer Res 46:3183, 1986.

Epstein MA. Vaccination against Epstein-Barr virus: current progress and future strategies. Lancet 1:1425, 1986.

Feramisco JR, Clark R, Wong G, et al. Transient reversion of ras oncogene-induced cell transformation by antibodies specific for amino acid 12 of ras protein. Nature 314:639, 1985.

FitzGerald DJ, Willingham MC, Pastan I. Antitumor effects of an immunotoxin made with Pseudomonas exotoxin in a nude mouse model of human ovarian cancer. Proc Natl Acad Sci USA 83:6627, 1986.

Garnett MC, Baldwin RW. An improved synthesis of a methotrexate-albumin-791T/36 monoclonal antibody conjugate cytotoxic to human osteogenic sarcoma cell lines. Cancer Res 46:2407, 1986.

Haspel MV, McCabe RP, Pomato N, et al. Generation of tumor cell-reactive human monoclonal antibodies using peripheral blood lymphocytes from actively immunized colorectal carcinoma patients. Cancer Res 45:3951, 1985.

Hoover HC Jr, Surdyke M, Dangel RB, et al. Delayed cutaneous hypersensitivity to autologous tumor cells in colorectal cancer patients immunized with an autologous tumor cell: bacillus Calmette-Guérin vaccine. Cancer Res 44:1671, 1984.

Houghton AN, Mintzer D, Cordon-Cardo C, et al. Mouse monoclonal IgG3 antibody detecting G_{D3} ganglioside: a phase I trial in patients with malignant melanoma. Proc Natl Acad Sci USA 82:1242, 1985.

Jung G, Honsik CJ, Reisfeld RA, et al. Activation of human peripheral blood mononuclear cells by anti-T3: killing of tumor target cells coated with anti-target-anti-T3 conjugates. Proc Natl Acad Sci USA 83:4479, 1986.

Kartner N, Evernden-Porelle D, Bradley G, et al. Detection of P-glycoprotein in multidrug-resistant cell lines by monoclonal antibodies. Nature 316:820, 1985.

Kennedy RC, Dreesman GR, Kohler H. Vaccines utilizing internal image anti-idiotypic antibodies that mimic antigens of infectious organisms. Biotechniques 3:404, 1985.

Koprowski H, Herlyn D, Lubeck M, et al. Human anti-idiotype antibodies in cancer patients: is the modulation of the immune response beneficial for the patient? Proc Natl Acad Sci USA 81:216, 1984.

Levy R, Stratte P, Link M, et al. Monoclonal antibodies to human lymphocytes: clinical application in therapy of leukemia. In: Kennett RH, Bechtol KB, McKearn TJ (eds), Monoclonal Antibodies and Functional Cell Line. New York: Plenum Press, 1984, p. 193.

Lindmo T, Boven E, Mitchell JB, et al. Specific killing of human melanoma cells by ^{125}I-labeled 9.2.27 monoclonal antibody. Cancer Res 45:5080, 1985.

Mach J-P, Chatal J-F, Lumbroso J-D, et al. Tumor localization in patients by radiolabeled monoclonal antibodies against colon carcinoma. Cancer Res 43:5593, 1983..

Maxwell IH, Maxwell F, Glode LM. Regulated expression of a diphtheria toxin α-chain gene transfected into human cells: possible strategy for inducing cancer cell suicide. Cancer Res 46:4660, 1986.

Mitsuya H, Broder S. Inhibition of the in vitro infectivity and cytopathic effect of human T-lymphotrophic virus type III/lymphodenopathy-associated virus (HTLV-III/LAV) by 2′,3′-dideoxynucleosides. Proc Natl Acad Sci USA 83:1911, 1986.

Perez P, Titus JA, Lotze MT, et al. Specific lysis of human tumor cells by T cells coated with anti-T3 cross-linked to anti-tumor antibody. J Immunol 137:2069, 1986a.

Perez P, Hoffman R, Titus J, Segal D. Specific targeting of human peripheral blood T cells by heteroaggregates containing anti-T3 crosslinked to anti-target cell antibodies J Exp Med 163:166, 1986b.

Perez P, Titus J, Lotze MT, et al. Specific lysis of human tumor cells by T cells coated with anti-T3 cross-linked to anti-tumor antibody. J Immunol 137:2069, 1986.

Rosenberg SA, Spiess P, Lofreniere R. A new approach to the adoptive immunotherapy of cancer with tumor infiltrating lymphocytes. Science 233:1318, 1986.

Segal DM, Perez P, Karpovsky B, et al. Targeting of cytotoxic cells with cross-linked antibody heteroaggregates. Mol Immunol 23:1211, 1986.

Smith CC, Aurelion L, Reddy MP, et al. Antiviral effect of an oligonucleoside methylphosphonate complementary to the splice junction of herpes simplex virus type 1 immediate early pre-mRNAs 4 and 5. Proc Natl Acad Sci USA 83:2787, 1986.

Staerz UD, Bevan MJ. Hybrid hybridoma producing a bispecific monoclonal antibody that can focus effector T-cell activity. Proc Natl Acad Sci USA 83:1453, 1986.

Sullivan LM, Quigley JP. An anticatalytic monoclonal antibody to avian plasminogen activator: its effect on behavior of RSV-transformed chick fibroblasts. Cell 45:905, 1986.

Taetle R, Castagnola J, Mendelsohn J. Mechanisms of growth inhibition by anti-transferrin receptor monoclonal antibodies. Cancer Res 46:1759, 1986.

Vitetta ES, Korlick KA, Miyama-Inaba M, et al. Immunotoxins: a new approach to cancer therapy. Science 219:644, 1983.

Zamecnik PS, Goodchild J, Taguchi Y, et al. Inhibition of replication and expression of human T-cell lymphotropic virus type III in cultured cells by exogenous synthetic oligonucleotides complementary to viral RNA. Proc Natl Acad Sci USA 83:4143, 1986.

15
Oncogene Paradigm: Contribution to a Fundamental Understanding of Malignancy

We have explored how the oncogene paradigm can help explain what a cancer is and the cascade of events that might induce the neoplastic phenotype. This concept provides an intellectual framework within which to create tools that will advance our ability to diagnose and treat cancer. We now understand that normal control of growth, development, and differentiation is determined by molecular genetic mechanisms residing ultimately in the cellular DNA. Because cancer is characterized by loss of this normal control, we can at least begin to define and understand malignancy in molecular terms. The oncogene concept provides us with a reasonable molecular "handle" with which to deal with cancer.

We know that most human cancers are not due to the introduction of foreign genetic material, e.g., through viral infection. The oncogene paradigm predicts that such foreign genes are not necessary, as our own genetic material carries all the elements necessary for malignant transformation. Even in animal models of viral oncogenesis, transduction or insertional activation of endogenous host proto-oncogenes is the usual mode of neoplastic induction.

We also know that in the cancerous state normal growth control and differentiation has gone awry. Cancer cells continue to divide; they live longer than normal cells and appear not to undergo the programmed cell death characteristic of normal cells; in addition, they fail to undergo normal differentiation to nondividing, terminally differentiated cells. Again, our new knowledge helps explain these phenomena, as proto-oncogenes appear to be involved in normal growth and differentiation (e.g., *src* in neuronal differentiation, *myc* in the cell cycle), and perturbations that change proto-oncogenes to oncogenes alter their normal functions. Therefore oncogenes help us define what cancer is.

Once we have identified those genes strongly associated with malignancy, we then may ask what converts a normal gene to a transforming gene. Answers here may help explain what causes cancer. In the discussions of each class of oncogenes, we examined the genetic events that activate proto-oncogenes to their transforming counterparts. We have shown that point mutations within genes (*ras, neu*), rearrangements within coding sequences (*bcr/abl*) and outside of coding sequences (*myc, bcl*-1 and 2), overexpression and amplification (c-*myc*, N-*myc, ras*), and deletions of portions of genes (EGF-r, *src*) or of whole genes (Rb-1) may be responsible for triggering the onset of malignancy or allowing cancer to progress.

Finally, once genetic markers of malignant transformation are identified, strategies to better diagnose and treat cancer can be rationally devised. Diagnostic tools based on the differences between proto-oncogenes and oncogenes are, in fact, already being formulated; and some of these approaches are being used. However, the oncogene concept suggests that specific anticancer treatment will not be easily found. We already know from our empirical experience that when treating a tumor we often sacrifice many normal cells. This destruction of normal cells is the reason for the significant side effects associated with current cancer therapies. Our knowledge of differences between proto-oncogenes and oncogenes again points to the difficulty of distinguishing "normal" from "malignant." Unlike antibiotic therapy, where differences between microbial and host enzymes (and other cellular features) can be exploited, oncogenes and proto-oncogenes (and their products) may differ on a truly miniscule level, such as one base alteration in DNA sequence or minor quantitative alterations of mRNA level or protein half-life. Though we have touched on possible therapies based on the oncogene paradigm, we concede that our knowledge poses more therapeutic challenges than answers.

We believe, however, that the oncogene concept, like the cell theory and the germ theory that preceded it, represents a major contribution to scientific thought. For this reason it is essential that both scientists and educated lay people familiarize themselves with it. Someday, hopefully, we will be able to befriend and convert our "enemies within."

Appendix: Oncogenes and Related Genes

Oncogene	Species of origin	Virus or method of detection	Viral/original tumor	Viral gene product	Cellular homologue
abl	Murine	Abelson leukemia virus	Lymphoid leukemia	p120 (*gag-abl*)	p150 (c-*abl*)
A-*raf*-1, A-*raf*-2	Human Human	DNA homologous to raf probe in human cDNA library	–	–	*raf*-1 related probably same as pks *raf*-1 related
(*bas*)	Murine	BALB/c murine sarcoma virus	Sarcoma	p21	p21
bcl 1, *bcl* 2	Human	Sequences adjacent to chromosome translocation	B cell lymphoma/leukemia	–	–
bcr	Human	Sequences adjacent to chromosome translocation	Chronic myelogenous leukemia	–	pp160 pp190
dbl	Human	Transfection of 3T3 cells with tumor DNA	B cell lymphoma	–	p66
erb A	Avian	Avian erythroblastosis virus	Erythroleukemia	pp75	–
erb B-1	Avian	Avian erythroblastosis virus	Sarcoma	gp65 (*erb*-B)	gp175 truncated EGF receptor
erb B-2 (*neu*)	Human	Human analogue of *neu*	Various	–	p138
ets	Avian	E26 chick erythroblastosis virus	Mixed erythroid-myeloid leukemia	p135 (*gag-myc-ets*)	1. p54 2. p60, p62, and p64
fes (homologous to *fps*)	Feline	Snyder-Theilen feline sarcoma virus	Sarcoma	p85 (*gag-fes*)	p92 (c-*fes*)

Oncogene	Activity/ function	Subcellular localization	Human chromosome location	Reference
abl	Tyrosine kinase	Cytoplasm, plasma membrane	9q34	De Klein et al. (Chapt. 7)
A-raf-1, A-raf-2	? serine-threonine kinase Pseudogene		Xp21-g11 7p14-g-21	Heubner et al. Bonner et al.
(*bas*)	GTP binding/ GTPase	Plasma membrane	Thought to be same as Ha-*ras*	See Ha-*ras*
bcl 1, *bcl* 2	–	–	*bcl* 1: 11g13 *bcl* 2: 18q21	Bakhshi et al.; Tsujimoto et al. (Chapt. 11)
bcr	–	–	22	Groffen et al.; Heisterkamp et al. (Chapt. 7)
dbl	Serine protein kinase	Cytoplasm	–	Srivastava et al. (Chapt. 7)
erb A	Thyroid hormone receptor	–	17q21→q22	Weinberger et al.; Graf & Beug; Spurr et al. (Chapt. 8)
erb B-1	Tyrosine kinase; receptor for EGF	Cell surface plasma membrane	7p11→p13	Downward et al.; Spurr et al. (Chapt. 8)
erb B-2 (*neu*)	Related to *erb* B_1	Cell surface plasma membrane	17p11-q21	Akiyama et al. (Chapt. 8)
ets	–	Nucleus	1. 11q23→ q24 2. 21q23	Chen, J.H.; Rovigatti et al. (Chapt. 11)
fes (homologous to *fps*)	Tyrosine kinase	Plasma membrane	1.15q24→ q26 2. q25→q21	See *fps*

Oncogene	Species of origin	Virus or method of detection	Viral/original tumor	Viral gene product	Cellular homologue
fgr	Feline	Gardner-Rasheed feline sarcoma virus	Sarcoma	p70 (*gag-fgr*)	–
fim-1	Murine	DNA adjacent to integrated retrovirus	Myeloblastic leukemia	–	–
fim-2	Murine	DNA adjacent to integrated retrovirus	Myeloblastic leukemia	–	–
fms	Feline	McDonough feline sarcoma virus	Sarcoma	gp140 (v-*fms*)	gp170 (c-*fms*) CSF-1
fos	Murine	FJB osteosarcoma virus	Sarcoma	pp55	p55/60 (c-*fos*)
fps (homologous to *fes*)	Avian	Fuginami sarcoma virus	Sarcoma	p140 (*gag-fps*)	p98 (c-*fps*)
Ha-*ras* (1)	Murine, rat	Harvey rat sarcoma virus	Sarcoma	p21 (v-H-*ras*)	p21 (c-H-*ras*)
Ha-*ras* (2) (pseudogene)	Murine, rat	Harvey rat sarcoma virus	Sarcoma	–	–
hst	Human	Transfection of 3T3 cells with tumor DNA	Stomach carcinoma	–	–
int-1	Murine	DNA adjacent to integrated retrovirus	Mammary carcinoma	–	gp (370 aa)
int-2	Murine	DNA adjacent to integrated retrovirus	Mammary carcinoma	–	–
Jun	Avian	Sarcoma virus 17	Fibrosarcoma	–	–
Ki-*ras* (1) (pseudogene)	Murine, rat	Kirsten rat sarcoma virus	Sarcoma		
Ki-*ras* (2)	Murine, rat	Kirsten rat sarcoma virus	Sarcoma	p21	p21
kit	Feline	Hardy-Zuckerman-4 feline sarcoma virus	Sarcoma	p80 (*gag-kit*)	–

Oncogene	Activity/ function	Subcellular localization	Human chromosome location	Reference
fgr	Tyrosine kinase	Cytoplasm	1p36	Naharro et al. (Chapt. 7)
fim-1	–	–	–	Sola et al. (Chapt. 11)
fim-2	–	–	–	Sola et al. (Chapt. 11)
fms	Tyrosine kinase; receptor for M-CSF	Plasma membrane	5q34	Anderson et al.; Manger et al.; Sherr et al. (Chapt. 7)
fos	Binds to *jun*	Nucleus	14q24→q25	Curran et al.; Van Beveren et al.; Franza et al. (Chapt. 10)
fps (homologous to *fes*)	Tyrosine kinase	Cytoplasm, plasma membrane	25q24→q21 15q24→q21	Shibuya & Hanafusa (Chapt. 7)
Ha-*ras* (1)	GTP binding/ GTPase	Plasma membrane	11p15.1→ p15	Chang et al.; Manne et al. (Chapt. 9)
Ha-ras (2)	(pseudogene)		Xpter-q28	See Ha-*ras* (1)
hst	Resembles FGF	–	–	Sakamoto et al. (Chapt. 11)
int-1	–	–	12q14→pter	Nusse & Varmus; Casey et al. (Chapt. 11)
int-2	Resembles basic FGF	–	–	Moore et al. (Chapt. 11)
Jun	Transacting factor AP-1	–	1p31→p32	Bohman et al.; Maki et al. (Chapt. 11)
Ki-*ras* (1) (pseudogene)	(pseudogene)		6p11→p12	Chang et al. (Chapt. 9)
Ki-*ras*(2)	GTP binding/ GTPase	Plasma membrane	12p12	See Ki-*ras*(1)
kit	Tyrosine kinase	–		Besmer et al. (Chapt. 7)

Oncogene	Species of origin	Virus or method of detection	Viral/original tumor	Viral gene product	Cellular homologue
K53	Human	Transfection of 3T3 cells with tumor DNA	Kaposi's sarcoma	–	p23 (206 aa)
L-*myc*	Human	Homology to c-*myc*	SCCL	–	–
mas	Human	Transfection of 3T3 cells, then injected into nude mouse	Epidermoid tumor	–	(325 aa)
mel	Human	Transfection of 3T3 cells with tumor DNA	Melanoma cell line	–	–
met	Human	Osteosarcoma cell line treated with carcinogen	Osteosarcoma	–	*nml*: p140 *onc*: p65
mil (*mht*)	Avian	Mill-Hill 2 carcinoma virus	Carcinoma Leukemia	p100 (*gag-mht*) or (*gag-mil*)	–
mos	Murine	Moloney sarcoma virus	Sarcoma	p37 (*mos*)	–
myb	Avian	E26 (see *ets*) avian myeloblastosis	Myeloid leukemia	p48 (v-*myb*)	p75 (c-*myb*) p110 (c-*myb*)
myc	Avian	MC29 carcinoma myelocytomatosis virus	Sarcoma Leukemia	p110 (*gag-myc*)	p58 (c-*myc*)
neu (*erb* B_2)	Rat	Transfection of tumor DNA into 3T3 cells	Neuroglioblastoma	–	p185
N-*myc*	Human	Transfection studies	Neuroblastoma	–	p62-64 doublet
N-*ras*	Human	Transfection studies	Neuroblastoma	–	p21 (c-N-*ras*)
p53	Human	Transfection with activated c-Ha-*ras*	Cellular protein; various tumors	–	p53

Oncogene	Activity/ function	Subcellular localization	Human chromosome location	Reference
K53	Homologous to FGF			Dell-Bovi et al. (Chapt. 22)
L-*myc*	–	–	1p32	Kirsch et al. (Chapt. 10)
mas	–	Membrane hydrophobic protein	–	Young et al. (Chapt. 11)
mel	–	–	19p13.2-q13.2	Padua et al.; Spurr et al. (Chapt. 9)
met	Tyrosine kinase related		7q21→q31	Dean et al.; Park et al.; White et al. (Chapt. 7)
mil (*mht*)	Serine, threonine kinase (homologue of murine v-*raf*)	Cytoplasm	–	Graf et al.; Flordellis et al. (Chapt. 7)
mos	Serine, threonine kinase	Cytoplasm	8q22	Baldwin (Chapt. 8)
myb	–	Nucleus, associated with nuclear matrix	6q22-q24	Klemprauer et al. (Chapt. 10)
myc	Possible role in rRNA processing	Nucleus co-localizes with small nuclear ribonuclear proteins	8q24	Colby et al.; Dalla-Favera et al.; Ralston & Bishop (Chapt. 10)
neu (*erb* B_2)	Related to *erb* B_1	Cell surface plasma membrane	17p11→q21 (*erb* B_2)	Bargmann et al.; Schechter et al. (Chapt. 8)
N-*myc*	Related to c-*myc*	Nucleus	2p23→p24	Kohl et al.; Slamon et al. (Chapt. 10)
N-*ras*	GTP binding/ GTPase	–	1p11→p13	Davis et al.; Taparowsky et al. (Chapt. 9)
p53	–	Nucleus, cytoplasm	17p	Miller et al.; Eliyahu et al. (Chapt. 10)

Oncogene	Species of origin	Virus or method of detection	Viral/original tumor	Viral gene product	Cellular homologue
pim	Murine	Cellular gene activated by retroviral insertion	T cell lymphoma	–	(313 aa)
pks	Human	Probe of leukocyte DNA with *raf*-1 sequences	Angioimmunoblastic lymphadenopathy	–	*raf*-1 related (probably same as A-*raf*-1)
pro 1, *pro* 2	Murine	Cloned from mouse cell line JB6	Sensitivity of tumor promotion by phorbol esters	–	–
pvt-1	Murine	DNA involved in translocation	B lymphoma	–	–
R-*ras*	Human	DNA homologous to ras probe human genomic DNA	–	–	218 aa
raf-1	Murine	3G11 murine sarcoma virus	Fibrosarcoma	p75 (*gag*-*raf*)	(human) p73c-*raf* p75c-*raf* (rat)
raf-2	Human	Related to *raf*-1		Pseudogene	–
ral	Simian	DNA homologous to *ras* probe	Simian B lymphocyte cell line	–	p23 (206 aa)
rel	Avian (turkey)	Reticuloendotheliosis strain T	Lymphoid leukemia	p59 (*env-rel*)	–
ret	Human	Transfection of 3T3 cells with tumor DNA	Lymphoma	–	–
ros	Avian	(UR-2) Rochester-2 sarcoma virus	Sarcoma	p68 (*gag*-*ros*)	–
sis	Woolly monkey	Simian sarcoma virus	Sarcoma	p28 (*sis*)	PDGF β-chain

Oncogene	Activity/ function	Subcellular localization	Human chromosome location	Reference
pim	Protein kinase homology	–	6pter-q12	Nagarcijan et al.; Selten et al.; Cuyper et al. (Chapt. 11)
pks	–	–	X	Mark et al. (Chapt. 7)
pro 1, *pro* 2	–	–	–	– (Chapt. 11)
pvt-1	–	–	–	Cory et al.; Webb et al. (Chapt 11)
R-*ras*	? GTP binding protein		19	Lowe, D et al. (Chapt. 9)
raf-1	Serine, threonine kinase (homologue of avian *mil* (*mht*)	Cytoplasm	3p25	Bonner et al.; Huebner et al. (Chapt. 7)
raf-2	–	–	4	Bonner et al. (Chapt. 7)
ral	? GTP binding protein	Plasma membrane	–	Chardin & Tantian (Chapt. 9)
rel		–	2p11→ q14	Gilmore & Temin; Wilhelmson et al. (Chapt. 11)
ret	–	–	–	Takahashi et al. (Chapt. 11)
ros	Tyrosine kinase related to insulin receptor	–	6q16→q22	Shibuya et al.; Ullrich et al. (Chapt. 7)
sis	PDGF agonist	Cytoplasm	22q12.3→ 13.1	Chiu et al.; Dalla-Favera et al.; Swan et al. (Chapt. 8)

Oncogene	Species of origin	Virus or method of detection	Viral/original tumor	Viral gene product	Cellular homologue
ski	Avian (chicken)	SKV 770 virus	Sarcoma	?p110 ?p125	–
src	Avian	Rous sarcoma virus	Sarcoma	pp60 (v-*src*)	pp60 (c-*src*)
tck	Murine	LSTRA M-MuLV-induced thymoma cell line	Thymoma	–	$p56^{tck}$
tcl	Human	Sequences adjacent to chromosome translocation	T cell lymphoma	–	–
trk (*onc* D)	Human	Transfection of 3T3 cells with tumor DNA	Colon	–	–
yes	Avian	Yamaguchi sarcoma virus	Sarcoma	p90 (*gag-yes*)	p62 (c-*yes*) p59 (c-*yes*)

Oncogene	Activity/ function	Subcellular localization	Human chromosome location	Reference
ski	–	Nucleus	1q12→ter	Staraczar et al.; Li et al. (Chapt. 10)
src	Tyrosine kinase	Plasma membrane	20g12→q13 and 1p34→ter p36	Hunter; LeBeau et al.; Swanstrom et al.; Willingham et al. (Chapter 7)
tck	Protein kinase	–	–	Voronova & Sefton (Chapt. 7)
tcl	–	–	7q35	Isobe et al.; Morton et al. (Chapt. 11)
trk (*onc* D)	Nonmuscle tropomycin ⊕ protein kinase	Plasma membrane	1q	Martin-Zanca et al.; Reinach & MacLeod (Chapt. 7)
yes	Tyrosine kinase	–	18q21.3	Shibuya et al.; Yoshida et al. (Chapt. 10)

Index